IET ENERGY ENGINEERING SERIES 186

Synchronous Reluctance Machines

Other volumes in this series:

Volume 1	**Power Circuit Breaker Theory and Design** C.H. Flurscheim (Editor)
Volume 4	**Industrial Microwave Heating** A.C. Metaxas and R.J. Meredith
Volume 7	**Insulators for High Voltages** J.S.T. Looms
Volume 8	**Variable Frequency AC Motor Drive Systems** D. Finney
Volume 10	**SF_6 Switchgear** H.M. Ryan and G.R. Jones
Volume 11	**Conduction and Induction Heating** E.J. Davies
Volume 13	**Statistical Techniques for High Voltage Engineering** W. Hauschild and W. Mosch
Volume 14	**Uninterruptible Power Supplies** J. Platts and J.D. St Aubyn (Editors)
Volume 15	**Digital Protection for Power Systems** A.T. Johns and S.K. Salman
Volume 16	**Electricity Economics and Planning** T.W. Berrie
Volume 18	**Vacuum Switchgear** A. Greenwood
Volume 19	**Electrical Safety: A guide to causes and prevention of hazards** J. Maxwell Adams
Volume 21	**Electricity Distribution Network Design, 2nd Edition** E. Lakervi and E.J. Holmes
Volume 22	**Artificial Intelligence Techniques in Power Systems** K. Warwick, A.O. Ekwue and R. Aggarwal (Editors)
Volume 24	**Power System Commissioning and Maintenance Practice** K. Harker
Volume 25	**Engineers' Handbook of Industrial Microwave Heating** R.J. Meredith
Volume 26	**Small Electric Motors** H. Moczala *et al.*
Volume 27	**AC–DC Power System Analysis** J. Arrillaga and B.C. Smith
Volume 29	**High Voltage Direct Current Transmission, 2nd Edition** J. Arrillaga
Volume 30	**Flexible AC Transmission Systems (FACTS)** Y.-H. Song (Editor)
Volume 31	**Embedded generation** N. Jenkins *et al.*
Volume 32	**High Voltage Engineering and Testing, 2nd Edition** H.M. Ryan (Editor)
Volume 33	**Overvoltage Protection of Low-Voltage Systems, Revised Edition** P. Hasse
Volume 36	**Voltage Quality in Electrical Power Systems** J. Schlabbach *et al.*
Volume 37	**Electrical Steels for Rotating Machines** P. Beckley
Volume 38	**The Electric Car: Development and future of battery, hybrid and fuel-cell cars** M. Westbrook
Volume 39	**Power Systems Electromagnetic Transients Simulation** J. Arrillaga and N. Watson
Volume 40	**Advances in High Voltage Engineering** M. Haddad and D. Warne
Volume 41	**Electrical Operation of Electrostatic Precipitators** K. Parker
Volume 43	**Thermal Power Plant Simulation and Control** D. Flynn
Volume 44	**Economic Evaluation of Projects in the Electricity Supply Industry** H. Khatib
Volume 45	**Propulsion Systems for Hybrid Vehicles** J. Miller
Volume 46	**Distribution Switchgear** S. Stewart
Volume 47	**Protection of Electricity Distribution Networks, 2nd Edition** J. Gers and E. Holmes
Volume 48	**Wood Pole Overhead Lines** B. Wareing
Volume 49	**Electric Fuses, 3rd Edition** A. Wright and G. Newbery
Volume 50	**Wind Power Integration: Connection and system operational aspects** B. Fox *et al.*
Volume 51	**Short Circuit Currents** J. Schlabbach
Volume 52	**Nuclear Power** J. Wood
Volume 53	**Condition Assessment of High Voltage Insulation in Power System Equipment** R.E. James and Q. Su
Volume 55	**Local Energy: Distributed generation of heat and power** J. Wood
Volume 56	**Condition Monitoring of Rotating Electrical Machines** P. Tavner, L. Ran, J. Penman and H. Sedding
Volume 57	**The Control Techniques Drives and Controls Handbook, 2nd Edition** B. Drury
Volume 58	**Lightning Protection** V. Cooray (Editor)
Volume 59	**Ultracapacitor Applications** J.M. Miller
Volume 62	**Lightning Electromagnetics** V. Cooray
Volume 63	**Energy Storage for Power Systems, 2nd Edition** A. Ter-Gazarian
Volume 65	**Protection of Electricity Distribution Networks, 3rd Edition** J. Gers
Volume 66	**High Voltage Engineering Testing, 3rd Edition** H. Ryan (Editor)
Volume 67	**Multicore Simulation of Power System Transients** F.M. Uriate
Volume 68	**Distribution System Analysis and Automation** J. Gers
Volume 69	**The Lightening Flash, 2nd Edition** V. Cooray (Editor)
Volume 70	**Economic Evaluation of Projects in the Electricity Supply Industry, 3rd Edition** H. Khatib
Volume 72	**Control Circuits in Power Electronics: Practical issues in design and implementation** M. Castilla (Editor)
Volume 73	**Wide Area Monitoring, Protection and Control Systems: The enabler for smarter grids** A. Vaccaro and A. Zobaa (Editors)

Volume 74 **Power Electronic Converters and Systems: Frontiers and applications** A.M. Trzynadlowski (Editor)
Volume 75 **Power Distribution Automation** B. Das (Editor)
Volume 76 **Power System Stability: Modelling, analysis and control** A.A. Sallam and Om P. Malik
Volume 78 **Numerical Analysis of Power System Transients and Dynamics** A. Ametani (Editor)
Volume 79 **Vehicle-to-Grid: Linking electric vehicles to the smart grid** J. Lu and J. Hossain (Editors)
Volume 81 **Cyber-Physical-Social Systems and Constructs in Electric Power Engineering** S. Suryanarayanan, R. Roche and T.M. Hansen (Editors)
Volume 82 **Periodic Control of Power Electronic Converters** F. Blaabjerg, K. Zhou, D. Wang and Y. Yang
Volume 86 **Advances in Power System Modelling, Control and Stability Analysis** F. Milano (Editor)
Volume 87 **Cogeneration: Technologies, optimisation and implementation** C.A. Frangopoulos (Editor)
Volume 88 **Smarter Energy: From smart metering to the smart grid** H. Sun, N. Hatziargyriou, H.V. Poor, L. Carpanini and M.A. Sánchez Fornié (Editors)
Volume 89 **Hydrogen Production, Separation and Purification for Energy** A. Basile, F. Dalena, J. Tong and T.N. Veziroǧlu (Editors)
Volume 90 **Clean Energy Microgrids** S. Obara and J. Morel (Editors)
Volume 91 **Fuzzy Logic Control in Energy Systems with Design Applications in MATLAB®/Simulink®** I.H. Altaş
Volume 92 **Power Quality in Future Electrical Power Systems** A.F. Zobaa and S.H.E.A. Aleem (Editors)
Volume 93 **Cogeneration and District Energy Systems: Modelling, analysis and & optimization** M.A. Rosen and S. Koohi-Fayegh
Volume 94 **Introduction to the Smart Grid: Concepts, technologies and evolution** S.K. Salman
Volume 95 **Communication, Control and Security Challenges for the Smart Grid** S.M. Muyeen and S. Rahman (Editors)
Volume 96 **Industrial Power Systems with Distributed and Embedded Generation** R. Belu
Volume 97 **Synchronized Phasor Measurements for Smart Grids** M.J.B. Reddy and D.K. Mohanta (Editors)
Volume 98 **Large Scale Grid Integration of Renewable Energy Sources** A. Moreno-Munoz (Editor)
Volume 100 **Modeling and Dynamic Behaviour of Hydropower Plants** N. Kishor and J. Fraile-Ardanuy (Editors)
Volume 101 **Methane and Hydrogen for Energy Storage** R. Carriveau and D.S.-K. Ting
Volume 104 **Power Transformer Condition Monitoring and Diagnosis** A. Abu-Siada (Editor)
Volume 106 **Surface Passivation of Industrial Crystalline Silicon Solar Cells** J. John (Editor)
Volume 107 **Bifacial Photovoltaics: Technology, applications and economics** J. Libal and R. Kopecek (Editors)
Volume 108 **Fault Diagnosis of Induction Motors** J. Faiz, V. Ghorbanian and G. Joksimović
Volume 110 **High Voltage Power Network Construction** K. Harker
Volume 111 **Energy Storage at Different Voltage Levels: Technology, integration, and market aspects** A.F. Zobaa, P.F. Ribeiro, S.H.A. Aleem and S.N. Afifi (Editors)
Volume 112 **Wireless Power Transfer: Theory, technology and application** N. Shinohara
Volume 114 **Lightning-Induced Effects in Electrical and Telecommunication Systems** Y. Baba and V.A. Rakov
Volume 115 **DC Distribution Systems and Microgrids** T. Dragičević, F. Blaabjerg and P. Wheeler
Volume 116 **Modelling and Simulation of HVDC Transmission** M. Han (Editor)
Volume 117 **Structural Control and Fault Detection of Wind Turbine Systems** H.R. Karimi
Volume 119 **Thermal & Power Plant Control and Instrumentation: The control of boilers and HRSGs, 2nd Edition** D. Lindsley, J. Grist and D. Parker
Volume 120 **Fault Diagnosis for Robust Inverter Power Drives** A. Ginart (Editor)
Volume 121 **Monitoring and Control Using Synchrophasors in Power Systems with Renewables** I. Kamwa and C. Lu (Editors)
Volume 123 **Power Systems Electromagnetic Transients Simulation, 2nd Edition** N. Watson and J. Arrillaga
Volume 124 **Power Market Transformation** B. Murray
Volume 125 **Wind Energy Modeling and Simulation, Volume 1: Atmosphere and plant** P. Veers (Editor)

Volume 126 **Diagnosis and Fault Tolerance of Electrical Machines, Power Electronics and Drives** A.J.M. Cardoso
Volume 128 **Characterization of Wide Bandgap Power Semiconductor Devices** F. Wang, Z. Zhang and E.A. Jones
Volume 129 **Renewable Energy from the Oceans: From wave, tidal and gradient systems to offshore wind and solar** D. Coiro and T. Sant (Editors)
Volume 130 **Wind and Solar Based Energy Systems for Communities** R. Carriveau and D.S.-K. Ting (Editors)
Volume 131 **Metaheuristic Optimization in Power Engineering** J. Radosavljević
Volume 132 **Power Line Communication Systems for Smart Grids** I.R.S. Casella and A. Anpalagan
Volume 134 **Hydrogen Passivation and Laser Doping for Silicon Solar Cells** B. Hallam and C. Chan (Editors)
Volume 139 **Variability, Scalability and Stability of Microgrids** S.M. Muyeen, S.M. Islam and F. Blaabjerg (Editors)
Volume 143 **Medium Voltage DC System Architectures** B. Grainger and R. De Doncker (Editors)
Volume 145 **Condition Monitoring of Rotating Electrical Machines** P. Tavner, L. Ran and C. Crabtree
Volume 146 **Energy Storage for Power Systems, 3rd Edition** A.G. Ter-Gazarian
Volume 147 **Distribution Systems Analysis and Automation, 2nd Edition** J. Gers
Volume 151 **SiC Power Module Design: Performance, robustness and reliability** A. Castellazzi and A. Irace (Editors)
Volume 152 **Power Electronic Devices: Applications, failure mechanisms and reliability** F. Iannuzzo (Editor)
Volume 153 **Signal Processing for Fault Detection and Diagnosis in Electric Machines and Systems** M. Benbouzid (Editor)
Volume 155 **Energy Generation and Efficiency Technologies for Green Residential Buildings** D. Ting and R. Carriveau (Editors)
Volume 156 **Lithium-ion Batteries Enabled by Silicon Anodes** C. Ban and K. Xu (Editors)
Volume 157 **Electrical Steels, 2 Volumes** A. Moses, K. Jenkins, P. Anderson and H. Stanbury
Volume 158 **Advanced Dielectric Materials for Electrostatic Capacitors** Q. Li (Editor)
Volume 159 **Transforming the Grid Towards Fully Renewable Energy** O. Probst, S. Castellanos and R. Palacios (Editors)
Volume 160 **Microgrids for Rural Areas: Research and case studies** R.K. Chauhan, K. Chauhan and S.N. Singh (Editors)
Volume 161 **Artificial Intelligence for Smarter Power Systems: Fuzzy logic and neural networks** M.G. Simoes
Volume 166 **Advanced Characterization of Thin Film Solar Cells** N. Haegel and M. Al-Jassim (Editors)
Volume 167 **Power Grids With Renewable Energy: Storage, integration and digitalization** A.A. Sallam and Om P. Malik
Volume 169 **Small Wind and Hydrokinetic Turbines** P. Clausen, J. Whale and D. Wood (Editors)
Volume 170 **Reliability of Power Electronics Converters for Solar Photovoltaic Applications** F. Blaabjerg, A.I. Haque, H. Wang, Z. Abdin Jaffery and Y. Yang (Editors)
Volume 171 **Utility-scale Wind Turbines and Wind Farms** A. Vasel-Be-Hagh and D.S.-K. Ting
Volume 172 **Lighting Interaction with Power Systems, 2 Volumes** A. Piantini (Editor)
Volume 174 **Silicon Solar Cell Metallization and Module Technology** T. Dullweber (Editor)
Volume 180 **Protection of Electricity Distribution Networks, 4th Edition** J. Gers and E. Holmes
Volume 182 **Surge Protection for Low Voltage Systems** A. Rousseau (Editor)
Volume 184 **Compressed Air Energy Storage: Types, systems and applications** D. Ting and J. Stagner
Volume 193 **Overhead Electric Power Lines: Theory and practice** S. Chattopadhyay and A. Das
Volume 194 **Offshore Wind Power Reliability, Availability and Maintenance, 2nd Edition** P. Tavner
Volume 198 **Battery Management Systems and Inductive Balancing** A. Van den Bossche and A. Farzan Moghaddam
Volume 199 **Model Predictive Control for Microgrids: From power electronic converters to energy management** J. Hu, J.M. Guerrero and S. Islam
Volume 204 **Electromagnetic Transients in Large HV Cable Networks: Modeling and calculations** Ametani, Xue, Ohno and Khalilnezhad
Volume 212 **Battery State Estimation: Methods and Models** S. Wang
Volume 905 **Power System Protection, 4 Volumes** The Electricity Training Association (Editor)

Synchronous Reluctance Machines

Analysis, optimization and applications

Nicola Bianchi, Cristian Babetto and Giacomo Bacco

The Institution of Engineering and Technology

Published by The Institution of Engineering and Technology, London, United Kingdom

The Institution of Engineering and Technology is registered as a Charity in England & Wales (no. 211014) and Scotland (no. SC038698).

First published 2021

The Institution of Engineering and Technology
Michael Faraday House
Six Hills Way, Stevenage
Herts, SG1 2AY, United Kingdom

www.theiet.org

British Library Cataloguing in Publication Data
A catalogue record for this product is available from the British Library

ISBN 978-1-83953-263-4 (hardback)
ISBN 978-1-83953-264-1 (PDF)

Typeset in India by MPS Limited

Contents

About the authors

Nicola Bianchi is a full professor at the Department of Industrial Engineering of the University of Padova, Italy. He is the head of the Electric Drive Laboratory, where the research activity is in the field of electrical motors and their control. His research activity focuses on the analysis of the performance, determination of the parameters, design, optimization and control of innovative machines. He is an IEEE Fellow Member. He is coauthor of more than two hundred papers and some books. He is recipient of some awards for best conference and journal papers.

Cristian Babetto is a design specialist at Marelli Motorsport, Corbetta, Italy. His activities focus on the design and analysis of electromagnetic devices and high-performance electric motors for motorsport applications. He received a PhD in industrial engineering from the University of Padova (Italy) in 2020. During his PhD studies, he mainly worked on the analysis and optimization of synchronous reluctance machines for fault-tolerant and high-speed applications.

Giacomo Bacco is an E-Motor Engineer in the Advanced Engineering group at Magna Powertrain. His work includes the analysis and design of electric drives for automotive applications. He received the BSc, MSc and PhD degrees from the University of Padova in 2013, 2016 and 2020, respectively. His research interests covered the analysis, design and optimization of electric machines, in particular anisotropic synchronous machines (IPMM and SyRM).

List of acronyms

BEMF	Back Electromotive Force	bemf
BM	Bonded Magnet	BM
CPSR	Constant Power Speed Range	CPSR
DE	Differential Evolution	DE
EA	Evolutionary Algorithm	EA
FE	Finite Element	FE
FEA	Finite-Element Analysis	FEA
FEO	Finite-Element-based Approach	FEO
FW	Flux-Weakening	FW
GA	Genetic Algorithm	GA
GT	Game Theory–based Optimization	GT
HF	High-Frequency	hf
HS	High-Speed	hs
IM	Induction Motor	IM
IPM	Interior Permanent Magnet	IPM
IRN	Improved Reluctance Network	IRN
MMF	Magnetomotive Force	mmf
MST	Maxwell Stress Tensor	MST
MTPA	Maximum Torque Per Ampere	MTPA
MTPV	Maximum Torque Per Voltage	MTPV
MTPVA	Maximum Torque Per Volt-Ampere	MTPVA
PM	Permanent Magnet	PM
PMaSyR	Permanent Magnet assisted Synchronous Reluctance	PMaSyR
PMSM	Permanent Magnet Synchronous Machine	PMSM
PSO	Particle Swarm Optimization	PSO
RE	Rare-Earth	re
RHS	Right-Hand Side	rhs
RMS	Root Mean Square	RMS
SA	Simulated Annealing	SA
SADA	Semi-Analytical Design Approach	SADA
SPM	Surface-mounted Permanent Magnet	SPM
SRN	Simple Reluctance Network	SRN
SwR	Switched Reluctance	SwR
SyR	Synchronous Reluctance	SyR
SyRM	Synchronous Reluctance Machine	SyRM

List of symbols

$\boldsymbol{A}$	The incidence matrix	
α_i^{e}	The current electric angle	[rad]
α_{s}	The slot angle	[rad]
B_{g}	The air-gap peak flux density	[T]
$\mathscr{B}_{\mathrm{g}}$	The air-gap flux-density spatial distribution	[T]
B_{rem}	The remanence of a permanent magnet	[T]
B_{y}	The yoke flux density	[T]
D_{e}	The stator outer diameter	[m]
$\Delta\vartheta$	The position estimation error	[rad]
D_{r}	The rotor diameter	[m]
D_{s}	The stator inner diameter	[m]
D_{sh}	The shaft diameter	[m]
e	Euler's number	
ℓ_x	The generic differential inductance	[H]
ℓ_d	The d-axis differential inductance	[H]
ℓ_Δ	The half-difference of the self-differential inductances along the q- and d-axis	[H]
ℓ_{dq}	The cross-saturation differential inductance	[H]
ℓ_q	The q-axis differential inductance	[H]
ℓ_{qd}	The second cross-saturation differential inductance	[H]
ℓ_Σ	The half-sum of the self-differential inductances along the q- and d-axis	[H]
$\Gamma_{\tilde{q}\tilde{d}}$	The inverse of the inductance that relates the estimated q-axis current with the estimated d-axis voltage	[1/H]
H_{c}	The magnetic coercive force	[Am]
$H_{\mathrm{c}B}$	The coercive force of the magnetic flux-density field	[Am]
$H_{\mathrm{c}J}$	The coercive force of the magnetization field	[Am]
h_{s}	The slot height	[m]
h_{y}	The stator yoke height	[m]
$\mathbf{I}$	The identity matrix	
I_{w}	The phase winding current	[A]
j	The imaginary unit	
k_{air}	The rotor magnetic insulation ratio	
k_{fill}	The slot fill factor	

k_{pack}	The lamination packing factor	
$\mathscr{K}_s$	The stator electric loading spatial distribution	[A/m]
F_{c}	The centrifugal force	[Wb]
λ_{mg}	The PM flux linkage	[Wb]
l_{b}	The flux-barrier length	[m]
L_{stk}	The length of the lamination stack	[m]
μ_0	The magnetic permeability of free space	[H/m]
μ_{rec}	The permanent magnet recoil relative permeability	
N_{b}	The number of flux-barriers	
N_c	The number of series conductor per phase	
n_c	The number of actual conductors in the slot	
n_{pp}	The number of the winding parallel paths	
N_s	The number of series conductor per phase	
ω	The angular frequency	[rad/s]
p_{s}	The slot pitch	[m]
Q	The number of stator slots	
$\Re$	The generic reluctance	[H^{-1}]
Res	The number of points used to discretize the domain	
S_{c}	The actual cross-section area of the conductors	[m^2]
$S_{\text{c,eq}}$	The equivalent cross-section area of the conductors	[m^2]
S_{slot}	The surface of the slot cross section	[m^2]
t_{b}	The flux-barrier thickness	[m]
ϑ_{b}	The flux-barrier angle	[rad]
$\vartheta_{\text{b}}^{\text{e}}$	The flux-barrier electrical angle	[rad]
ϑ_{e}	The generic electrical angular coordinate	[rad]
ϑ_{m}	The rotor angular position	[rad]
ϑ_{r}	The generic angular coordinate in the rotor reference	[rad]
ϑ_{s}	The generic angular coordinate in the stator reference frame	[rad]
THD	The Total Harmonic Distortion	
t_{m}	The permanent magnet thickness	[m]
$\langle T_{\text{m}} \rangle$	The average torque	[Nm]
T_{m}	The electromechanical torque	[Nm]
ΔT_{m}	The torque ripple	[Nm]
U_{r}	The rotor magnetic scalar potential	[A]
$\mathscr{U}_{\text{r}}$	The rotor magnetic scalar potential spatial distribution	[A]
$\mathscr{U}_{\text{s}}$	The stator magnetic scalar potential spatial distribution	[A]
w_{c}	The flux-carrier width	[m]
w_{m}	The permanent magnet width	[m]
w_{rib}	The width of iron ribs	[m]
w_{s}	The slot width	[m]
w_{t}	The tooth width	[m]

Chapter 1
Introduction

In recent years, there is growing interest for Synchronous Reluctance (SyR) machines. Even if such a machine has been proposed several years ago, only recently it is more and more adopted in various applications due to its robustness, high efficiency and the absence of excitation winding or Permanent Magnets (PMs) (Figure 1.1).

SyR motors are becoming attractive in many fields and applications requiring high dynamics, high torque density and fault-tolerant capabilities such as traction and household appliances [1,2]. A substantial advantage is also the virtual absence of rotor losses [3], leading to lower rotor temperatures if compared with induction motors.

However, if not properly designed, SyR machine presents some drawbacks, such as high torque ripple and low-power factor. The design methodology is well established since works have been carried out in the field of torque ripple reduction [4,5] or power factor improvement [6].

Both with or without PM content, the SyR machine is one of the most promising candidate that can meet the requirements of efficient and low-cost drive [7,8]. The key benefits of this technology are a rotor structure made of cutouts and iron parts only, as in Figure 1.2. A rotor with neither excitation coils nor PMs leads to a very cost-effective structure.

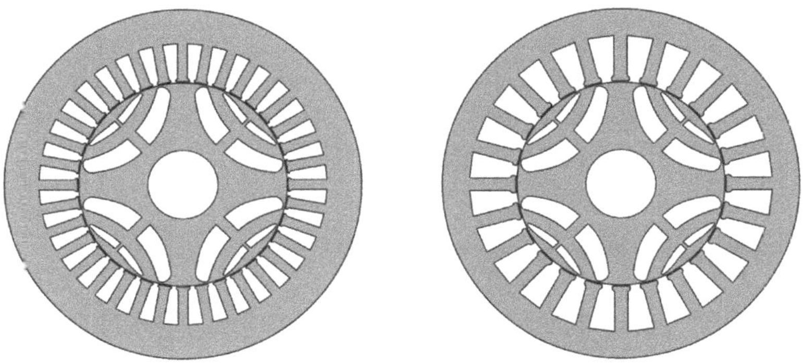

Figure 1.1 SyR motor structures, characterized by two flux-barriers per pole

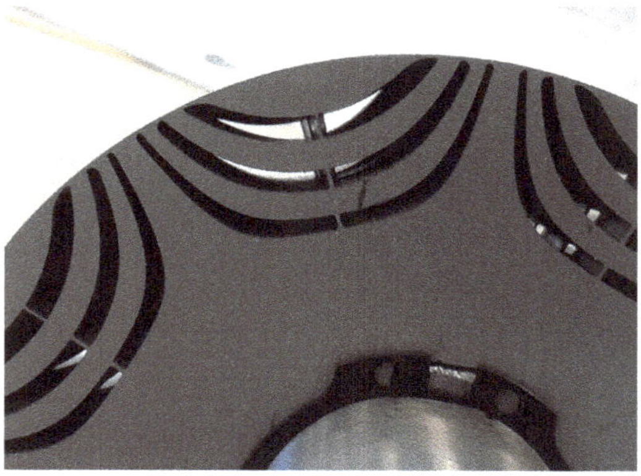

Figure 1.2 SyR motor rotor

1.1 A brief history

The SyR machine topology was first introduced in the 1920s [9]; however, it was not adopted into industrial applications as other machines until the middle of the twentieth century. The induction motors dominated the electrical machine market in applications such as industrial fans, pumps and mills [10]. In the 1960s, PM motors started to be used in high-performance applications. They were commercialized in the early 1980s. SyR motors gained interest later on, as an alternative to induction motors [11,12]. They were proposed as a commercial product around 2010.

The most recent revolution of electrical machines was due to the introduction of Rare-Earth (RE) permanent magnets. They are characterized by a very high magnetic energy density compared to previous magnetic materials.

Samarium–Cobalt (SmCo) PMs represent a family of high-energy hard-magnetic materials. They had an important development in the 1970s and are still extensively used in applications characterized by high operating temperatures or high armature reaction. The RE Neodymium–Iron–Boron (NdFeB) magnets were presented in the 1980s [13]. They were proposed as interesting candidates to substitute SmCo magnets thanks to the more abundant primary resources. NdFeB PMs are mainly used in PM motors designed for high-performance applications.

1.2 Drawbacks in using rare-earth PMs

PMs are an essential component of modern electric motors and generators. Key properties of PMs are coercivity and the remanent flux density. One of the most widely used PMs for traction motors and power generators contains Neodymium (Nd) and

Dysprosium (Dy). Dy is used to sustain the NdFeB PMs coercivity at higher temperatures [14]. Both Dy and Nd are considered Rare-Earth Elements (REE) and are listed as critical materials by US Department of Energy as well as other international institutes due to the high risk of the supply chain.

The main drawback of RE materials is their high cost and its extremely variable price, as it happened in August 2011, when the price of NdFeB PMs rose rapidly by a factor of 20 compared to the beginning of 2010. Let us remember that other materials, in smaller percentage, are used in these magnets. Neodymium (Nd) and Praseodymium (Pr) are the primary RE PMs component, and components as Cerium (Ce) and Lanthanum (La) are almost always present in the PM compound [15,16]. In particular, Ref. [15] showed that the synergic action of light RE as Cerium (Ce) and Lanthanum (La) enables an enhancement of coercivity. Alternatively, Ref. [16] demonstrated that magnets with excellent properties can be obtained always using both light and heavy RE. However, most of the proposed compounds require the presence of very expensive heavy REE, like Terbium (Tb) or Dysprosium (Dy), in order to increase significantly the magnet coercivity and the temperature stability. The high prices of the PM components, and their relevant variability, have led to a price ratio between NdFeB PMs and ferrite PMs roughly in the range between 7 and 12.

The average price of the Dy oxide has spiked from 245 USD/kg in 2010 to 1 410 USD/kg in 2011 then gradually went down to 185 USD/kg in the following 5 years. The prices of the Nd oxide also have increased from 88 USD/kg in 2010 to 195 USD/kg in 2011 and then went also down to 39 USD/kg.

The increase in the price of NdFeB magnets was mainly caused by huge increase in demand in REE as well as the monopoly of the critical REE mines in specific areas of the globe. China and big European countries are currently forcing the Electric Vehicles (EVs) market to replace the internal combustion engine ones. With the increase in number of EVs, NdFeB magnet price will possibly continue to grow in the coming years.

Other alternatives are non-Rare-Earth (non-RE) PMs with lower magnetic performance. They are still attractive for many applications where cost saving is a priority. The demand for NdFeB PMs are significantly increasing due to the demanding needs of modern drives applications: higher energy efficiency, performance and power density. The most pragmatic approach to reduce the usage of RE materials is the development of non-RE magnets that can fill the magnetic performance gap between ferrite and RE magnets.

To reduce the demand in RE, non-RE or less-RE PM technologies are being investigated intensively. Significant advancements in the reduction of Dy content in NdFeB, keeping required coercivity, were achieved through the reduction of the grains size. The development of EM technologies that do not require PMs excitation is one of the key approaches to the described problem.

The current trends toward higher efficiency as well as RE-free technologies raise the interest toward reluctance machines. The rapid industrial acceptance of the SyR is a matter of time, as the biggest motor manufacturers and R&D institutions are working toward the goals aforementioned. There are more and more studies of adopting such a machine in various types of applications such as EVs, wind turbines and actuators.

1.3 Increasing interest in Synchronous Reluctance machines

As mentioned earlier, SyR is a promising technology which has features that are aligned with the previously mentioned industrial needs, that is, high efficiency and low cost. Currently, world leading manufacturers and research and development institutions are constantly investigating the possibility of increasing the efficiency using cost-effective solutions. In industrial applications, in order to comply with new standards and regulations, growing interest in higher efficiency or superpremium efficiency, EMs took place. This was driven by global trends toward a better utilization of the energy produced and consumed, in order to minimize the human footprint and minimize the greenhouse gas emissions.

However, despite its advantages, there are still a number of challenges that are still subjects of research. From the machine design perspective, the main challenges come from the complex anisotropic structure of the rotor that is held in place by thin iron ribs, serving mainly for mechanical retention. High torque ripple, low power factor and other secondary effects such as rotor iron losses, vibration and noise are the main issues that need to be carefully considered during the design of an SyR motor. These are conventionally addressed by means of comprehensive analyses and optimization loops using both analytical and finite-element tools.

In this manuscript, some special applications of the SyR motor are dealt with, so as to give a contribution to the development of such a motor. After comparing the operating principles with other synchronous machine topologies, outlining the working principle of SyR machine, advantages and drawbacks are highlighted when the motor is designed for special applications.

At first, design techniques to achieve SyR motors with a reduced torque ripple are described, with the aim of providing the reader proper tools to achieve satisfactory performance. Innovative techniques are introduced by the Authors' and the scientific community with the aim of significantly reduce time and effort required to design SyR machines, from a preliminary sizing to a final design stage. In some cases, design optimization by means of finite-element analysis can be bypassed, applying new sizing methods, quick and effective tools for the design of SyR machines in a wide power range.

Special attention is given to high-speed applications such as vacuum pumps, turbo-compressors and machining spindles, requiring compact and efficient drive systems [17] (Figure 1.3). Both magnetic and thermal aspects are taken into account in the design. Mechanical aspect represents an additional design key point to guarantee the structural integrity against the high centrifugal forces.

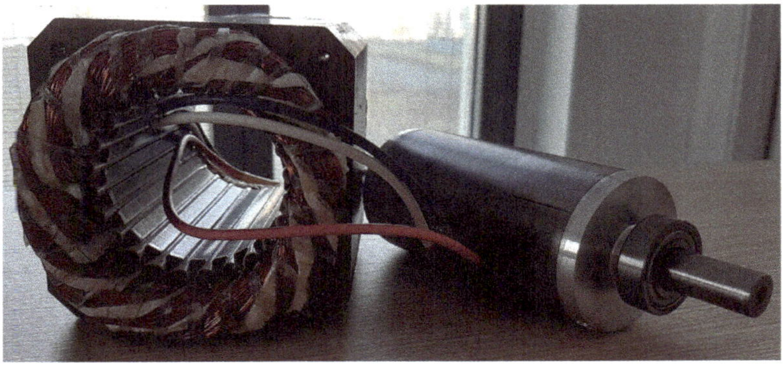

Figure 1.3 SyR motor deigned for high-speed application

A second deep investigation is in the regard of fault-tolerant SyR motor drives. The reliability of an electric drive system is a mandatory requirement in applications (automotive, aeronautic, mining) where failures cause economic costs or could affect the safety of operation. Fault-tolerance is the ability to ensure proper operations even in or after an event of failure. A dual-three-phase system is applied to an SyR motor to increase the reliability of the overall drive. In addition, it is shown that such a motor is suitable when a short-circuit failure occurs.

Chapter 2
Magnetic materials

All materials can be classified in terms of their magnetic behavior falling into one of five categories, depending on their magnetic susceptibility.

The two most common types of magnetism are diamagnetism and paramagnetism, which account for most of the periodic table of elements at room temperature (see Figure 2.1). These elements are usually referred to as nonmagnetic, whereas those that are referred to as magnetic are actually classified as ferromagnetic. The only other type of magnetism observed in pure elements at room temperature is antiferromagnetism. Finally, magnetic materials can also be classified as ferrimagnetic, although this is not observed in any pure element but can only be found in compounds, such as the mixed oxides, known as ferrites, from which ferrimagnetism derives its name [18] (Table 2.1).

When a field, $\boldsymbol{H}$, is applied to a material, it responds by producing a magnetic field, called magnetization and referred to as $\boldsymbol{M}$. This magnetization is a measure of the magnetic moment per unit volume of material. The applied field is the one that would be present if the field were applied to a vacuum. However, due to magnetization, the total flux of magnetic field lines is taken into account by the magnetic induction, $\boldsymbol{B}$, also called magnetic flux density. It counts exactly the magnetic flux crossing a unit area of the material. The flux is produced by both the external applied field and the internal reaction field. Therefore, these quantities are related through

$$\boldsymbol{B} = \mu_0(\boldsymbol{H} + \boldsymbol{M}) \tag{2.1}$$

Then the magnetic susceptibility, χ_m, can be defined, measuring this reaction and identifying the type of magnetic material:

$$\chi_m = \frac{\boldsymbol{M}}{\boldsymbol{H}}$$

The other parameter that relates the magnetic induction and the magnetic field is the permeability of the material

$$\mu = \frac{\boldsymbol{B}}{\boldsymbol{H}}$$

All these relationships are summarized in the following equations:

$$\boldsymbol{B} = \mu\boldsymbol{H} = \mu_0\mu_r\boldsymbol{H} = \mu_0\boldsymbol{H} + \boldsymbol{J} = \mu_0(\boldsymbol{H} + \boldsymbol{M}) = \mu_0(1 + \chi_m)\boldsymbol{H} \tag{2.2}$$

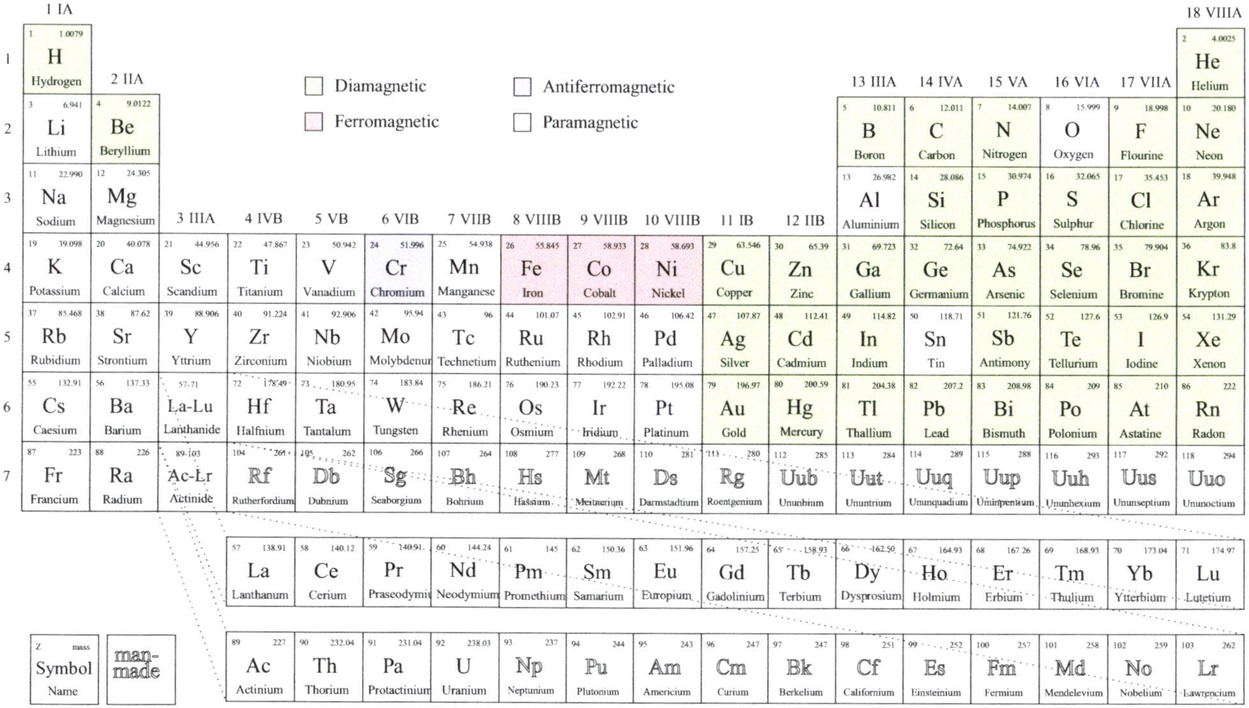

Figure 2.1 Periodic table. Elements are classified according to their magnetic behavior.
The original figure code comes from https://texample.net/tikz/examples/periodic-table-of-chemical-elements/

Table 2.1 Susceptibility reference value at room temperature for each type of magnetic material [18]

Type of magnetism	Susceptibility	Example	
Diamagnetism	Small and negative	Cu	$-0.77 \cdot 10^{-6}$
Paramagnetism	Small and positive	Mn	$66.10 \cdot 10^{-6}$
Ferromagnetism	Large and positive, function of the applied field	Fe	Up to ~100 000
Antiferromagnetism	Small and positive	Cr	
Ferrimagnetism	Large and positive, function of the applied field	Ba Ferrite	Up to ~3

where

$$\mu_r = \frac{\mu}{\mu_0} = \chi_m + 1$$

and $\boldsymbol{J}$ is the magnetic polarization, also referred to as the intensity of magnetization.

2.1 Ferromagnetic materials

Ferromagnetic materials are magnetic even without the external applied field. They exhibit spontaneous magnetization as a result of the ordered internal magnetic moments.

Ferromagnetism is only possible when atoms are arranged in a lattice and the atomic magnetic moments can interact to align parallel to each other. In 1907, Weiss postulated the presence of magnetic domains within the material, which are regions where the atomic magnetic moments are aligned. The movement of these domains determines how the material responds to a magnetic field. Consequently the susceptibility is a function of applied magnetic field and usually ferromagnetic materials are compared in terms of saturation magnetization rather than susceptibility.

In Figure 2.1, it can be seen that the only ferromagnetic materials at room temperature are iron, cobalt and nickel.

The relation between the magnetizing force $\boldsymbol{H}$ and the flux density or magnetic induction $\boldsymbol{B}$ which it produces in a ferromagnetic material is of considerable importance. It is most conveniently expressed by means of characteristic curves.

For purposes of explanation, a ferromagnetic material is considered to be placed in a region where the magnetic field intensity can be varied. The magnetization force is varied through changing the current in the coil. The material is originally demagnetized. If the flux density is measured as a function of the magnetizing force, and the relation is plotted, the *first magnetization* curve is obtained (the path 0–1 in Figure 2.2).

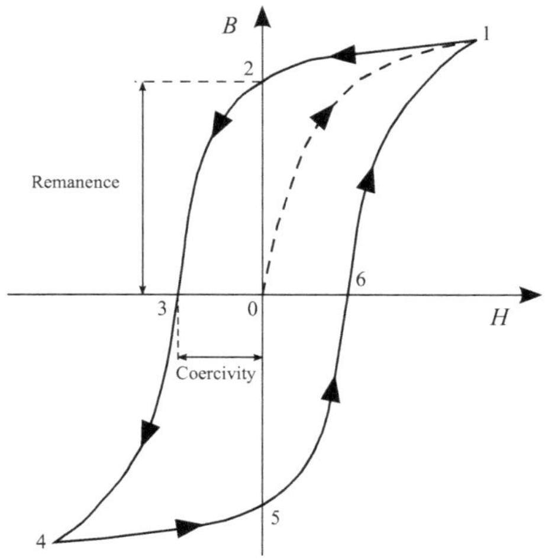

Figure 2.2 Example of a ferromagnetic material B–H curve

This curve is sometimes referred to as the *rising* magnetization curve. If H is now decreased, a different relation between B and H is found, as is shown by the curve 1–2, which lies above the rising curve. A flux density, given by the 1–2 on the plot, remains when H is made zero. This flux density is called the *remanence* or *remanent magnetism*. To reduce the flux density to zero, a magnetizing force 2–3 must be applied in the direction opposite to that of the force formerly applied. For certain conditions of magnetization, as discussed later, this magnetizing force is called the *coercive force*.

As H becomes more negative until it reaches the minimum value, the relation between B and H follows the curve 3–4. Then if H is increased from the minimum value through 0, to the maximum value, the curve follows a path such as 4–5–6–1.

The points do not coincide exactly after the first turns but, after many cycles, the path becomes a closed loop.

So the values of B on the falling curve are greater than those on the rising curve. Thus, the material has the property of tending to oppose a change in the value of the flux density. This property is known as *hysteresis*, which is a term meaning a *lagging behind*. The obtained closed loop is known as a *hysteresis loop*.

When the properties of different magnetic materials are compared, some of the properties are denoted by special terms. Those of major interest are as follows:

Remanence is the flux density, or magnetic induction, which remains in a magnetic material after the removal of an applied magnetizing force.

Coercive force H_c for a magnetic material is the magnitude of the magnetizing force at which the flux density is zero when the material is being symmetrically magnetized.

In many magnetic problems, the history of the material is unknown. Many magnetic calculations are therefore performed using a magnetization curve, called the *normal magnetization curve*, which is obtained by drawing a single-valued curve through the tips of a series of increasingly larger symmetrical hysteresis loops.

Diamagnetic materials
Diamagnetism is present in every material but empty space. The materials oppose the external magnetic field. Generally, other magnetic properties hide it due to their greater effect.

Paramagnetic materials
Paramagnetic materials are nonmagnetic without an applied field, while they are magnetic once the field is applied. The susceptibility is positive but small.

2.2 Permanent magnets

The sustained success of Permanent Magnet (PM) use in industry was leaded by the improved magnetic characteristics in terms of energy, remanence and coercivity, which permitted marked reductions in motor frame size for the same output compared with motors using ferrite (ceramic) magnets. However, ceramic magnets are considerably cheaper. Both ceramic and NdFeB (Neodymium, Iron and Boron) magnets are sensitive to temperature and special care must be taken in design for working temperature above 100°C. For very high temperature applications, alnico or rare-earth cobalt magnets must be used (samarium–cobalt is usable up to 200°C or even more) (Table 2.2).

For lowest cost, ferrite or ceramic magnets are the universal choice. This class of magnet materials has been steadily improved and is now available with remanence of 0.38 T or more and almost straight demagnetization characteristic throughout the second quadrant. The temperature characteristics of ferrite magnets can be tailored to the application requirements so that maximum performance is obtained at the normal operating temperature, which may be as high as 100°C.

Table 2.2 Magnetic, mechanical and thermal properties of magnetic materials [19]

Property	Unit	Alnico 5–7	Ceramic	Sm_2Co_{17}	NdFeB
B_r	T	1.35	0.405	1.06	1.12
$\mu_0 H_c$	T	0.074	0.37	0.94	1.06
$(BH)_{max}$	MGOe	7.5	3.84	26.0	30.0
μ_{rec}		1.9	1.1	1.03	1.1
Specific gravity		7.31	4.8	8.2	7.4
Resistivity	$\mu\Omega$ cm	47	$>1 \cdot 10^4$	86	150
Thermal expansion	10^{-6}/°C	11.3	13	9	3.4
B_r temp. coeff.	%/°C	−0.02	−0.2	−0.025	−0.1
Saturation H	kOe	3.5	14.0	>40	>30

2.2.1 B–H Loop and magnetization characteristic

The starting point for understanding magnet characteristics is the *B–H* loop or "hysteresis loop," report in Figure 2.3(a). The *x*-axis measures the magnetizing force or field intensity H in the material. The *y*-axis is the magnetic flux density in the material. An unmagnetized sample has $B = 0$ and $H = 0$ and therefore starts out at the origin. If it is subjected to a magnetic field, as, for example, in a magnetizing fixture (an electromagnet with specially shaped pole pieces to focus flux into the magnet), B and H in the magnet will follow the curve OA as the external ampere-turns are increased. If the external ampere-turns are switched off, the magnet relaxes along AB. Its operating point (H, B) will depend on the shape of the magnet and the permeance of the surrounding "magnetic circuit." If the magnet is surrounded by a highly permeable magnetic circuit, that is, if it is "keepered," its poles are effectively shorted together so that $H = 0$ and the flux density is then the value at point B, the remanence B_r. The remanence is the maximum flux density that can be retained by the magnet at a specified temperature after being magnetized to saturation. External ampere-turns applied in the opposite direction cause the magnet's operating point to follow the curve from B through the second quadrant to C.

A *hard* PM material is one in which the hysteresis loop is straight throughout the second quadrant, where the magnet normally operates in service. In this case, the recoil line is coincident with the second quadrant section of the hysteresis loop. This is the characteristic of ceramic, rare-earth/cobalt and NdFeB magnets, and the recoil permeability is usually between 1.0 and 1.1. *Soft* PM materials are those with a "knee" in the second quadrant, such as alnico. While alnico magnets have very

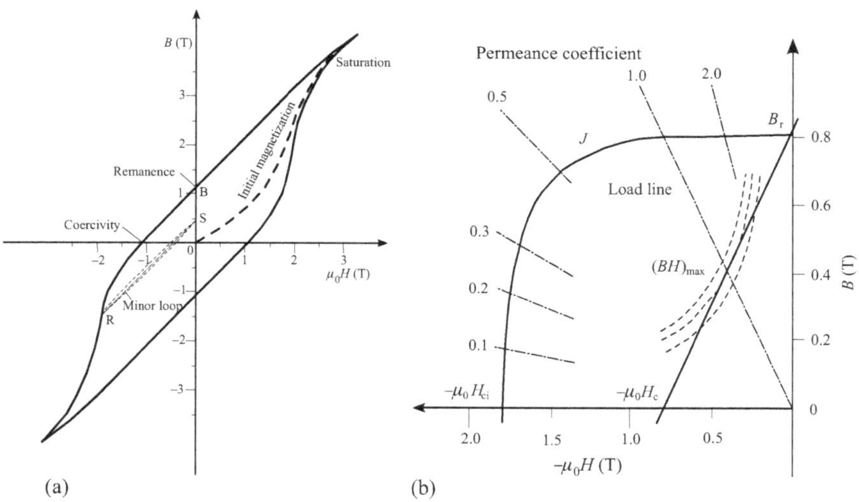

Figure 2.3 Permanent magnet characteristic B–H curves: (a) example of a magnet B–H curve and (b) example of a magnet second quadrant characteristic with load lines

high remanence and excellent mechanical and thermal properties, they are limited in the demagnetizing field they can withstand. It should be noted that compared with lamination steels, even the soft PM materials are very hard: in other words, the hysteresis loop of a typical nonoriented electrical steel is very narrow compared with that of even the alnico magnets.

The most important part of the *B–H* loop is the second quadrant. This is called the demagnetization curve. In the absence of externally applied ampere-turns, the magnet's operating point is at the intersection of the demagnetization curve and the *load line*, slope of which is the product of μ_0 and the permeance coefficient of the external circuit.

Recalling (2.2), the first term is the flux density that would exist if the magnet were removed and the magnetizing force remained at the value H. Therefore the second term can be regarded as the contribution of the magnet to the flux density within its own volume. Clearly, if the demagnetization curve is straight, and if its relative slope and therefore the recoil permeability are both unity, J is constant. In most hard magnets, the recoil permeability is slightly greater than 1 and there is a slight decrease of J as the negative magnetizing force increases, but this is reversible down to the "knee" of the *B–H* loop (which may be in either the second or the third quadrant, depending on the material and its grade).

Evidently, the magnet can recover or recoil back to its original flux density as long as the magnetization is constant. The coercive force required to permanently demagnetize the magnet is called the intrinsic coercivity and this is shown as H_{cJ}.

Another parameter often calculated is the magnet energy product, which is simply the product of B and H in the magnet. This is not the actual stored magnet energy, which depends on the history or trajectory by which the magnet arrived at its operating point and usually cannot be calculated except under very artificial conditions. The energy product is a measure of the stored energy but, more importantly, it gauges how hard the magnet is working to provide flux against the demagnetizing influence of the external circuit.

2.2.2 Temperature effects

2.2.2.1 High-temperature effects

Exposure to sufficiently high temperatures for long enough periods produce metallurgical changes that may impair the ability of the material to be magnetized and may even render it nonmagnetic. There is also a temperature, called the Curie temperature, at which all magnetization is reduced to zero. After a magnet has been raised above the Curie temperature, it can be remagnetized to its prior condition provided that no metallurgical changes have taken place. The temperature at which significant metallurgical changes begin is lower than the Curie temperature in the case of the rare-earth/cobalt magnets, NdFeB and alnico; but in ceramic ferrite magnets, it is the other way round. Therefore, ceramic magnets can be safely demagnetized by heating them just above the Curie point for a short time. This is useful if it is required to demagnetize them for handling or finishing purposes. Table 2.3 shows these temperatures for some of the important magnets used in motors.

Table 2.3 Metallurgical change and Curie temperature

	Metallurgical change [°C]	Curie temperature [°C]
Alnico 5	550	890
Ceramics	1 080	450
$SmCo_5$	300	700
Sm_2Co_{17}	350	800
NdFeB	200	310

Table 2.4 Reversible temperature coefficients [%/°C]

	B_r	H_c	H_{cJ}
Alnico 5–7	−0.02		
Ceramics	−0.19	0.20	
$SmCo_5$	−0.04		−0.25
Sm_2Co_{17}	−0.02		−0.20
NdFeB	−0.11		−0.60

2.2.2.2 Reversible losses

The *B–H* loop changes shape with temperature. Over a limited range, the changes are reversible and approximately linear, so that temperature coefficients for the remanence and coercivity can be used. Table 2.4 gives some typical data.

Ceramic magnets have a positive coefficient of H_c, whereas the high-energy magnets lose coercivity as temperature increases. In ceramic magnets, the knee in the demagnetization curve moves down toward the third quadrant, and the permeance coefficient at the knee decreases. Thus, ceramic magnets become better able to resist demagnetization as the temperature increases up to about 120°C. The greatest risk of demagnetization is at low temperatures when the remanent flux density is high and the coercivity is low; in a motor, this results in the highest short-circuit current when the magnet is least able to resist the demagnetizing ampere-turns.

All the magnets lose remanence as temperature increases. For a working temperature of 50°C above an ambient of 20°C, for instance, a ceramic magnet will have lost about 10%. This is spontaneously recovered as the temperature falls back to ambient.

2.2.3 Simple magnetic circuits

Consider the example of Figure 2.4(a): let the iron be ideal, and let the magnet and the gap have the same section area:

$$S_m = S_g = S$$

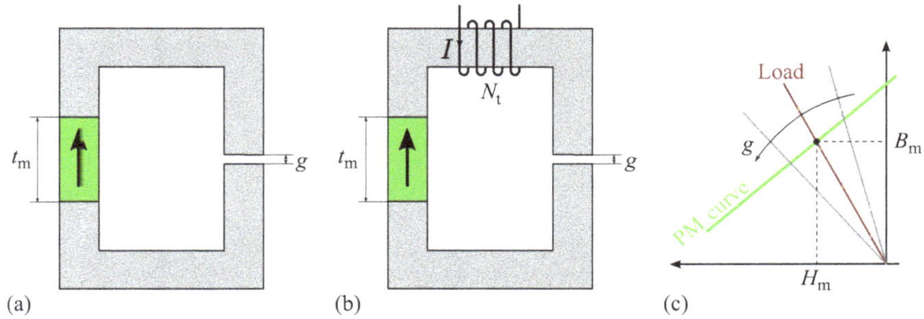

Figure 2.4 Examples of simple magnetic circuits and magnet characteristic: (a) simple magnetic circuit with the inset of a PM, (b) simple magnetic circuit with a PM and a coil, (c) magnet characteristic at load

Then we can write

$$\begin{cases} H_m\, t_m + H_g\, g = 0 & \text{Ampère's Law} \\ B_m\, S_m = B_g\, S_g & \text{Gauss's Law} \\ B_g = \mu_0\, H_g & \\ B_m = B_{rem} + \mu_{rec}\mu_0 H_m & \text{magnet second quadrant characteristic} \end{cases}$$

Substituting the second and the third equation into the first one

$$H_m\, t_m + \frac{B_g}{\mu_0}\, g = 0 \qquad \Longrightarrow \qquad B_m = -\mu_0 \frac{t_m}{g}\, H_m$$

This equation expresses the load line of the magnetic circuit. Moving onto the last equation of the system

$$B_m = B_{rem} - \mu_{rec}\mu_0 \frac{g}{\mu_0 t_m}\, B_m$$

and finally

$$B_m = \frac{B_{rem}}{1 + \mu_{rec} \frac{g}{t_m}} \tag{2.3}$$

So, through this equation, one can see the effect of changing the air gap length or the magnet thickness.

If we add a coil around the ferromagnetic core, like shown in Figure 2.4(b), Ampère's Law becomes

$$H_m\, t_m + H_g\, g = N_t I \qquad \Longrightarrow \qquad B_m = -\mu_0 \frac{t_m}{g}\, H_m + \mu_0 \frac{N_t I}{g}$$

This causes a shift of the load line rightward or leftward, depending on the sign of the current. A high-valued positive current could saturate the iron, while a negative one could demagnetize the magnet. It is important to stay sufficiently far away from the knee of the demagnetization curve.

2.2.4 Finite element simulations of PM

As explained in Section 2.2.1 and shown in Figure 2.3(a), a hard magnet characteristic is almost a straight line in the second quadrant. This feature simplifies their simulation in Finite Element (FE) programs.

Typically, the magnet supplier makes available

- remanence at 20°C
- recoil permeability
- coercive field at 20°C

Furthermore, for simulations at a higher temperature, we also require

- temperature coefficient for B_{rem}
- temperature coefficient for H_{cJ}

The recoil permeability could be obtained through

$$\mu_{\text{rec}} = \frac{B_{\text{rem}}}{\mu_0 |H_{cB}|}$$

Thanks to the temperature coefficients, it is possible to get the residual flux density $B_{\text{rem}}^{@T}$ and the coercive force $H_{cJ}^{@T}$ at the required temperature T (Figure 2.5). Then the value of the coercivity to use for the FE program is

$$H_c^* = \frac{-B_{\text{rem}}^{@T}}{\mu_{\text{rec}}\mu_0}$$

At this point, we have to verify where the knee of the curve is. If it is below the x-axis (which means $|H_c^*| < |H_{cJ}|$), the coercivity H_c^* can be safely used. Otherwise,

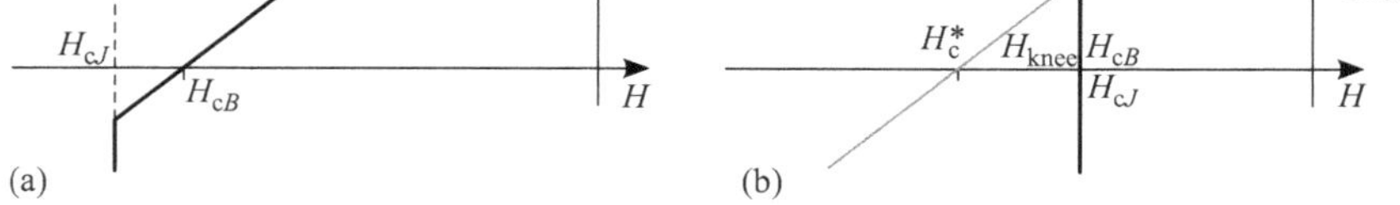

Figure 2.5 Illustration of the effect of temperature on the demagnetization characteristic of hard magnetic material: (a) demagnetization characteristic at standard room temperature and (b) demagnetization characteristic at 120°C

the magnet flux density should be checked in order not to irreversibly damage the magnet. In fact, it has to be true

$$B_{m} > B_{knee}$$

where

$$B_{knee} = B_{rem} + \mu_{rec}\mu_0 H_{knee}$$

2.3 Bonded magnets

The Bonded Magnets (BMs) represent a valid solution since they can easily be shaped or directly injected, at fluid state, into the final volume to fill [20–22]. Figure 2.6(b) shows how the BMs can be effectively used for the realization of a Permanent Magnet assisted Synchronous Reluctance (PMASYR) rotor and highlights the higher amount of PM materials with respect to the classical solution in Figure 2.6(a). The BMs are constituted by a mixture of a magnetic powder and polymeric binder [23]. The magnetic powder may be hard ferrite, NdFeB, SmCo, while the binders depend on the typology of the molding process. The normally used polymeric materials are Polyamide (PA) for injection molding and thermoset resins for compression molding [24]. The BMs can be made in very complex shapes using both molding technologies [25]. Compression molding is suitable for relatively simple geometries and small production volumes compared to injection molding; on the other hand, the tool and mold costs are lower and higher magnetic properties are reached. Finally, the magnetic characteristics of the BMs can be "tuned" through the percentage of the plastic binder. The NdFeB BMs have better magnetic characteristics with respect to ferrites, mainly for the high intrinsic coercivity.

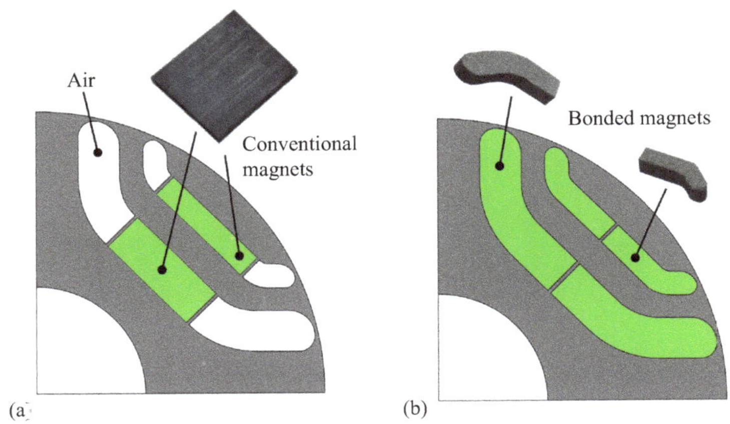

Figure 2.6 Permanent Magnet assisted Synchronous Reluctance (PMASYR) motors with different magnet types: (a) sintered magnets and (b) Bonded Magnets (BMs)

2.4 BMs preparation

2.4.1 Compression molding

In the compression technique, the phenolic resin has been used as polymeric binder and mixed with NdFeB powder. For this process, the proper amount of binder content has to be carefully selected on the basis of the operating pressure and resin typology. In general, it ranges between 0.8% and 5% in weight (wt%). After some mechanical tests carried out in laboratory, a value of 4% has been set to ensure a proper mechanical strength.*

For the BMs mixture, a phenolic resin and MQP 14-12 NdFeB powder have been selected. The latter is a magnetic powder for high-temperature applications (up to 150°C). At first, the obtained magnetic mixture is compressed by a hydraulic press at an operating pressure of 600 MPa (Figure 2.7(a)). Then, the obtained sample is heat treated at 150°C for 45 min. The processed discoid is finally machined (Figure 2.7(b)). to obtain the final samples shown in Figure 2.7(c).

2.4.2 Injection molding

For injection molding, the BMs are prepared using the commercial product MQIP-M260, which is NdFeB compound with polyamide 12 (PA12) in 8 wt% in pellet form as reported in Figure 2.8(a). Such value usually represents the lowest limit for the plastic part because the extrusion process with twin-screw needs a certain viscosity. The pellets are first heated at a temperature of about 275°C to reach the fluid state. Once the mixture is completely fluid, the temperature is reduced to 80°C and kept constant during the injection process. In this study, the magnetic mixture has been injected in shaped molds and not directly in the rotor lamination. The final product is reported in Figure 2.8(b).

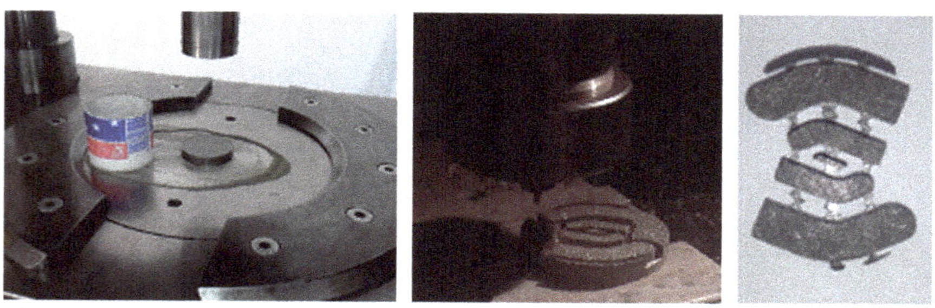

(a) (b) (c)

Figure 2.7 Production steps: (a) discoid sample after the process molding, (b) working process of the compacted disk, (c) final product

*A sample with a polymer content of 5 wt% exhibited structural problems after the thermal treatment.

(a) (b)

Figure 2.8 Magnet pellets and final samples: (a) MQIP-M260 pellets used for the injection molding and (b) final BMs samples

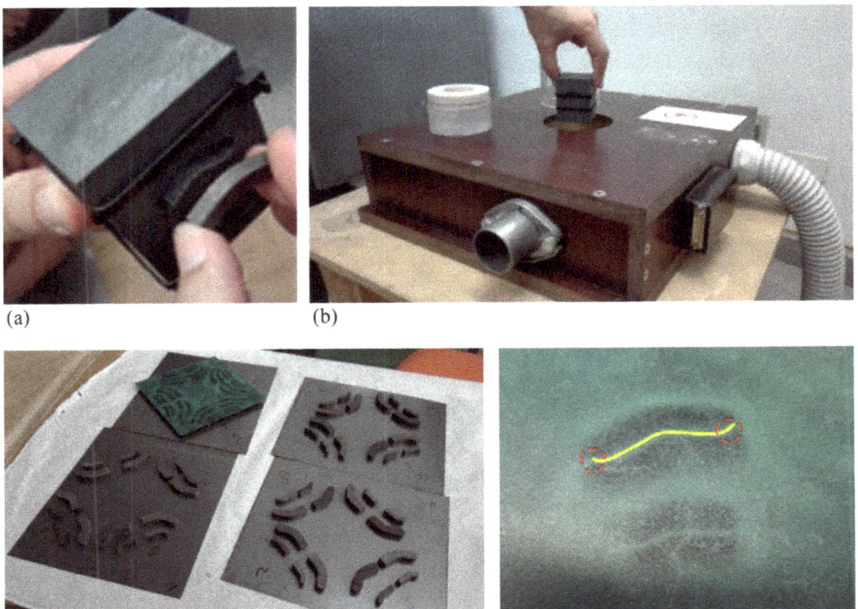

(a) (b) (c) (d)

Figure 2.9 BMs magnetization steps: (a) Step 1, (b) Step 2, (c) Step 3 (d) Step 4. The photos refer to the bonded PMs realized by Emir Poskovic.

2.5 BMs magnetization and characterization

2.5.1 Magnetization

The custom magnet holder in Figure 2.9(a) has been manufactured to keep the BMs pieces in position during the magnetization impulse. Such a magnet holder has been also designed to obtain a magnetization direction as much as possible perpendicular to the flux-barrier edges with the standard magnetizer shown in Figure 2.9(b). The BMs magnetization in situ using the assembled prototype was not possible for the high current (over 50 A) required for the process. This operation has been carried out for all the BMs pieces required to fill the PMASYR rotors. In this step, the magnetized BMs were organized as reported in Figure 2.9(c) in order to keep trace of their polarity. By means of the green magnetic paper test, in Figure 2.9(d), it has been observed a negligible local flux density reduction near the BMs extremities as confirmation of the effective magnet holder design.

2.5.2 Magnetic characterization

Some magnetized BM samples have been characterized by means of a hysteresisgraph as reported in [21]. The resulting averaged demagnetization curves of the compressed and injected BMs are shown in Figure 2.10(a) and (b), respectively. From a comparison between the two graphs, it is evident that the compression molding allows to obtain the highest remanence flux, while the injection molding yields BMs with higher coercive field. Such magnetic properties have been used in FE analysis to estimate the PMASYR machines performance.

2.6 Losses in magnetic materials

The iron losses in the magnetic parts of rotating electrical machines are also referred to as core losses or magnetic losses. They are generated by the changing magnetic field in the stator and in the rotor cores. They are usually divided in two components: hysteresis and eddy currents losses.

It should be kept in mind that the engineering approach of iron loss separation into different loss types (e.g. hysteresis losses and eddy current losses) and related models represents an empirical approach that tries to separate the different physical influences due to frequency and flux density variations in electromagnetic systems into different components. However, it does not explain the real physical phenomena of the iron losses directly. From a physical point of view, both the hysteresis and the eddy current losses in the conducting ferromagnetic materials are based on Joule heating. They are caused by the same physical phenomena: every change in magnetization (which also occurs at DC magnetization) is a movement of magnetic domain walls and creates (microscopic and macroscopic) eddy currents which, in turn, create Joule heating. The fact that hysteresis losses also arise at almost zero frequency is due to the fact that even if the macroscopic magnetization change is very slow, the local magnetization inside the domains is changing rapidly and discrete in time, which generates eddy current losses. Nevertheless, the loss separation shows in most cases good correlation with measurements and has therefore its justification in the engineering science.

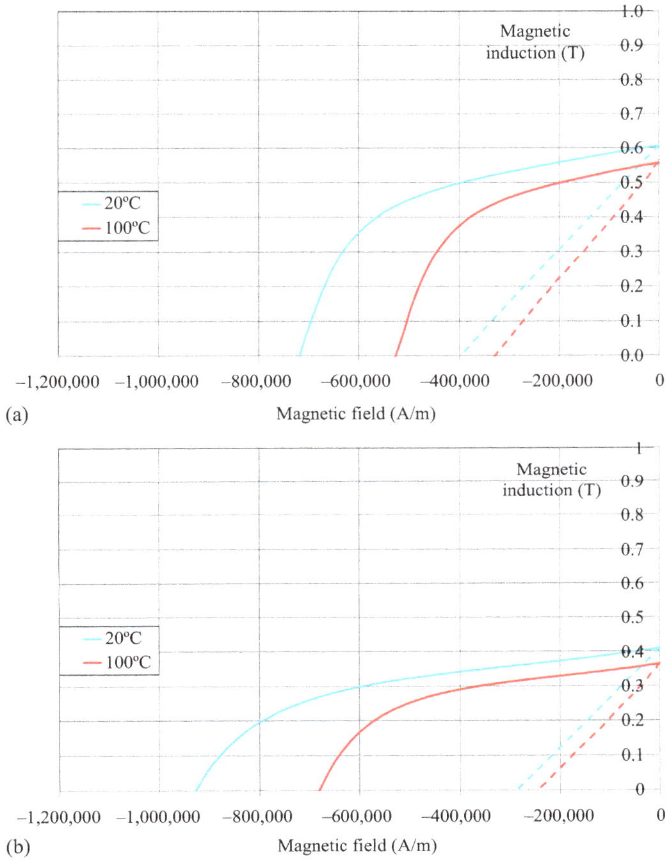

Figure 2.10 Measured demagnetization curve of NdFeB BMs prepared with different processes: (a) compression and (b) moulding

2.6.1 Hysteresis losses

The hysteresis cycle explained in Section 2.1 causes an effective dissipation of energy, called *hysteresis loss*, within the material when cyclic variations of magnetizing force are considered. The area enclosed by the hysteresis loop is equal to the energy loss density (expressed in [J/m^3]). The power loss density can be determined by

$$p_{\mathrm{hy}} = f \oint H(B)\,\mathrm{d}B \quad [\mathrm{W/m^3}]$$

where f is the frequency of the magnetic field H. From now on, the subscript hy denotes the quantity related to hysteresis effect.

2.6.2 Eddy current losses

Eddy currents are created by the induced voltages in the conducting magnetic materials due to the changing magnetic flux, leading to dynamic iron losses. These eddy currents counteract the variations (time and direction) of the magnetic fields. They lead to a broadening of the hysteresis curve and thus increase the magnetic coercivity H_c, as it can be seen in Figure 2.11.

2.6.3 Measuring the iron losses

There are several international standards (e.g. IEC 60404-1, ASTM A677) describing magnetic materials for electrical machines. In these standards, the materials are usually classified by their lamination thickness and iron losses. The typical material M800-50A, e.g. is a non-oriented 0.5-mm-thick lamination sheet with maximum iron losses of 8 W/kg at 1.5 T and 50 Hz sinusoidal flux density excitation. However, these values are only the maximum guaranteed values of the iron losses. Therefore, manufacturers provide often also typical average values in their catalogs, which are closer to the real and expected values, and thus more suitable to use in iron loss calculations.

However, the manufacturing process of the magnetic machine parts is usually the main reason for discrepancies between the data sheet values and measurement results of the final assembled machine (Figure 2.12).

The challenging point about determining the value of iron losses is that it is not possible to measure them directly. Indirect thermal measurements (inverse thermal models and calorimetric measurements), magnetic field measurements (search coil windings, hall sensors, etc.) or loss separation calculations (subtracting all known other losses from the total losses) are the applied possibilities to determine indirectly iron losses in ferromagnetic materials for electrical machines.

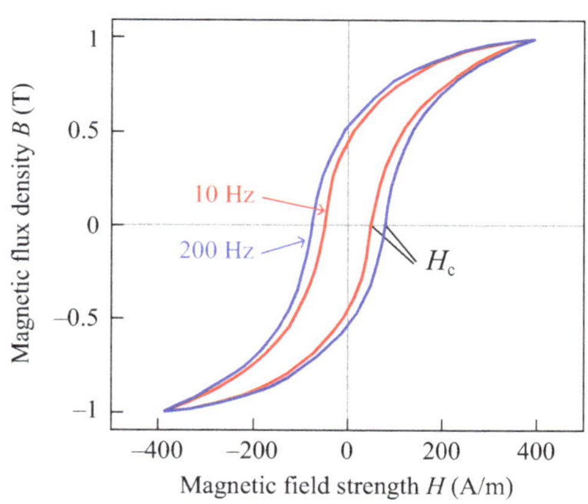

Figure 2.11 B–H hysteresis curve of a ferromagnetic material at 10 Hz (red) and 200 Hz (blue)

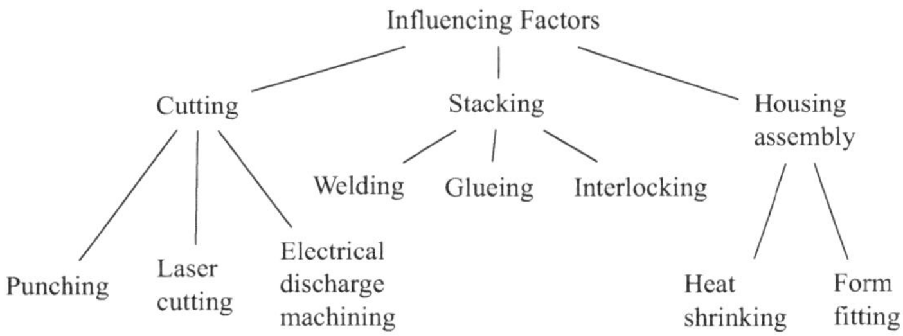

Figure 2.12 Influencing factors of the manufacturing process of electrical machine cores on the iron losses

2.6.3.1 Mathematical formulation

The most used mathematical formulation of iron losses is an extension of the classical Steinmetz equation, made by Jordan. One term depends linearly on the variation of the frequency, the other one depends on the frequency squared. This means that the losses are separated into (static) hysteresis losses and (dynamic) eddy current losses:

$$p_{\mathrm{Fe}} = p_{\mathrm{hy}} + p_{\mathrm{ec}} = c_{\mathrm{hy}} f \hat{B}^2 + c_{\mathrm{ec}} f^2 \hat{B}^2 \tag{2.4}$$

In (2.4), c_{hy} and c_{ec} are the hysteresis and eddy currents coefficient, respectively. In Jordan's approach, it is assumed that the hysteresis losses are proportional to the hysteresis loop area of the material at low frequencies ($f \rightarrow 0$ Hz).

One of the approaches to improve (2.4) is to introduce an additional loss term p_{ex} to take into account the excess losses as a function of the flux density and frequency. It separates the iron loss formula p_{Fe} into three terms, the static hysteresis losses p_{hy}, dynamic eddy current losses p_{ec} and the excess losses p_{ex}:

$$p_{\mathrm{Fe}} = p_{\mathrm{hy}} + p_{\mathrm{ec}} + p_{\mathrm{ex}} = g_{\mathrm{hy}} f \hat{B}^2 + g_{\mathrm{ec}} f^2 \hat{B}^2 + g_{\mathrm{ex}} f^{1.5} \hat{B}^{1.5} \tag{2.5}$$

It has to be noted that the presented loss separation approaches do not hold if the skin effect is not negligible, which means at high frequency (Table 2.5).

Table 2.5 Influence factors on SiFe steel sheets properties

Factor	p_{hy}	p_{ec}	p_{ex}	J_s	H_c
Grain size (+)	−		+		−
Impurities (+)	+			−	+
Thickness (t+)	−	+			−
Internal stress (+)	+				+
Cutting/punching process	+			−	
Pressing process		+			
Welding process	+				
Alloy content (%Si+)		−		−	

2.6.3.2 Determination of the coefficients

The Steinmetz formulation described by (2.4) needs only two coefficients to approximate the iron losses. And since both the terms depend on B^2, we need to change the frequency to obtain the coefficients. So c_{hy} and c_{ec} could be obtained in many ways, the two most common being interpolation through two points and least squares method. Here, the former is described. The data points may be obtained from the material datasheet, or through experimental measurements on a sample of the ferromagnetic material.

Let us pick two pairs of flux density and frequency $(B_i, f_i)_{i=\{1,2\}}$, and their respective power loss density p_i. The following system can be written:

$$\begin{cases} p_1 = c_{hy} B_1^2 f_1 + c_{ec} B_1^2 f_1^2 \\ p_2 = c_{hy} B_2^2 f_2 + c_{ec} B_2^2 f_2^2 \end{cases} \tag{2.6}$$

Since all the terms but c_{hy}, c_{ec} are known, this system can be solved directly:

$$c_{hy} = \frac{p_1}{B_1^2 f_1} - c_{ec} f_1 \qquad [\mathrm{W/(kgT^2Hz)}]$$

$$p_2 = p_1 \frac{B_2^2 f_2}{B_1^2 f_1} - c_{ec} B_2^2 f_1 f_2 + c_{ec} B_2^2 f_2^2$$

$$c_{ec} = \frac{p_2 - p_1 \frac{B_2^2 f_2}{B_1^2 f_1}}{B_2^2 f_2 (f_2 - f_1)} = \frac{p_2 B_1^2 f_1 - p_1 B_2^2 f_2}{B_1^2 f_1 B_2^2 f_2 (f_2 - f_1)} \qquad [\mathrm{W/(kgT^2Hz^2)}]$$

Alternatively, Bertotti's formulation, shown in (2.5), could be used. In this equation, g_{ec} is computed directly from the material properties, according to†

$$g_{ec} = \frac{\pi^2 t^2}{6 \rho \gamma} \qquad [\mathrm{W/(kgT^2Hz^2)}] \tag{2.7}$$

where t is the thickness of the lamination material, ρ is the material electrical resistivity and γ is the material density. After that, only the two coefficients g_{hy} and g_{ex} are unknown, so a system of two equations is needed, similarly to before. That is:

$$\begin{aligned} p_1 &= g_{hy} B_1^2 f_1 + g_{ec} B_1^2 f_1^2 + g_{ex} B_1^{1.5} f_1^{1.5} \\ p_2 &= g_{hy} B_2^2 f_2 + g_{ec} B_2^2 f_2^2 + g_{ex} B_2^{1.5} f_2^{1.5} \end{aligned} \tag{2.8}$$

†For the derivation of this coefficient, see Appendix A.

And so

$$g_{\text{ex}} = \frac{p_1 - g_{\text{hy}} B_1^2 f_1 - g_{\text{ec}} B_1^2 f_1^2}{B_1^{1.5} f_1^{1.5}} \qquad [\text{W}/(\text{kgT}^{1.5}\text{Hz}^{1.5})]$$

$$p_2 = g_{\text{hy}} B_2^2 f_2 + g_{\text{ec}} B_2^2 f_2^2 + \frac{p_1 - g_{\text{hy}} B_1^2 f_1 - g_{\text{ec}} B_1^2 f_1^2}{B_1^{1.5} f_1^{1.5}} B_2^{1.5} f_2^{1.5}$$

$$\frac{p_2}{B_2^2 f_2} = g_{\text{hy}} + g_{\text{ec}} f_2 + p_1 \frac{B_2^{-0.5} f_2^{0.5}}{B_1^{1.5} f_1^{1.5}} - g_{\text{hy}} \frac{B_1^2 f_1 B_2^{-0.5} f_2^{0.5}}{B_1^{1.5} f_1^{1.5}} - g_{\text{ec}} \frac{B_1^2 f_1^2 B_2^{-0.5} f_2^{0.5}}{B_1^{1.5} f_1^{1.5}}$$

With some more algebraic steps

$$g_{\text{hy}} \left(1 - \frac{B_1^2 f_1 B_2^{-0.5} f_2^{0.5}}{B_1^{1.5} f_1^{1.5}} \right) = \frac{p_2}{B_2^2 f_2} - p_1 \frac{B_2^{-0.5} f_2^{0.5}}{B_1^{1.5} f_1^{1.5}} - g_{\text{ec}} f_2 + g_{\text{ec}} \frac{B_1^2 f_1^2 B_2^{-0.5} f_2^{0.5}}{B_1^{1.5} f_1^{1.5}}$$

$$g_{\text{hy}} = \frac{\frac{p_2}{B_2^2 f_2} - p_1 \frac{B_2^{-0.5} f_2^{0.5}}{B_1^{1.5} f_1^{1.5}} + g_{\text{ec}} \left(\frac{B_1^2 f_1^2 B_2^{-0.5} f_2^{0.5}}{B_1^{1.5} f_1^{1.5}} - f_2 \right)}{1 - \frac{B_1^2 f_1 B_2^{-0.5} f_2^{0.5}}{B_1^{1.5} f_1^{1.5}}} \qquad [\text{W}/(\text{kgT}^2\text{Hz})]$$

Chapter 3
Synchronous reluctance motor geometry drawing

This chapter reports the design rules adopted to draw different Synchronous Reluctance (SYR) rotor geometries.

At first, the *fluid* and *segment* shaped flux-barriers rotor drawings are described. Then, a design methodology to choose the proper SYR rotor parameters is presented.

3.1 Fluid flux-barrier rotor geometry

The criteria to draw the fluid flux-barriers are to align their edges to determined flux lines that would appear in the case of a d-axis magnetization of a solid rotor geometry [26]. The concept is clearly explained in Figure 3.1 that shows how the flux-barrier edges overlap the flux lines obtained with a solid rotor. This design strategy allows one to minimize the d-axis flux lines distortion and effectively block the q-axis ones with beneficial effect on the rotor anisotropy.

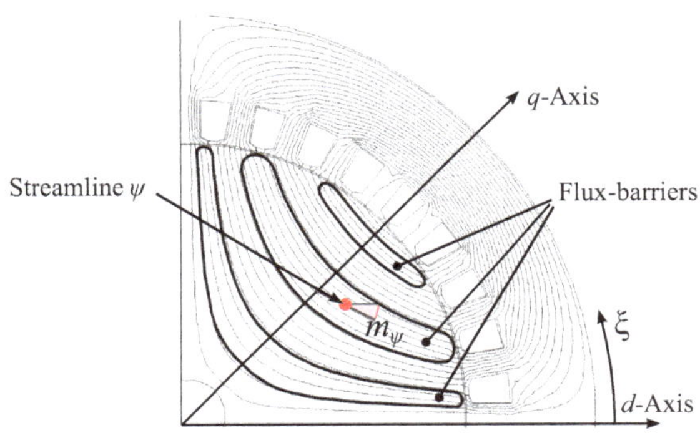

Figure 3.1 d-Axis flux lines and relative flux-barriers geometry

3.1.1 Flow past a cylinder

Let ρ_0 be the radius of the cylinder, (ρ, ξ) the polar coordinate system in use (Figure 3.2). One possible solution of this problem has these potential and streamline functions:

$$\phi(\rho, \xi) = \left(\rho + \frac{\rho_0^2}{\rho}\right) \cos \xi$$

$$\psi(\rho, \xi) = \left(\rho - \frac{\rho_0^2}{\rho}\right) \sin \xi$$

Although these equations are deeply coupled, the radius ρ and the phase ξ can be obtained as a function of the other quantities. For our purposes, we use ψ:

$$\rho(\psi, \xi) = \frac{\psi + \sqrt{\psi^2 + 4\rho_0^2 \sin^2 \xi}}{2 \sin \xi}$$

$$\xi(\psi, \rho) = \arcsin\left(\frac{\rho\,\psi}{\rho^2 - \rho_0^2}\right)$$

The velocity field can also be derived through

$$v_\rho(\rho, \xi) = \frac{\partial \phi}{\partial \rho} = \left(1 - \frac{\rho_0^2}{\rho^2}\right) \cos \xi$$

$$v_\xi(\rho, \xi) = \frac{1}{\rho}\frac{\partial \phi}{\partial \xi} = -\left(1 + \frac{\rho_0^2}{\rho^2}\right) \sin \xi$$

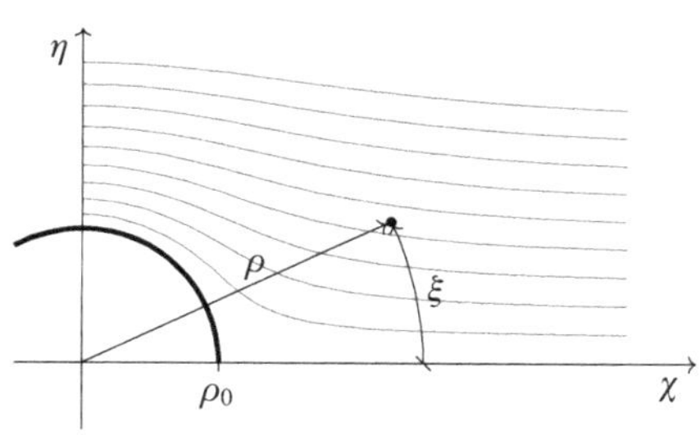

Figure 3.2 Sketch of fluid flow past a cylinder

3.1.2 Conformal mapping

From the reference plane, which is equivalent to a two-pole machine, we use a complex map to obtain the quantities in the actual plane. Let p be the number of pole pairs. Then,

$$\begin{aligned} \zeta &\xrightarrow{\mathscr{M}} z = \sqrt[p]{\zeta} \\ \rho\, e^{j\xi} &\xrightarrow{\mathscr{M}} r\, e^{j\vartheta} = \sqrt[p]{\rho}\, e^{j\xi/p} \\ \chi + j\eta &\xrightarrow{\mathscr{M}} x + jy \end{aligned}$$

It is easy to find the inverse map as

$$\mathscr{M} : \sqrt[p]{\cdot} \qquad \mathscr{M}^{-1} : (.)^p$$

In the transformed plane, the velocities have a different expression as follows:

$$\begin{aligned} v_r(r,\vartheta) &= \frac{\partial \phi}{\partial r} = \frac{\partial \phi}{\partial \rho}\frac{\partial \rho}{\partial r} = p\left(r^{p-1} - \frac{R_0^{2p}}{r^{p+1}}\right)\cos p\vartheta \\ v_\vartheta(r,\vartheta) &= \frac{1}{r}\frac{\partial \phi}{\partial \theta} = \frac{1}{r}\frac{\partial \phi}{\partial \rho}\frac{\partial \rho}{\partial \theta} = -p\left(r^{p-1} + \frac{R_0^{2p}}{r^{p+1}}\right)\sin p\vartheta \end{aligned} \tag{3.1}$$

This vector field is tangent to the streamlines in every point in the transformed plane. In order to work with this field in (x, y) coordinates, we need a rotational map as follows:

$$\begin{aligned} v_x(r,\vartheta) &= v_r \cos\vartheta - v_\vartheta \sin\vartheta \\ v_y(r,\vartheta) &= v_r \sin\vartheta + v_\vartheta \cos\vartheta \end{aligned} \tag{3.2}$$

3.1.3 Computation of flux-barrier base points

Refer to Figure 3.3 for the points naming scheme. Keep in mind that A′ is not simply the projection of A onto the q-axis, but it represents the original starting point for the barrier sideline, so it lies on the flux-barrier streamline. The same is true for points B′, B, C′, C, and D′, D.

Let the flux-barrier and flux-carrier thicknesses and widths be given. Then the base points for the flux-barriers can be computed easily. Let D_r be the rotor outer

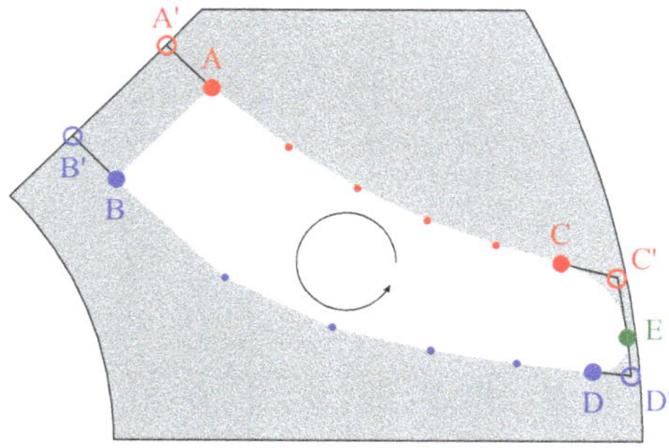

Figure 3.3 Flux-barrier base points description

diameter, $w_{\mathrm{rib,t}}$ the tangential iron rib width, w_{ck} the kth flux-carrier width, and t_{bk} the kth flux-barrier thickness.* Then

$$\begin{aligned} R_{\mathrm{rib}} &= \frac{D_{\mathrm{r}}}{2} - w_{\mathrm{rib,t}} \\ R_{\mathsf{A}'_1} &= R_{\mathrm{rib}} - w_{\mathrm{c}1} \\ R_{\mathsf{B}'_1} &= R_{\mathsf{A}'_1} - t_{\mathrm{b}1} \\ &\vdots \end{aligned}$$

where R represents the radius from the origin. So, in general,

$$\begin{aligned} R_{\mathsf{A}'_k} &= R_{\mathsf{B}'_{k-1}} - w_{ck} \\ R_{\mathsf{B}'_k} &= R_{\mathsf{A}'_{k-1}} - t_{bk} \end{aligned}$$

with the exception $R_{\mathsf{B}'_0} = R_{\mathrm{rib}}$.

Now we know both the radii and the angle—always $\pi/(2p)$—of the flux-barrier internal points. So we can compute their respective streamline value.

3.1.3.1 Magnet insertion

$$w_{\mathrm{rib},k} \leftarrow w_{\mathrm{rib},k} + w_{\mathrm{pm},k}$$

where $w_{\mathrm{pm},k}$ is the kth magnet width.

3.1.3.2 Central base points

We refer to points A and B. If the rib width is zero $\mathsf{A} \equiv \mathsf{A}'$ and $\mathsf{B} \equiv \mathsf{B}'$.

*The main dimensions of the flux-carrier and flux-barrier differ in the names (width versus thickness). This is because we refer to width when the flux flows perpendicularly to the dimension, and to thickness when it flows in parallel.

The line describing the q-axis is

$$y = mx + q$$
$$m = \tan \frac{\pi}{2p}$$
$$q = \frac{w_{\text{rib}}}{2\cos\frac{\pi}{2p}}$$

and the following system can be set up to find point A

$$\begin{cases} y_{\mathsf{A}} - mx_{\mathsf{A}} - q = 0 \\ x_{\mathsf{A}} - r_{\mathsf{A}}(\psi_{\mathsf{A}'}, \vartheta_{\mathsf{A}}) \cos \vartheta_{\mathsf{A}} = 0 \\ y_{\mathsf{A}} - r_{\mathsf{A}}(\psi_{\mathsf{A}'}, \vartheta_{\mathsf{A}}) \sin \vartheta_{\mathsf{A}} = 0 \end{cases}$$

where ϑ_{A} is used as the third degree of freedom and r_{A} is then a function of it. The solution of such system can be determined solving the single equation

$$r_{\mathsf{A}}(\psi_{\mathsf{A}'}, \vartheta_{\mathsf{A}})(\sin \vartheta_{\mathsf{A}} - m \cos \vartheta_{\mathsf{A}}) - q = 0$$

in the unknown ϑ_{A}. The function $r(\psi, \vartheta)$ is simply

$$r(\psi, \vartheta) = \sqrt[p]{\rho(\psi, \vartheta/p)}$$

The same equation can be written for point B with the proper substitutions and repeated for all the flux-barriers (Figure 3.4).

3.1.4 Outer base points

We refer to points C, D and E of Figure 3.3. If the flux-barrier angle, ϑ_{b}, is given, then

$$x_{\mathsf{E}} = R_{\text{rib}} \cos\left(\tfrac{\pi}{2p} - \vartheta_{\text{b}}\right)$$
$$y_{\mathsf{E}} = R_{\text{rib}} \sin\left(\tfrac{\pi}{2p} - \vartheta_{\text{b}}\right)$$

Points C and D result from the connection of the flux-barrier sidelines and point E. This connection should be as smooth as possible in order to avoid dangerous mechanical stress concentrations. We are going to use circular arcs to make this connection. So we impose the tangency between the flux-barrier sideline and the arc, between

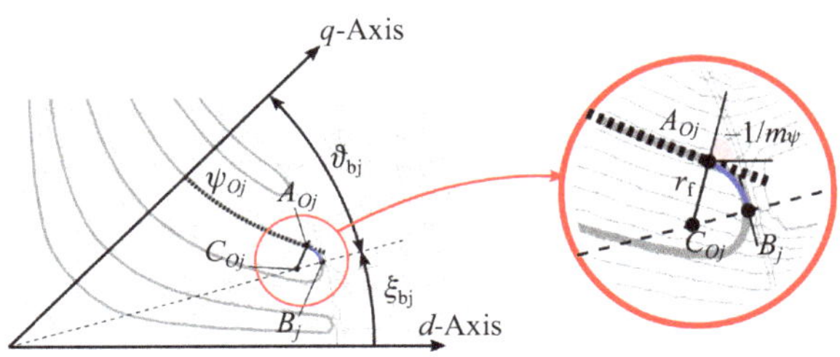

Figure 3.4 Computation of the fillet point of the jth flux-barrier outer edge

the arc and the radius through point E. The tangent to the sideline can be obtained through the velocity field described with (3.1) and (3.2).

Then we want point C to lay on the flux-barrier sideline. These conditions represent a nonlinear system of four equations, in six unknowns. So we need two more equations, which are that points C and E belong to the fillet circle with radius $R^{\dagger}$.

$$\begin{cases} x_C - r_C(\psi_C, \vartheta_C)\cos\vartheta_C = 0 \\ y_C - r_C(\psi_C, \vartheta_C)\sin\vartheta_C = 0 \\ (x_C - x_{O_C})^2 + (y_C - y_{O_C})^2 - R_{EC}^2 = 0 \\ (x_E - x_{O_C})^2 + (y_E - y_{O_C})^2 - R_{EC}^2 = 0 \\ (x_{O_C} - x_E)y_E - (y_{O_C} - y_E)x_E = 0 \\ (x_{O_C} - x_C)v_x(r_C, \vartheta_C) + (y_{O_C} - y_C)v_y(r_C, \vartheta_C) = 0 \end{cases}$$

The very same system can be written and solved for point D.

3.1.4.1 Choice of initial position

For the good convergence of the nonlinear system, we have to choose a proper initial position for the points of interest, namely C and O_C for the top part of the flux-barrier.

Since point C should be close to E and C′, a good initial guess could be

$$x_{C^{(0)}} = \frac{x_E + x_{C'}}{2}, \qquad y_{C^{(0)}} = \frac{y_E + y_{C'}}{2}$$

A slightly better guess shifts the points a bit to the left, in this way:

$$x_{C^{(0)}} = \frac{x_E + x_{C'} + 0.1x_A}{2.1}, \qquad y_{C^{(0)}} = \frac{y_E + y_{C'}}{2}$$

On the other hand, point O_C lies on one edge of the triangle of vertices E, C, O, where O represents the origin and where we are going to use $C^{(0)}$ instead of C because it is still unknown. Then,

$$x_{O_C}^{(0)} = \frac{x_E + x_{C^{(0)}} + 0}{3}, \qquad y_{O_C}^{(0)} = \frac{y_E + y_{C^{(0)}} + 0}{3}$$

Similar considerations can be made for point D, with slight changes:

$$x_{D^{(0)}} = \frac{x_E + x_{D'}}{2}, \qquad y_{D^{(0)}} = \frac{y_E + y_{D'}}{2}$$

$$x_{O_D}^{(0)} = \frac{x_E + x_{D^{(0)}} + x_C}{3}, \qquad y_{O_D}^{(0)} = \frac{y_E + y_{D^{(0)}} + x_C}{3}$$

Notice that here we use point C that has already been found.

†Actually a system of five equations and five unknowns may be written but the simpler solution was preferred.

3.1.5 Flux-barrier sideline points

Let us consider the top flux-barrier sideline, so the one going from point A to point C. We want to create such sideline using a predetermined number of steps, N_{step}. From now on, let us call this N_k for the k-th flux-barrier.

One of the best way to distribute the points along the streamline is to use the potential function, ϕ, defined in Section 3.1.1. We start computing the potential for points A and C:

$$\phi_{\mathsf{A}} = \phi(\rho_{\mathsf{A}}, \xi_{\mathsf{A}})$$
$$\phi_{\mathsf{C}} = \phi(\rho_{\mathsf{C}}, \xi_{\mathsf{C}})$$

Then, we want to find $N - 1$ points along the streamline between points A and C with a uniform distribution of the potential function. We define

$$\Delta\phi_{\mathsf{AC}} = \frac{\phi_{\mathsf{C}} - \phi_{\mathsf{A}}}{N}$$

So we can compute the potentials we are looking for

$$\phi_i = \phi_{\mathsf{A}} + i\Delta\phi_{\mathsf{AC}}\,, \quad i = 1, \ldots, N-1$$

and finally the location of the point with this potential value and the streamline function value required to lie on the flux-barrier sideline. This translates to the following system of equations:

$$\begin{cases} \psi_{\mathsf{AC}} - \psi(\rho, \xi) = 0 \\ \phi_i - \phi(\rho, \xi) = 0 \end{cases}$$

The system is well defined because there are two unknowns and two independent equations. This system must be solved for every flux-barrier sideline point, for the two sides and for every flux-barrier.

3.1.6 Examples

Figure 3.5 sketches four different fluid flux-barriers rotor geometries obtained with the drawing procedure. Figure 3.5(a) and (b) presents the same flux-barrier inner

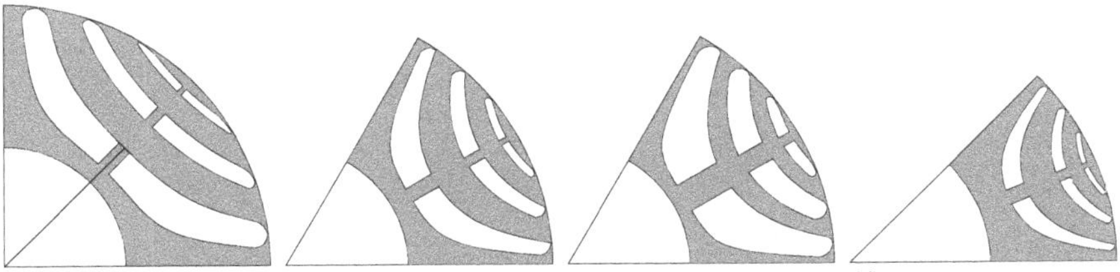

(a) (b) (c) (d)

Figure 3.5 Fluid flux-barrier rotors with different number of poles, flux-barrier geometry and radial ribs size: (a) fluid four-pole, (b) fluid six-pole, (c) fluid six-pole $k_{air} = 0.6$, (d) fluid eight-pole

and outer middle radii but different number of poles. Figure 3.5(b) and (c) has the same number of poles but the second one presents larger radial ribs and thicker flux-barriers. Figure 3.5(d) shows an eight-pole rotor with a bigger rotor shaft diameter and thin flux-barriers.

3.2 Segmented flux-barrier rotor

The segmented flux-barriers rotor presents a simple geometry and, for this reason, it is often preferred to the fluid flux-barriers one even though it yields a lower rotor anisotropy. A further advantage of this geometry is that it allows one to achieve a high flux-barriers filling with the common sintered Permanent Magnet (PM) without affecting the rotor geometry.

A sketch of a segmented flux-barriers rotor is shown in Figure 3.6. As can be noted, the simplicity of the present rotor geometry is due to the fact that, except the fillets, the flux-barrier edges are mainly straight lines. In particular, all the lateral flux-barrier sides are parallel to the d-axis. Moreover, the flux-barrier thicknesses are constant while those of the iron channel are not. It is worth noticing that an alternative rotor geometry with constant iron channel thickness and variable flux-barrier thicknesses can be drawn as well. However, it is not here described hereafter since it has not been used in the following studies.

3.2.1 *Flux-barrier edges identification*

The computation of the jth flux-barrier starts from the points B_j, O_j and I_j reported in Figure 3.7(a). The first one is given by

$$\begin{cases} x_{Bj} = -\left(\frac{D_r}{2} - w_{\text{rib,t}}\right) \sin \vartheta_{bj} \\ y_{Bj} = \left(\frac{D_r}{2} - w_{\text{rib,t}}\right) \cos \vartheta_{bj} \end{cases} \tag{3.3}$$

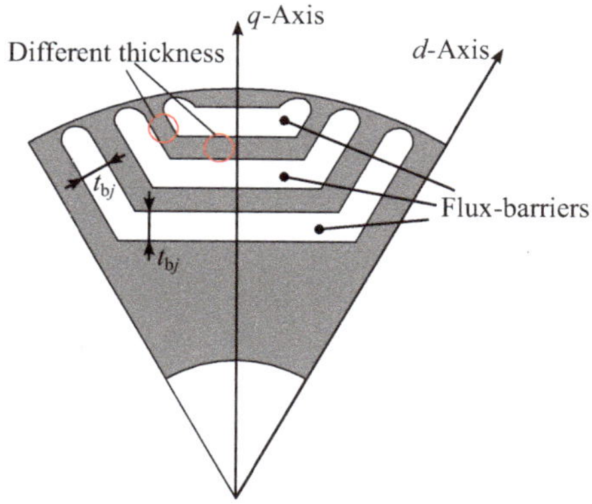

Figure 3.6 Sketch of segmented flux-barriers rotor

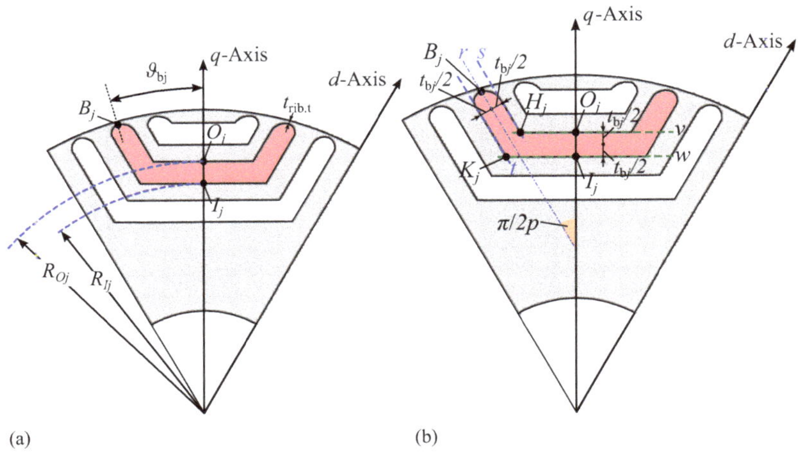

Figure 3.7 Identification of the jth flux-barrier edges: (a) Depiction of the main flux-barrier parameters and (b) Depiction of the auxiliary flux-barrier parameters

the slope of line which connects B_j to the origin is

$$m_{Bj} = \tan \frac{y_{Bj}}{x_{Bj}} \tag{3.4}$$

The points O_j and I_j lie on the vertical q-axis at height R_{Oj} and R_{Ij}, respectively.

Then the straight lines r, s, t, v and w, reported in Figure 3.7(b), are determined. The slope and intercept of the first one are

$$m_r = \tan\left(\frac{\pi}{2p} + \frac{\pi}{2}\right)$$
$$q_r = y_{Bj} - m_r x_{Bj}$$

the straight lines s and t are distant from r of $+t_{bj}/2$ and $-t_{bj}/2$, respectively. Their intercepts are given by

$$q_s = q_r + \frac{t_{bj}}{2\cos\frac{\pi}{2p}}$$
$$q_t = q_r - \frac{t_{bj}}{2\cos\frac{\pi}{2p}}$$

The straight lines v and w are simply defined by their intercepts that are $q_v = R_{Oj}$ and $q_w = R_{Ij}$, respectively. Finally, the flux-barrier edge equations are

$$\begin{aligned} s: &\quad y = m_r x + q_s \\ t: &\quad y = m_r x + q_t \\ v: &\quad y = q_v \\ w: &\quad y = q_w \end{aligned}$$

The point coordinates of H_j, intersection between s and v, are

$$\begin{cases} x_{Hj} = \frac{q_v - q_s}{m_r} \\ y_{Hj} = q_v \end{cases} \tag{3.5}$$

similarly, the point K_j is given by

$$\begin{cases} x_{Kj} = \frac{q_w - q_t}{m_r} \\ y_{Kj} = q_w \end{cases} \tag{3.6}$$

3.2.2 *Flux-barrier end fillet*

The procedure to compute a circular fillet that joints the tangentially the flux-barrier edges to the point B_j is similar to the one described in Section 3.1.4. Regarding the fillet of the jth outer flux-barrier edge reported in Figure 3.8, the following conditions have to be honored:

$$y_{CO_j} = \tan\left(\frac{\pi}{2} + \vartheta_{bj}\right) x_{CO_j} \tag{3.7}$$

$$y_{CO_j} = -\frac{1}{m_r} x_{CO_j} + q \tag{3.8}$$

$$y_{A_j} = -\frac{1}{m_r} x_{A_j} + q \tag{3.9}$$

$$y_{Aj} = m_r x_{Aj} + q_{Aj} \tag{3.10}$$

$$\left(x_{Aj} - x_{C_{Oj}}\right)^2 + \left(y_{Aj} - y_{C_{Oj}}\right)^2 - r_f^2 = 0 \tag{3.11}$$

$$\left(x_{Bj} - x_{C_{Oj}}\right)^2 + \left(y_{Bj} - y_{C_{Oj}}\right)^2 - r_f^2 = 0 \tag{3.12}$$

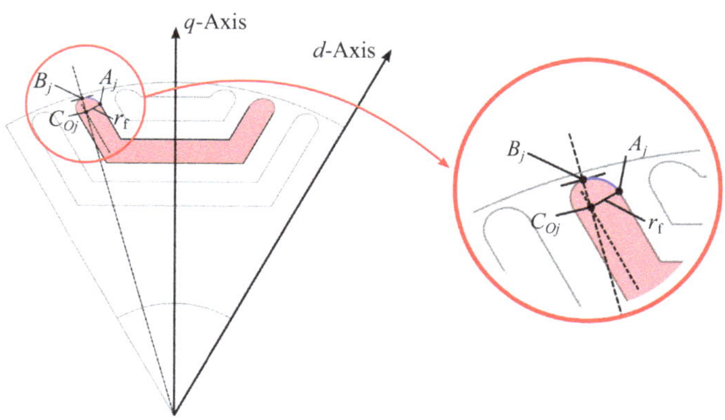

Figure 3.8 Computation of the fillet in a segmented flux-barrier rotor

Equation (3.7) fixes the point x_{CO_j} on the straight line that connects the origin to point B_j. Equations (3.8) and (3.9) impose that the point C_{Oj} must lie on the straight line, perpendicular to r, that pass honored the point A_j. Equation (3.10) fixes the point A_j on the straight line r. Equations (3.11) and (3.12) set the same distance $\overline{AC_j}$ and $\overline{BC_j}$ to be equal. Unlike the flux-barrier geometry, the system can be solved directly. After some equation manipulation and substitutions, it results in

$$\begin{aligned}(1 + 2k_3 + m_r^2 + 2k_5 m_r)x_{Aj}^2 \\ + (-2k_2 + 2m_r q_r - 2k_4 m_r + 2k_5 q_r - 2x_{Bj}k3 - 2y_{Bj}k_5)x_{Aj} \\ + (q_r^2 - 2q_r k_4 - x_{Bj}^2 + 2x_{Bj}k2 - y_{Bj}^2 + 2y_{Bj}k_4) = 0 \end{aligned} \tag{3.13}$$

where

$$\begin{cases} k_1 = m_r + \frac{1}{m_r} \\ k_2 = \dfrac{q_r}{\frac{1}{m_r} + m_{B_j}} \\ k_3 = -\frac{k_1}{\frac{1}{m_r} + m_{B_j}} \\ k_4 = k_2 m_{Bj} \\ k_5 = k_3 m_{Bj} \end{cases} \tag{3.14}$$

Equation (3.13) provides two solutions, but only the one that meets the following constraint is selected:

$$\sqrt{x_{Aj}^2 + y_{Aj}^2} < \frac{D_\mathrm{r}}{2} \tag{3.15}$$

which means that the joint point must be within the rotor diameter. The other joint point, D_j, on the inner flux-barrier edge can be found with the same computation with the obvious substitutions.

3.2.3 Radial ribs

As shown in Figure 3.9, the computation of the radial rib points is quite simple. In particular, their coordinates are

$$\begin{cases} x_{Mj} = -t_{\mathrm{rib},j}/2 \\ y_{Mj} = R_{Oj} \end{cases} \quad \text{and} \quad \begin{cases} x_{Nj} = -t_{\mathrm{rib},j}/2 \\ y_{Nj} = R_{Ij} \end{cases} \tag{3.16}$$

3.2.4 Middle PM slot

A PM rectangular slot can be also added in the middle of the flux-barrier as reported in Figure 3.10. In this case, the radial rib points are as follows:

$$\begin{cases} x_{Mj} = -t_{\mathrm{rib},j} - w_{\mathrm{pm}j}/2 \\ y_{Mj} = R_{Oj} \end{cases}, \quad \begin{cases} x_{Nj} = -t_{\mathrm{rib},j}/2 - w_{\mathrm{pm}j}/2 \\ y_{Nj} = R_{Ij} \end{cases} \tag{3.17}$$

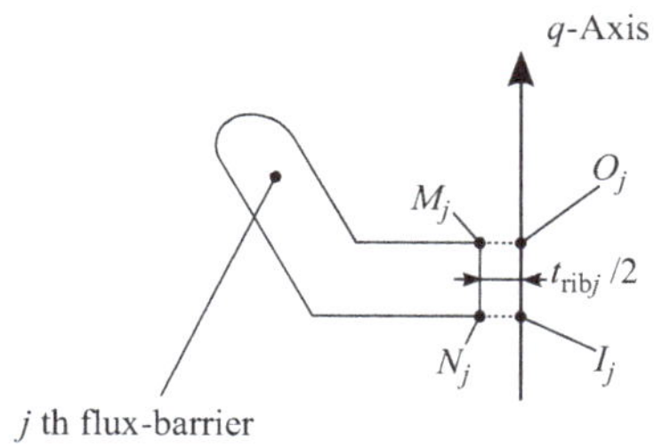

Figure 3.9 Computation of the radial rib nodes of the jth flux-barrier

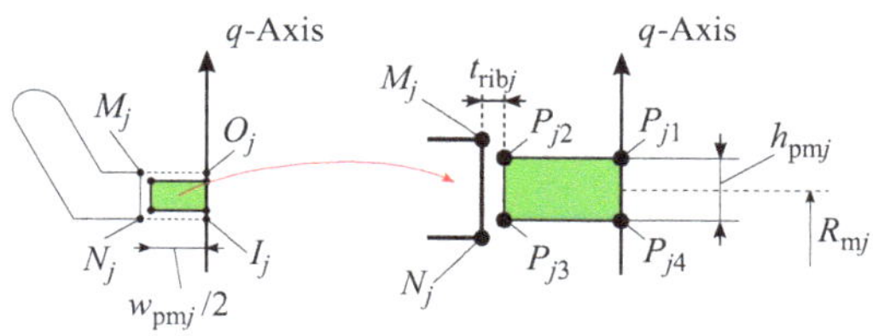

Figure 3.10 Computation of the PM nodes of the jth flux-barrier

while the PM points are as follows:

$$P_{j1}: \begin{cases} x_{Pj1} = 0 \\ y_{Pj1} = R_{mj} + t_{pmj}/2 \end{cases}, \quad P_{j2}: \begin{cases} x_{Pj2} = -w_{pmj}/2 \\ y_{Pj2} = R_{mj} + t_{pmj}/2 \end{cases}$$

$$P_{j3}: \begin{cases} x_{Pj3} = -w_{pmj}/2 \\ y_{Pj3} = R_{mj} - t_{pmj}/2 \end{cases}, \quad P_{j4}: \begin{cases} x_{Pj4} = 0 \\ y_{Pj4} = R_{mj} - t_{pmj}/2 \end{cases}$$

where $R_{mj} = (R_{Oj} + R_{Ij})/2$.

3.2.5 Flux-barrier edges fillet

The flux-barrier edges can be finally smoothed with a fillet, of radius R_{fbj}, between the horizontal and inclined edges as reported in Figure 3.11. In order to keep the flux-barrier thickness constant along the fillet, the same center is adopted for the outer flux-barrier fillet.

At first, the center of the arc fillet can be computed as follows:

$$\begin{cases} y_{C_{fbj}} = R_{Oj} + R_{fbj} \\ y_{C_{fbj}} = m_r x_{C_{fbj}} + q_s + \dfrac{R_{fbj}}{\cos \frac{\pi}{2p}} = m_r x_{C_{fbj}} + q_{C_{fbj}} \end{cases} \tag{3.18}$$

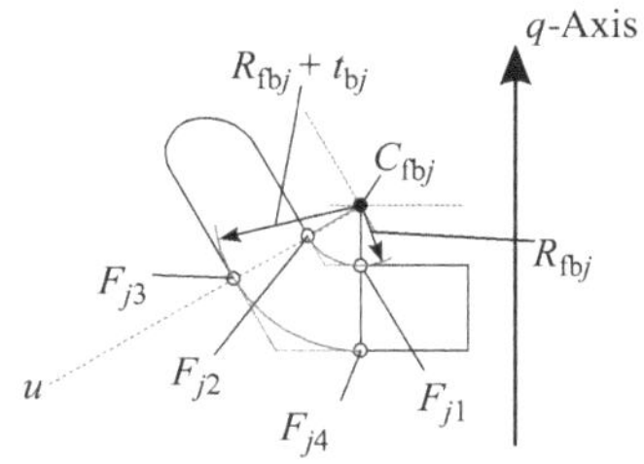

Figure 3.11 Computation of the PM nodes of the jth flux-barrier

then, the points F_{j1} and F_{j4} are easily determined by

$$F_{j1}: \begin{cases} x_{F_{j1}} = x_{C_{fbj}} \\ y_{F_{j1}} = R_{Oj} \end{cases} \quad \text{and} \quad F_{j4}: \begin{cases} x_{F_{j4}} = x_{C_{fbj}} \\ y_{F_{j4}} = R_{Ij} \end{cases}$$

while the other two points require some few steps more.
The intercept of the straight line u, perpendicular to r and passing through C_{fbj}, is

$$q_{fbj} = y_{C_{fbj}} + \frac{1}{m_r} x_{C_{fbj}} \tag{3.19}$$

so, the fillet joint point F_{j2} is the intersection between s and u, which results in

$$\begin{cases} x_{F_{j2}} = \dfrac{q_{C_{fbj}} - q_s}{m_r + \frac{1}{m_r}} \\ y_{F_{j2}} = m_r x_{F_{j2}} + q_s \end{cases} \tag{3.20}$$

while the point, F_{j3} is given by

$$\begin{cases} x_{F_{j3}} = \dfrac{q_{C_{fbj}} - q_t}{m_r + \frac{1}{m_r}} \\ y_{F_{j3}} = m_r x_{F_{j3}} + q_t \end{cases} \tag{3.21}$$

3.2.6 Automatic segmented flux-barrier drawing algorithm

Hereafter, the pseudocode to draw a segmented flux-barriers rotor is reported as

Input: Tangential rib thickness: $w_{\text{rib,t}}$
Input: Radial rib thicknesses: $t_{\text{rib},1}, \ldots, t_{\text{rib},N_\text{b}}$
Input: Flux-barrier angles: $\vartheta_{\text{b}1}, \ldots, \vartheta_{\text{b}N_\text{b}}$
Input: Magnet thicknesses: $t_{\text{pm}1}, \ldots, t_{\text{pm}N_\text{b}}$
Input: Magnet widths: $w_{\text{pm}1}, \ldots, w_{\text{pm}N_\text{b}}$
Input: Flux-barrier fillet radii: $R_{\text{fb}1}, \ldots, R_{\text{fb}N_\text{b}}$
Input: Middle flux-barrier outer radii: $R_{O1}, \ldots, R_{ON_\text{b}}$
Input: Middle flux-barrier inner radii: $R_{I1}, \ldots, R_{IN_\text{b}}$

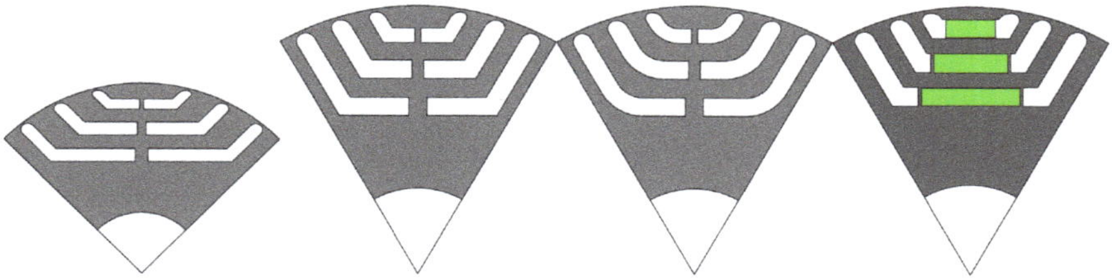

(a) (b) (c) (d)

Figure 3.12 Segmented flux-barrier rotors with different number of poles and fillets: (a) four-pole, (b) six-pole, (c) six-pole smooth, (d) six-pole smooth with PM

for $j = 1$ **To** N_b **do**
  Compute point B_j with (3.3)
  Compute points H_j and K_j with (3.5) and (3.6)
  Compute joint point fillets A_j and D_j system
  if $w_{pmj} > 0$ **then**
    Compute radial rib with (3.17)
    Compute PM points with (3.2.4)
  else
    Compute radial rib with (3.16)
  end if
  if $R_{fbj} > 0$ **then**
    Compute flux-barrier fillet with (3.2.5), (3.20) and (3.21)
  end if
end for

3.2.7 Examples

Figure 3.12 shows some segmented flux-barrier rotor pole sketches. All the rotors presents the same flux-barrier and iron channel thicknesses. The difference between Figure 3.12(a) and (b) is the number of poles: four and six, respectively. The rotor in Figure 3.12(c) is the smoothed version of the one in Figure 3.12(b). The rotor in Figure 3.12(d) presents both radial ribs and PMs in the middle of the flux-barriers.

3.3 Flux-barriers and iron channels design

Both the drawing procedures seen in the previous sections require, at least, the knowledge of the intercept points between the flux-barriers and q-axis and the flux-barrier angles.

The rotor diameter, the shaft diameter and the flux-barriers middle radii define, in turn, N_b flux-barrier thicknesses, $t_{bj}, \ldots, t_{bN_b}$ and $N_b + 1$ iron channel widths, $w_{cj}, \ldots, w_{cN_b+1}$, as reported in Figure 3.13.

In the following, an SyR machine design methodology to reduce the number of parameters on the basis of analytical considerations is presented [27,28].

3.3.1 Magnetic insulation ratio k_{air}

On the basis of the flux-barrier thicknesses and iron channel widths, it is possible to define the magnetic insulation ratio, k_{air}, as

$$k_{air} = \frac{\sum_{j=1}^{N_b} \sum t_{bj}}{\frac{D_r - D_{sh}}{2}} = \frac{\sum_{j=1}^{N_b} \sum t_{bj}}{\sum_{j=1}^{N_b} t_{bj} + \sum_{j=1}^{N_b+1} w_{cj}} = \frac{l_{air}}{l_{air} + l_{fe}} \tag{3.22}$$

which gives an indication of the amount of magnetic insulation (i.e. air) along the q-axis. This is clearly shown in Figure 3.14 where SyR rotors with different values of k_{air} and the same flux-barrier angles are reported. It can be noted that, for $k_{air} = 0.2$, the flux-barriers are quite thin while the iron channels are thick. For $k_{air} = 0.5$ the

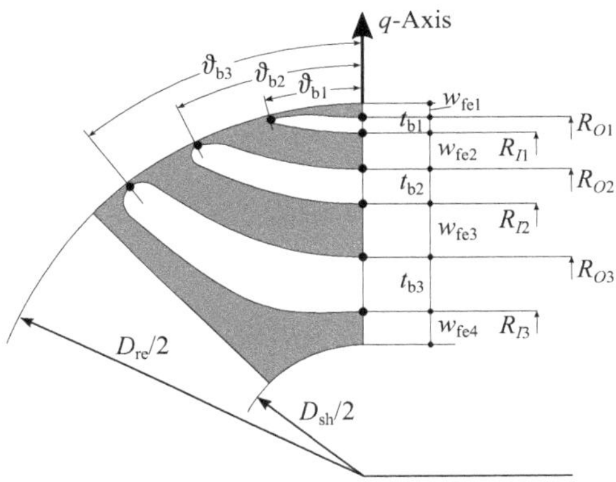

Figure 3.13 Sketch of an SyR rotor with three flux-barriers per pole with relative quotes

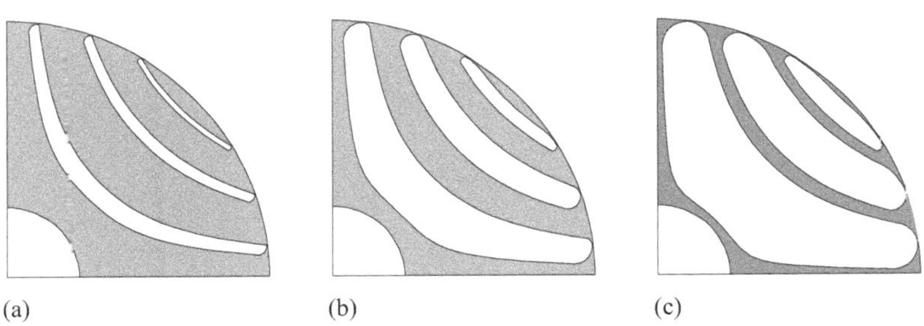

Figure 3.14 SyR rotors with different values of k_{air}: (a) 0.2, (b) 0.5 and (c) 0.8

amount of air and iron along the q-axis is the same. For $k_{\text{air}} = 0.8$, there are thick flux-barriers with thin iron channels.

3.3.2 Iron channel widths design

The aim of the criteria to select the optimal iron channel widths is to maximize the d-axis flux linkage by assigning to each iron channel a width proportional to the magnetomotive force (MMF) at the air gap [26].

The main assumptions that allow one to get an analytical design criterion are linear magnetic behavior of the iron and sinusoidal distribution of the stator scalar magnetic, potential within the air gap. The latter can be expressed in per unit (p.u.) by

$$\mathcal{U}_s(\vartheta^{\text{e}}) = \sin p\vartheta_{\text{m}} \tag{3.23}$$

where ϑ_{m} is the mechanical rotor coordinated whose origin lies on the q-axis. The average value of the MMF in front of the jth iron channel, f_j, can be computed as follows:

$$\begin{cases} f_1 = \frac{1}{\vartheta_{\text{b}1}} \int_0^{\vartheta_{\text{b}1}} \mathcal{U}_s \sin(p\vartheta_{\text{m}})\, d\vartheta_{\text{m}} \\ \quad\vdots \\ f_j = \frac{1}{\vartheta_{\text{b}j} - \vartheta_{\text{b}(j-1)}} \int_{\vartheta_{\text{b}(j-1)}}^{\vartheta_{\text{b}j}} \mathcal{U}_s \sin(p\vartheta_{\text{m}})\, d\vartheta_{\text{m}} \\ \quad\vdots \\ f_{N_{\text{b}}+1} = \frac{2}{\frac{\pi}{2p} - \vartheta_{\text{b}N_{\text{b}}}} \int_{\vartheta_{\text{b}N_{\text{b}}}}^{\frac{\pi}{2p}} \mathcal{U}_s \sin(p\vartheta_{\text{m}})\, d\vartheta_{\text{m}} \end{cases} \tag{3.24}$$

this operation is shown qualitatively in Figure 3.15.

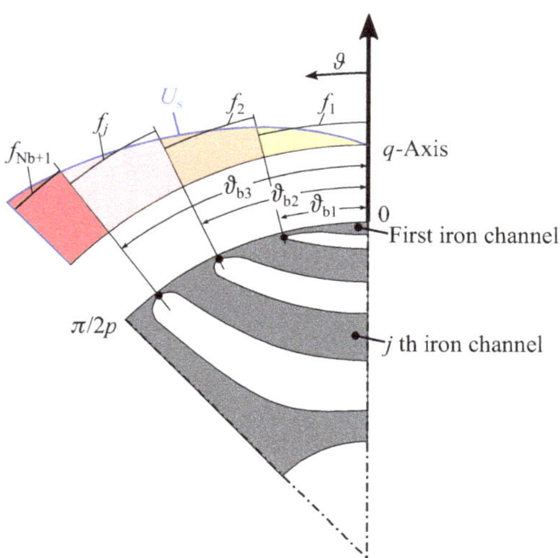

Figure 3.15 Computation of the average MMF in front of each iron channel

Then, the iron channel widths can be computed solving the following system:

$$\begin{cases} \sum_{j=1}^{N_b+1} w_{cj} = (1 - k_{air})\dfrac{D_r - D_{sh}}{2} \\ \dfrac{w_{ck}}{w_{ch}} = \dfrac{f_k}{f_h} \end{cases} \tag{3.25}$$

after some arrangements it results:

$$\begin{cases} w_{c1} = \dfrac{l_{fe}}{\sum_{j=1}^{N_b+1} \frac{f_j}{f_1}} \\ w_{cj} = w_{c1}\dfrac{f_j}{f_1} \end{cases} \tag{3.26}$$

finally, in order to consider the tangential rib's contribution, the first iron channel width has to be adjusted as follows:

$$w'_{c1} = w_{c1} - w_{rib,t} \tag{3.27}$$

3.3.3 Flux-barrier thicknesses design

The aim of the flux-barrier thicknesses design is to minimize the q-axis flux linkage [26,29]. The p.u. behavior of the MMF within the air gap with a q-axis magnetization is expressed by

$$\mathscr{U}_s = \cos p\vartheta_m \tag{3.28}$$

from which it is possible to compute the average stator scalar magnetic potential, f_j, in front of the jth flux-barrier:

$$\begin{cases} f_1 = \frac{1}{\vartheta_{b1}} \int_0^{\vartheta_{b1}} \mathscr{U}_s \cos(p\vartheta_m)\, d\vartheta_m \\ f_j = \frac{1}{\vartheta_{bj} - \vartheta_{b(j-1)}} \int_{\vartheta_{b(j-1)}}^{\vartheta_{bj}} \mathscr{U}_s \cos(p\vartheta_m)\, d\vartheta_m \end{cases} \tag{3.29}$$

from which it is possible to compute the voltage drops across each flux-barrier, Δf_j:

$$\begin{cases} \Delta f_j = f_j - f_{(j+1)} \qquad j = 1, \ldots, N_b - 1 \\ \Delta f_{N_b} = f_{N_b} \end{cases} \tag{3.30}$$

this computation is described qualitatively in Figure 3.16.

From the derivations given in [29], the relationship that yields the minimum q-axis flux linkage is

$$\frac{t_{bj}}{t_{bk}} = \frac{\Delta f_h}{\Delta f_k}\sqrt{\frac{l_{bh}}{l_{bk}}} \tag{3.31}$$

where l_{bj} is the length of the jth flux-barrier. The constraint

$$\sum_{j=1}^{N_b} t_{bj} = k_{air}\frac{D_r - D_{sh}}{2} \tag{3.32}$$

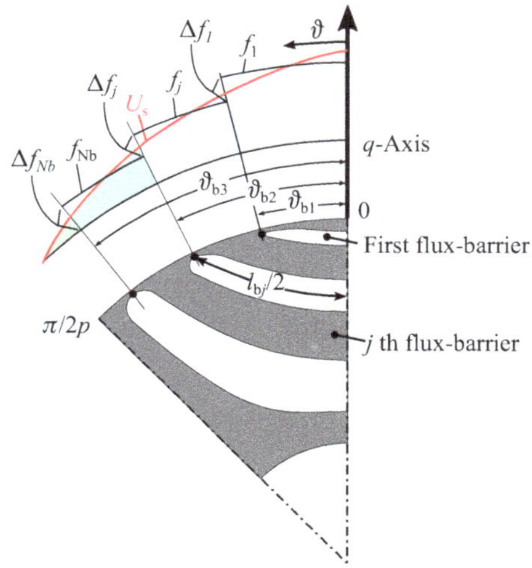

Figure 3.16 Computation of the average MMF in front of each flux-barrier

and the simplification $l_{bj} = D_r \vartheta_{bj}$ allow one to express all the flux-barrier thicknesses as follows:

$$\begin{cases} \frac{t_{b2}}{t_{b1}} = \frac{\Delta f_2}{\Delta f_1}\sqrt{\frac{\vartheta_{b2}}{\vartheta_{b1}}} \\ \vdots \\ \frac{t_{bj}}{t_{b1}} = \frac{\Delta f_j}{\Delta f_1}\sqrt{\frac{\vartheta_{bj}}{\vartheta_{b1}}} \\ \vdots \\ \frac{t_{bN_b}}{t_{b1}} = \frac{\Delta f_{N_b}}{\Delta f_1}\sqrt{\frac{\vartheta_{bN_b}}{\vartheta_{b1}}} \end{cases} \tag{3.33}$$

and, after some arrangements, it results in

$$t_{b1} = \frac{l_{air}}{\sum_{j=1}^{N_b} \frac{\Delta f_j}{\Delta f_1}\sqrt{\frac{\vartheta_{bj}}{\vartheta_{b1}}}} \qquad t_{bj} = \frac{\Delta f_j}{\Delta f_1}\sqrt{\frac{\vartheta_{bj}}{\vartheta_{b1}}}\, t_{b1} \qquad j = 2, \ldots, N_b \tag{3.34}$$

3.3.4 Flux-barrier inner and outer middle radii

Once all the flux-barrier and iron channel dimensions are known, it is possible to compute the jth inner and outer flux-barrier radii as follows:

$$R_{Oj} = D_r/2 - \sum_{i=1}^{j} w_{ci} - \sum_{i=1}^{j-1} t_{bi}$$
$$R_{Ij} = R_{Oj} - t_{bj}$$

3.3.5 Automatic drawing algorithm

In the following, the pseudocode to draw an SYR rotor with optimal flux-barrier and iron channel dimensions is reported.

Algorithm 1: Complete rotor parameters computation

Input: Rotor and shaft diameters: D_r and D_{sh}
Input: Tangential rib thickness: $w_{rib,t}$
Input: Flux-barrier angles: $\vartheta_{b1}, \ldots, \vartheta_{bN_b}$
Input: Magnetic insulation ratio: k_{air}
  Compute iron channel widths with (3.26)
  Compute flux-barrier thicknesses with (3.34)
  Compute middle inner and outer radii with (3.3.4) and (3.3.4)
  if Rotor topology = *fluid* **then**
    Draw fluid flux-barrier
  else if Rotor topology = *segmented* **then**
    Draw segmented flux-barrier
  end if

Chapter 4
Reluctance network model of high-speed synchronous reluctance machines

The aim of this chapter is to describe the magnetic behavior of the high-speed (HS)-Synchronous Reluctance (SyR) rotor through an Improved Reluctance Network (IRN) [30,31] that takes into account the radial ribs contribution on the rotor saturation and the cross coupling between d-axis and q-axis using nonlinear reluctances [32].

To prove the proposed method, some rotors are designed for different rated speeds. For the sake of simplicity, only the radial rib thickness is mechanically designed, while the rotor flux-barrier angles and insulation ratios are kept the same as reported in Table 4.1. The stator geometry is fixed and its main data are reported in Table 4.1. The motors performance are evaluated in maximum torque per ampere condition.

The motors are also analyzed using Finite Element (FE) magnetostatic simulations and the results compared with the ones coming from the IRN.

Table 4.1 Main data of the motors

Motor part	Symbol	Value
Outer stator diameter	D_e	300 mm
Inner stator diameter	D_s	190 mm
Stack length	L_{stk}	100 mm
Pole number	$2p$	4
Slot number	Q	48
Slot height	h_s	25 mm
Tooth width	w_t	6.76 mm
Air gap	g	1 mm
Flux-barrier angles	$\vartheta_{b1}, \vartheta_{b2}, \vartheta_{b3}$	14.8°, 26.7°, 38.7°
Magnetic insulation ratio	k_{air}	0.4
Series conductor per slot	n_{ncs}	1
Phase current	$\hat{I}$	572 A
Iron lamination		N020
Ultimate tensile strength	σ_r	500 MPa

4.1 Improved reluctance network

As previously mentioned, the nonlinear magnetic behavior of the iron parts (i.e. iron channels and radial ribs) and the cross-coupling effect, due to the higher q-axis flux, have to be considered in order to better describe the rotor saturation. To this purpose, a magnetic model based on an IRN is built [31].

In such an approach, all the motor parts are modeled either with a magnetic reluctance or with a magnetomotive force (MMF) generator. In particular, linear and nonlinear magnetic reluctances are used to model the air and iron regions, respectively. Therefore, the proposed IRN is solved in a iterative process in which the unknowns are the flux density values in each reluctance component [31,33]. In this context, there is a seed value for each magnetic flux density. Then, the equation system is solved and the new values of magnetic fluxes are obtained. If these values are different from the seed values, the process will start again using the solutions as seed values until convergence is reached.

The aim of the IRN is to model the iron saturation locally with particular care of the rotor radial ribs and channels. This allows to take into account the unusual flux density distribution in the rotor that cannot be described accurately with the Simple Reluctance Network (SRN).

4.1.1 Stator and air gap magnetic network

A sketch of the stator magnetic network is shown in Figure 4.1(a). Each stator slot is modeled by two magnetic reluctances (the tooth and the yoke one) and an MMF generator. The reluctances expression is

$$\Re = \frac{l}{\mu_{\text{steel}}(H)\, S} \tag{4.1}$$

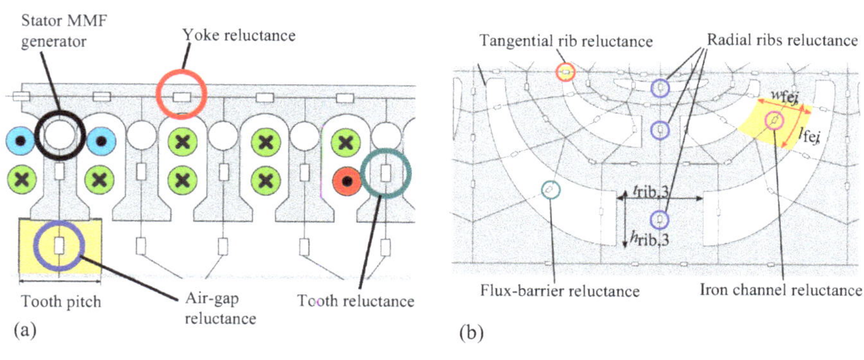

Figure 4.1 Improved magnetic reluctance magnetic network of an HS-SyR machine with three flux-barriers per pole: (a) stator network and (b) rotor network

where $\mu_{\text{steel}}(H)$, function of the magnetic field, is the absolute magnetic permeability of the lamination, l is the length in the flux direction and S is the surface normal to flux direction.

The value of each MMF generator depends on the winding distribution and on the time instant considered [30]. For example, Table 4.2 reports the MMF value of each generator when the winding is fed with only d-axis and q-axis current, respectively. The direct MMF is obtained with

$$I_c = n_{\text{cs}} I \qquad \text{and} \qquad I_b = I_c = \frac{-n_{\text{cs}} I}{2}, \tag{4.2}$$

while the q-axis MMF is obtained with

$$I_a = 0 \qquad \text{and} \qquad I_b = -I_c = \frac{\sqrt{3}}{2\, n_{\text{cs}} I} \tag{4.3}$$

in general, for a generic current vector amplitude, $\hat{I}$, and angle, α_i^{e}, the phase currents I_a, I_b and I_c are computed by means of Park's anti-transformation and used to determine the stator MMF generator values. The air gap in front of each tooth is modeled with a magnetic reluctance computed as follows:

$$\mathfrak{R}_g = \frac{g\, k_{\text{carter}} Q}{\mu_0 \pi D_{\text{si}}}, \tag{4.4}$$

while the slotting effect is taken into account with Carter's coefficient as for the SRN. Conversely, the saturation factor is not necessary since the reluctance network already models nonlinearity.

4.1.2 Rotor magnetic network

The rotor reluctance networks of one pole is shown in Figure 4.1(b). The iron channels and the iron ribs (i.e. radials and tangentials) are modeled by nonlinear magnetic reluctances to take into account the saturation, while the flux-barriers are modeled by linear magnetic reluctances as they contain air. It is worth noting that the proposed network is not unique but it is trade-off between the magnetic accuracy and the computational time [30].

Table 4.2 d-Axis and q-axis stator magnetomotive forces of one pole

Phase A	$2I_a$	$2I_a$	$2I_a$	$2I_a$	$1.5I_a$	I_a	0	$-I_a$	$-1.5I_a$	$-2I_a$	$-2I_a$	$-2I_a$
Phase B	$-1.5I_b$	$-I_b$	0	I_b	$1.5I_b$	$2I_b$	$2I_b$	$2I_b$	$2I_b$	$2I_b$	$2I_b$	$2I_b$
Phase C	$-1.5I_c$	$-2I_c$	$-2I_c$	$-2I_c$	$-2I_c$	$-2I_c$	$-2I_c$	$-2I_c$	$-1.5I_c$	$-I_c$	0	I_c
d-Axis MMF	3.5	3.5	3.0	2.5	1.75	1.0	0.0	−1.0	−1.75	−2.5	−3.0	−3.5
q-Axis MMF	0	−0.87	−1.73	−2.60	−3.03	−3.46	−3.46	−3.46	−3.03	−2.60	−1.73	−0.87

The d-axis MMF is computed with $I_a = n_{\text{cs}} I$ and $I_b = I_c = -n_{\text{cs}} I/2$. The q-axis MMF is computed with $I_a = 0$ and $I_b = -I_c = -\sqrt{3}/2 n_{\text{cs}} I$

4.2 Finite element analysis results

Figure 4.2 reports the flux density plots of the four different rotors designed for 500 rpm, 10 krpm, 20 krpm and 25 krpm with a safety factor $\nu_\sigma = 2.5$. It is quite evident the deep influence of the radial rib thickness on the rotor saturation. In particular, when the radial ribs are saturated, the flux density in the iron channels is lower than 1.64 T, as usual. This condition is achieved for speeds lower than 10 krpm. As the speed increases (i.e. near 20 krpm), the radial rib thickness of the third flux-barrier is such that the flux density level, in the same rib, is lower than the iron channel one. The saturation of the other radial ribs is higher with respect to the third one. At the highest speeds (i.e.25 krpm), the saturation of the rotor parts is reversed: the iron channels' flux density levels are about 1.9 T, while within the radial ribs, the flux density is lower than 1.7 T.

The FE inductances and the saliency ratio behaviors versus the speed are reported in Figure 4.3. It is evident that both the inductances depend on the speed. The d-axis

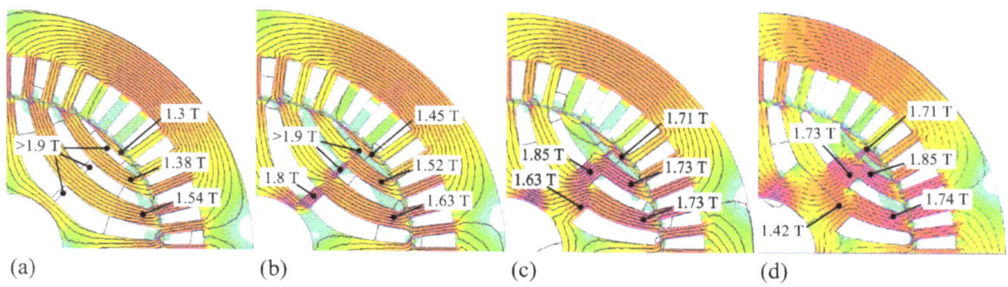

Figure 4.2 Flux density plots of different HS-SyR rotor design for different rated speeds: (a) 500 rpm, (b) 10 krpm, (c) 20 krpm, (d) 25 krpm

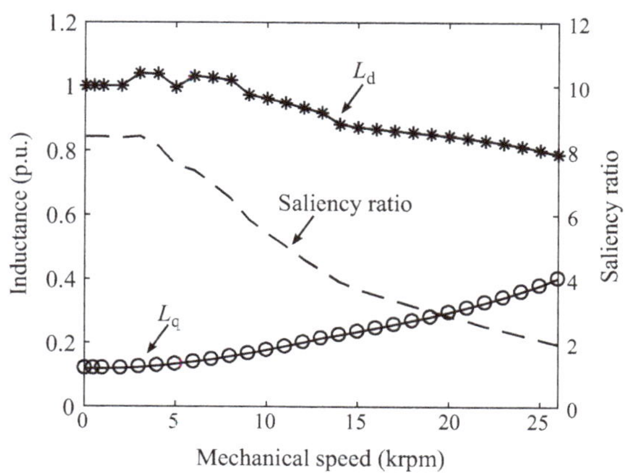

Figure 4.3 Finite Element Analysis (FEA) results: inductances and saliency ratio versus mechanical speed. The base value is the d-axis inductance at 0 rpm

inductance decreases as the speed increases because of the iron channels' saturation. The q-axis inductance increases with the speed because of the higher flux linkage that flows through the iron ribs.

4.3 Results comparison

This section reports the comparison between the FE and the proposed reluctance network results in terms of torque and power limit curves.

4.3.1 Simplified reluctance network results

Figure 4.4(a) and (b) shows the torque and the mechanical power behaviors versus the rotor speed obtained with SRN and FE analyses. For speeds lower than 10 krpm, there is a good agreement between the curves resulting from the two methods. This is because the analytical model hypotheses (i.e. radial ribs saturated and low iron channel saturation) are met. For higher speeds, the larger radial rib thicknesses affect the flux density distribution in the rotor channels so that the analytical model overestimates the electromagnetic torque with respect to the FE analysis results. In particular, at the highest speed (25 krpm), the torque computed analytically is 36% higher than the FE one.

4.3.2 Improved reluctance network results

The torque and power versus speed curves, obtained with IRN and FE analyses, are reported in Figure 4.5(a) and (b), respectively. There is a good agreement between the two in the whole speed range. However, at the highest speed, the IRN reports a torque 11% higher than FE one. The same observation is valid for the power curves, of course. This is due to the fact that, as reported in Figure 4.6, the IRN underestimates the q-axis inductance and so the relative flux linkage. Conversely, along the d-axis, there is a good agreement between FE analysis and IRN curves. The underestimation

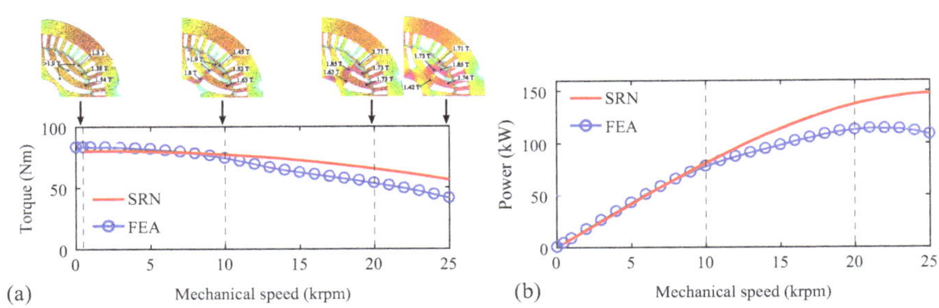

Figure 4.4 (a) Torque and (b) power versus mechanical speed: comparison between the simplified reluctance network and FE analysis results

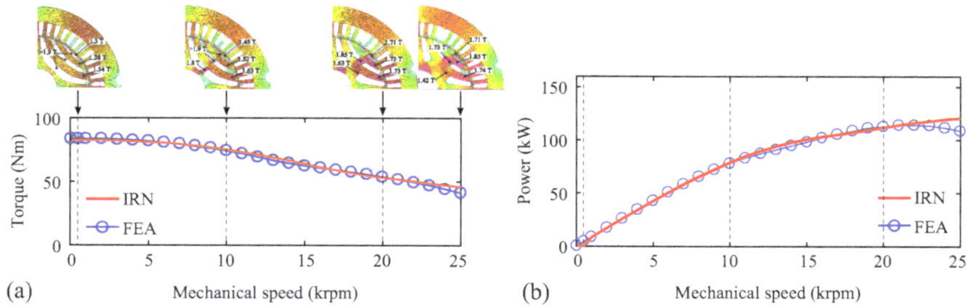

Figure 4.5 (a) Torque and (b) power versus mechanical speed: comparison between IRN and FE results

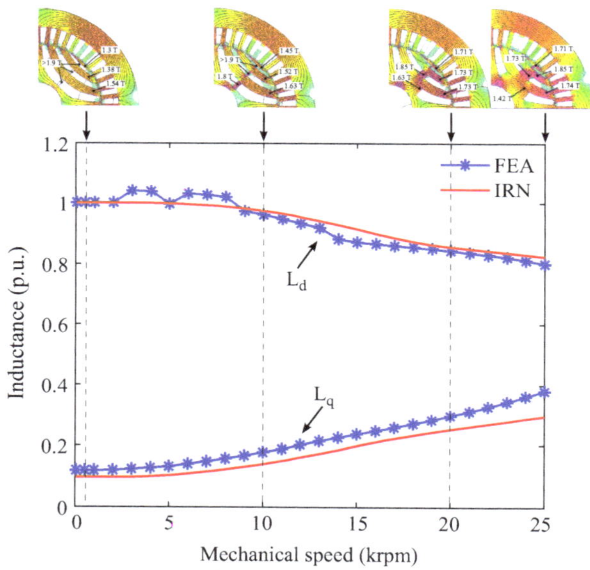

Figure 4.6 Inductances versus mechanical speed: comparison between FE analysis and IRN results. The base value is the FE analysis d-axis inductance at 0 rpm

of the q-axis flux linkage is related to the variable flux density distribution within the radial rib regions, as observable in Figure 4.2(d). A better estimation of the flux that flows through the radial ribs can be achieved increasing the number of local magnetic reluctances.

4.4 Discussion

This chapter dealt with the analysis of HS-SYR by means of two different reluctance networks: a simplified and an improved one. Some motors have been designed for different rated speeds taking into account both magnetic (flux-barriers shape) and structural aspects (radial rib thicknesses) to prove the suitability of the proposed methods for the analysis and design of HS-SYR. Comparing the results with FE, it was shown that the simplified reluctance network is not able to describe the motor's performance with accuracy in a wide speed range, since the rotor saturation in the radial ribs and in the iron channels is not modeled properly. This is because the simplified model does not describe locally the iron channels' saturation and considers the radial ribs to be saturated with a constant flux density regardless of the radial rib thicknesses. These simplifications are reasonable with the common SYR motors but are not met as the radial rib thicknesses and the saturation of the iron channels' increase with the speed as in the considered case study.

Conversely, a good agreement has been found between FE and the IRN results in the whole speed range considered. This is because the IRN considers the local saturation of stator and rotor as all the motor regions are modeled by magnetic reluctances, which are nonlinear for the iron parts.

Chapter 5
Nonlinear analytical model for synchronous reluctance machines

This chapter presents a nonlinear analytical model of the Synchronous Reluctance (SyR) machine at the air gap, used to derive both average and torque harmonics as a function of the rotor geometry.

At first, a linear model is described. Then it is extended to include the saturation of some core parts of the machine. Saturation is taken into account following all the main flux lines paths inside the machine. After that, a proper saturation coefficient is assigned to each of these paths and applied at the air gap. The analytical model proves to be fast and fairly accurate in any relevant calculation. Therefore, it is quick and easy to obtain the behavior of the average torque and torque ripple as functions of the rotor flux-barrier geometry. The result is presented using maps that are essential for finding proper combinations of barrier angles that maximize the average torque and minimize torque ripple. The torque maps are compared with those obtained from both linear analytical and Finite Element (FE) models. The maps computed analytically show good agreement with those derived by means of FE analysis, and they are obtained in a much shorter amount of time.

5.1 Development of the analytical model

Starting very generally, suppose to distribute a single-turn, non-chorded infinitesimal coil inside the slots of a given stator lamination stack (Figure 5.1). The conductor distribution function is equal to

$$n_{\mathrm{d}}(\vartheta^{\mathrm{e}}) = \hat{n}\left[\delta(\vartheta^{\mathrm{e}} - \tfrac{\pi}{2}) - \delta(\vartheta^{\mathrm{e}} - \tfrac{\pi}{2} - \pi)\right], \quad \vartheta^{\mathrm{e}} \in [0, 2\pi]$$

where $\delta(.)$ is the Dirac delta impulse and ϑ^{e} is the electrical angular coordinate whose origin is along the magnetic axis of this coil.

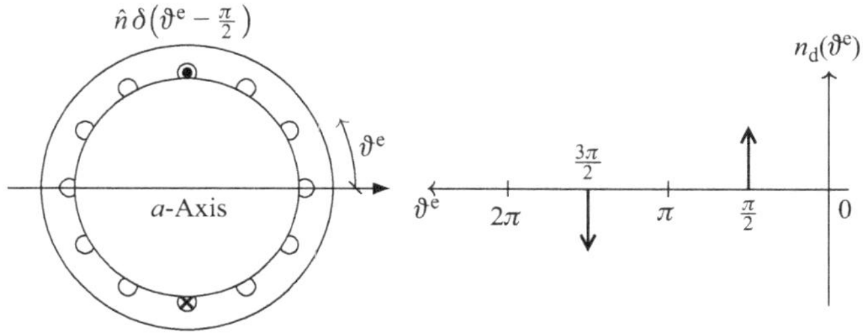

Figure 5.1 Sketch of a stator lamination with one infinitesimal coil and relative conductor distribution function plot

From the integration of n_d in one semi-period, it has to result in one conductor*:

$$\int_0^{\pi} n_d(\vartheta^e)\tfrac{D_s}{2}\,\mathrm{d}\vartheta^e = \int_0^{\pi} \hat{n}\,\delta(\vartheta^e - \tfrac{\pi}{2})\tfrac{D_s}{2}\,\mathrm{d}\vartheta^e = 1 \quad\Longrightarrow\quad \hat{n} = \tfrac{2}{D_s}$$

so

$$n_d(\vartheta^e) = \frac{2}{D_s}\left[\delta(\vartheta^e - \tfrac{\pi}{2}) - \delta(\vartheta^e - \tfrac{\pi}{2} - \pi)\right] \tag{5.1}$$

where the value 2 represents the total number of conductors in one electrical period. Note that this result is valid for any number of poles and conductors. In fact defining

$$n_d(\vartheta) = \frac{N_s}{D_s}\left[\delta(p\vartheta - \tfrac{\pi}{2}) - \delta(p\vartheta - \tfrac{\pi}{2} - \pi)\right], \quad \vartheta \in \left[0, \tfrac{2\pi}{p}\right] \tag{5.2}$$

and integrating over a mechanical semi-period, it results in

$$\int_0^{\pi/p} \frac{N_s}{D_s}\,\delta(p\vartheta - \tfrac{\pi}{2})\frac{D_s}{2}\,\mathrm{d}\vartheta = \frac{N_s D_s}{2pD_s} = \frac{N_s}{2p}$$

which is exactly the number of series conductors per phase in one pole. Note that (5.1) and (5.2) represent just the first pole pair, so they have to be repeated through the operator

$$\mathrm{rep}_T\{g(t)\} = \sum_{k=-\infty}^{+\infty} g(t - kT)$$

*Remember that the integral of the Dirac delta function is 1 and the property

$$\int_{\mathbb{R}} \delta(a\tau)\,\mathrm{d}\tau = \frac{1}{|a|}.$$

Then

$$n_{\mathrm{d}}(\vartheta^{\mathrm{e}}) = \frac{2}{D_{\mathrm{s}}}\,\mathrm{rep}_{2\pi}\left[\delta(\vartheta^{\mathrm{e}} - \tfrac{\pi}{2}) - \delta(\vartheta^{\mathrm{e}} - \tfrac{\pi}{2} - \pi)\right] \tag{5.3}$$

The conductor distribution function (5.3) is now periodic, odd and owns half-wave symmetry, so its Fourier series coefficients are

$$\begin{aligned} b_k &= \frac{1}{\pi}\int_0^{2\pi} n_{\mathrm{d}}(\vartheta^{\mathrm{e}}) \sin k\vartheta^{\mathrm{e}} \,\mathrm{d}\vartheta^{\mathrm{e}} \\ &= \frac{2}{\pi}\int_0^{\pi} \frac{2}{D_{\mathrm{s}}}\delta(\vartheta^{\mathrm{e}} - \tfrac{\pi}{2}) \sin k\vartheta^{\mathrm{e}} \,\mathrm{d}\vartheta^{\mathrm{e}} \\ &= \frac{4}{\pi D_{\mathrm{s}}} \sin k\tfrac{\pi}{2} \end{aligned}$$

and the series is

$$n_{\mathrm{d}}(\vartheta^{\mathrm{e}}) = \frac{4}{\pi D_{\mathrm{s}}} \sum_{\substack{k=1 \\ k \text{ odd}}}^{+\infty} \sin k\tfrac{\pi}{2} \sin k\vartheta^{\mathrm{e}}$$

Similarly, the conductor distribution function for the other phases can be written, specifically "b" and "c" for a three-phase machine:

$$\begin{aligned} n_{\mathrm{da}}(\vartheta^{\mathrm{e}}) &= \frac{4}{\pi D_{\mathrm{s}}} \sum_k \sin k\tfrac{\pi}{2} \sin k\vartheta^{\mathrm{e}} \\ n_{\mathrm{db}}(\vartheta^{\mathrm{e}}) &= \frac{4}{\pi D_{\mathrm{s}}} \sum_k \sin k\tfrac{\pi}{2} \sin k(\vartheta^{\mathrm{e}} - \tfrac{2\pi}{3}) \\ n_{\mathrm{dc}}(\vartheta^{\mathrm{e}}) &= \frac{4}{\pi D_{\mathrm{s}}} \sum_k \sin k\tfrac{\pi}{2} \sin k(\vartheta^{\mathrm{e}} - \tfrac{4\pi}{3}) \end{aligned}$$

Usually these windings are fed with a system of symmetric sinusoidal currents:

$$\begin{aligned} i_{\mathrm{a}}(t) &= \hat{I}\cos(\omega t + \alpha_i^{\mathrm{e}}) \\ i_{\mathrm{b}}(t) &= \hat{I}\cos(\omega t + \alpha_i^{\mathrm{e}} - \tfrac{2\pi}{3}) \\ i_{\mathrm{c}}(t) &= \hat{I}\cos(\omega t + \alpha_i^{\mathrm{e}} - \tfrac{4\pi}{3}) \end{aligned}$$

such that the electric loading of the machine is

$$\mathscr{K}_{\mathrm{s}}(\vartheta^{\mathrm{e}}, t) = n_{\mathrm{da}}(\vartheta^{\mathrm{e}})i_{\mathrm{a}}(t) + n_{\mathrm{db}}(\vartheta^{\mathrm{e}})i_{\mathrm{b}}(t) + n_{\mathrm{dc}}(\vartheta^{\mathrm{e}})i_{\mathrm{c}}(t)$$

The sinusoidal functions products result in

$$\begin{aligned}
&\sin k\vartheta^{\mathrm{e}}\cos(\omega t+\alpha_i^{\mathrm{e}})\\
&\quad+\sin k(\vartheta^{\mathrm{e}}-\tfrac{2\pi}{3})\cos(\omega t+\alpha_i^{\mathrm{e}}-\tfrac{2\pi}{3})\\
&\quad+\sin k(\vartheta^{\mathrm{e}}-\tfrac{4\pi}{3})\cos(\omega t+\alpha_i^{\mathrm{e}}-\tfrac{4\pi}{3})\\
&\qquad=\sin k\vartheta^{\mathrm{e}}\cos(\omega t+\alpha_i^{\mathrm{e}})\\
&\qquad\quad+\left(\sin k\vartheta^{\mathrm{e}}\cos k\tfrac{2\pi}{3}-\cos k\vartheta^{\mathrm{e}}\sin k\tfrac{2\pi}{3}\right)\\
&\qquad\qquad\cdot\left(\cos(\omega t+\alpha_i^{\mathrm{e}})\cos\tfrac{2\pi}{3}+\sin(\omega t+\alpha_i^{\mathrm{e}})\sin\tfrac{2\pi}{3}\right)\\
&\qquad\quad+\left(\sin k\vartheta^{\mathrm{e}}\cos k\tfrac{4\pi}{3}-\cos k\vartheta^{\mathrm{e}}\sin k\tfrac{4\pi}{3}\right)\\
&\qquad\qquad\cdot\left(\cos(\omega t+\alpha_i^{\mathrm{e}})\cos\tfrac{4\pi}{3}+\sin(\omega t+\alpha_i^{\mathrm{e}})\sin\tfrac{4\pi}{3}\right)\\
&\qquad=\sin \nu\vartheta^{\mathrm{e}}\cos(\omega t+\alpha_i^{\mathrm{e}})\\
&\qquad\quad+\tfrac{1}{2}\sin \nu\vartheta^{\mathrm{e}}\cos(\omega t+\alpha_i^{\mathrm{e}})-\tfrac{3}{2}\cos \nu\vartheta^{\mathrm{e}}\sin(\omega t+\alpha_i^{\mathrm{e}})
\end{aligned}$$

where $\nu = 6k+1,\ k\in\mathbb{Z}$ (see Appendix B for further insights). The final result is

$$\mathscr{K}_{\mathrm{s}}(\vartheta_{\mathrm{s}}^{\mathrm{e}},t)=\frac{6}{\pi D_{\mathrm{s}}}\hat{I}\sum_{\substack{\nu=6k+1\\ k\in\mathbb{Z}}}^{\infty}\sin \nu\tfrac{\pi}{2}\sin(\nu\vartheta_{\mathrm{s}}^{\mathrm{e}}-\omega t-\alpha_i^{\mathrm{e}})$$

where $\vartheta_{\mathrm{s}}^{\mathrm{e}}$ is the electrical angular coordinate fixed with the stator and with the origin on the magnetic axis of the first phase (a). Now it is possible to substitute the previous single-turn winding (with two conductors) with the single-coil equivalent winding of a distributed and chorded one. This equivalent winding has N_{s} series conductors per phase and its winding factor is k_{w}^{ν}. Then it is possible to define

$$\hat{K}_{\mathrm{s}}=\frac{3N_{\mathrm{s}}\hat{I}}{\pi D_{\mathrm{s}}}$$

and

$$K_\nu=k_{\mathrm{w}}^{\nu}\hat{K}_{\mathrm{s}}\sin \nu\tfrac{\pi}{2}=\frac{3\hat{I}\,k_{\mathrm{w}}^{\nu}N_{\mathrm{s}}}{\pi D_{\mathrm{s}}}\sin \nu\tfrac{\pi}{2}$$

so the electric loading of a symmetric three-phase winding is [34–36]

$$\mathscr{K}_{\mathrm{s}}(\vartheta_{\mathrm{s}},t)=\sum_{\substack{\nu=6k+1\\ k\in\mathbb{Z}}}^{\infty}K_\nu\sin(\nu p\vartheta_{\mathrm{s}}-\omega t-\alpha_i^{\mathrm{e}}) \tag{5.4}$$

This is the starting point of the following analysis thanks to the fact that (5.4) contains all the harmonics generated by the discretized winding.

5.2 Analytical model

The analytical model considers SYR machines with transversally laminated rotor. Furthermore, only integral-slot winding are considered. The electric loading of a symmetric three-phase distributed winding is

$$\mathscr{K}_s(\vartheta_s, \vartheta_m) = \sum_{\substack{\nu=6k+1 \\ k\in\mathbb{Z}}}^{\infty} \hat{K}_\nu \sin(\nu p\vartheta_s - p\vartheta_m - \alpha_i^e) \tag{5.5}$$

where ν is the space harmonic order belonging to the set $\{1, -5, 7, -11, 13, \ldots\}$, $\hat{K}_\nu$ is the amplitude of the electric loading νth harmonic, p is the number of pole pairs, ϑ_s is the angular coordinate fixed to the stator, ϑ_m is the rotor angular position, and α_i^e is the current electric angle. It is worth noticing that (5.5) produces every space harmonic generated by the discretized winding.

The electric loading generates the stator magnetic scalar potential, which is simply its integral, given by

$$\mathscr{U}_s(\vartheta_s, \vartheta_m) = \int \mathscr{K}_s(\vartheta_s, \vartheta_m)\frac{D_s}{2}\,\mathrm{d}\vartheta_s \tag{5.6}$$

where D_s is the stator diameter at the air gap. For a three-phase machine, it is

$$\mathscr{U}_s(\vartheta_s, \vartheta_m) = -\frac{D_s}{2}\sum_{\substack{\nu=6k+1 \\ k\in\mathbb{Z}}} \frac{\hat{K}_\nu}{p\nu}\cos(\nu p\vartheta_s - p\vartheta_m - \alpha_i^e) \tag{5.7}$$

Similarly, the rotor magnetic scalar potential, which reacts to the stator potential, can also be expressed by means of its Fourier series expansion:

$$\mathscr{U}_r(\vartheta_r, t) = \sum_{\xi=1}^{+\infty} \hat{U}_\xi \sin(\xi p\vartheta_r) \tag{5.8}$$

It is different from zero when the flux flows in the q-axis direction, crossing the flux-barriers. The rotor magnetic scalar potentials and the rotor barrier angles are grouped in vectors:

$$\begin{aligned} \boldsymbol{u}_r &= \{U_{r1}, U_{r2}, \ldots, U_{rn}\}^T, \\ \boldsymbol{\vartheta}_b &= \{\vartheta_{b1}, \vartheta_{b2}, \ldots, \vartheta_{bn}\}^T \end{aligned} \tag{5.9}$$

A quasi-diagonal matrix $\boldsymbol{G}$ can be built, given by

$$\boldsymbol{G} = \begin{bmatrix} +1 & -1 & & & \\ & +1 & -1 & & \\ & & \ddots & \ddots & \\ & & & +1 & -1 \\ & & & & +1 \end{bmatrix} \tag{5.10}$$

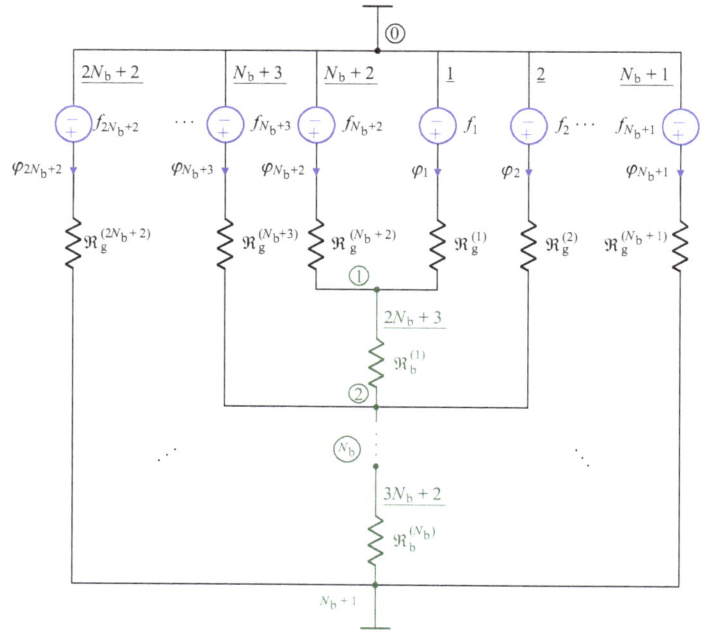

Figure 5.2 Magnetic circuit of one pole of the rotor of an SyR machine. Circled labels represent nodes while underlined labels represent edges of the circuit

with +1 in the main diagonal and −1 in the second right-hand diagonal, such that

$$\hat{U}_\xi = \frac{4}{\pi \xi} \sin(\xi \tfrac{\pi}{2}) \left[\sin\left(\xi p \boldsymbol{\vartheta}_{\mathrm{b}}^{\mathrm{T}}\right) \boldsymbol{G}\boldsymbol{u}_{\mathrm{r}}\right] \tag{5.11}$$

where $\sin(\xi \boldsymbol{\vartheta}_{\mathrm{b}}^{\mathrm{T}})$ is a row vector of sines, $\boldsymbol{G}\boldsymbol{u}_{\mathrm{r}}$ is the product of the matrix (5.10) and the vector $\boldsymbol{u}_{\mathrm{r}}$, defined in (5.9). The magnetic potentials of the islands can be calculated as the solution of the magnetic circuit shown in Figure 5.2.

5.3 Rotor magnetic potentials computation

Figure 5.2 represents the magnetic circuit of an SyR motor with N_{b} flux-barriers. There are N_{b} nodes, each one at its own magnetic potential $U_{\mathrm{r}j}$. The magnetomotive force (MMF) generators are defined between two distinct barrier angles and so are their reluctances. The remaining reluctances are the barriers' one.

The first $N_b + 1$ edges connects node 0 to the nodes of the islands, so the partial incidence matrix is

$$A' = \begin{array}{c} \begin{matrix} \underline{1} & \underline{2} & \cdots & \underline{N_b} & \underline{N_b+1} \end{matrix} \\ \left[\begin{matrix} +1 & +1 & \cdots & +1 & +1 \\ -1 & & & & \\ & -1 & & & \\ & & \ddots & & \\ & & & -1 & \\ & & & & -1 \end{matrix}\right] \begin{matrix} 0 \\ 1 \\ 2 \\ \vdots \\ N_b \\ N_b+1 \end{matrix} \end{array}$$

In these edges there are nonideal (real) MMF generator, so an ideal generator in series with a reluctance. In the following $N_b + 1$ nodes, there are the very same elements and structure, so the matrix $\boldsymbol{A}'$ is repeated.

The last part of the incidence matrix connects the following nodes between flux barrier reluctances, so it is the path followed by the flux that crosses all the barriers:

$$A'' = \begin{array}{c} \begin{matrix} \underline{2N_b+3} & \underline{2N_b+4} & \cdots & \underline{3N_b+1} & \underline{3N_b+2} \end{matrix} \\ \left[\begin{matrix} +1 & & & & \\ -1 & +1 & & & \\ & \vdots & \ddots & \vdots & \\ & & & -1 & +1 \\ & & & & -1 \end{matrix}\right] \begin{matrix} 1 \\ 2 \\ \vdots \\ N_b \\ N_b+1 \end{matrix} \end{array}$$

$$\boldsymbol{A}_c = \left[\boldsymbol{A}'_{(N_b+2,N_b+1)} \middle| \boldsymbol{A}'_{(N_b+2,N_b+1)} \middle| \begin{matrix} \boldsymbol{0}_{(1,N_b)} \\ \\ \boldsymbol{A}''_{(N_b+1,N_b)} \end{matrix} \right]$$

This is the scheme of the incidence matrix. To obtain the reduced one, we drop the *first* and *last* rows, the ones at "ground" potential. The remaining nodes are located inside the rotor island. Note that some columns are filled with zeros now, and they are the extreme edges corresponding to the main channel. This means that they are not dependent on the quantities of the grid and their fluxes can be solved immediately. Thanks to the order of construction, active components are present only in the first $2N_b + 2$ edges; the rest of the circuit is made of passive components. So the circuit can be solved with the *tableau method*, which is [37]

$$\underbrace{\begin{bmatrix} \boldsymbol{A} & \boldsymbol{0} & \boldsymbol{0} \\ \boldsymbol{0} & -\boldsymbol{I} & \boldsymbol{A}^{\mathrm{T}} \\ \boldsymbol{R} & \boldsymbol{G} & \boldsymbol{0} \end{bmatrix}}_{\boldsymbol{M}} \cdot \underbrace{\begin{pmatrix} \boldsymbol{\varphi} \\ \boldsymbol{\psi} \\ \boldsymbol{u} \end{pmatrix}}_{\boldsymbol{x}} = \underbrace{\begin{pmatrix} \boldsymbol{0} \\ \boldsymbol{0} \\ \boldsymbol{s} \end{pmatrix}}_{\boldsymbol{g}} \tag{5.12}$$

5.3.1 Reluctance of the flux barriers

The reluctance of the Hth flux barrier is given by

$$\mathfrak{R}_{bH} = \frac{t_{bH}}{\mu_0 L_{stk} \, l_{bH}}$$

5.3.2 Reluctance of partial air-gap

There are $2N_b + 2$ reluctances in front of each generator, representing part of the air-gap reluctance

$$\mathfrak{R}_{gH} = \frac{2g}{\mu_0 L_{stk} D_s \left[\vartheta_{bH} - \vartheta_{b(H-1)}\right]}$$

5.3.3 MMF generator expression

The first MMF generator comprehends the angles between $((\pi/2) - \vartheta^e_{b1})$ and $\pi/2$:

$$f_1 = \frac{1}{\vartheta^e_{b1}} \int_{(\pi/2)-\vartheta^e_{b1}}^{\pi/2} \mathscr{U}_s(\vartheta^e_r) \, d\vartheta^e_r$$

$$= -\sum_{\nu,k} \frac{K_\nu D_s}{2p\vartheta^e_{b1} \, \nu^2}\left[\sin\lambda_\nu (1 - \cos\nu\vartheta^e_{b1}) + \cos\lambda_\nu \sin\nu\vartheta^e_{b1}\right]$$

In general, given θ_H and θ_{H-1}[†],

$$f_H = \frac{1}{\theta_H - \theta_{H-1}} \int_{(\pi/2)-\theta_H}^{(\pi/2)-\theta_{H-1}} \mathscr{U}_s(\vartheta^e_r) \, d\vartheta^e_r$$

$$= \frac{-1}{\theta_H - \theta_{H-1}} \sum_{\nu,k} \frac{K_\nu D_s}{2p\nu^2}\Big[\sin\lambda_\nu (\cos\nu\theta_{H-1} - \cos\nu\theta_H) - \cos\lambda_\nu (\sin\nu\theta_{H-1} - \sin\nu\theta_H)\Big]$$

On the other hand, for the other half pole, it results in

$$f_H = \frac{1}{\theta_H - \theta_{H-1}} \int_{(\pi/2)+\theta_{H-1}}^{(\pi/2)+\theta_H} \mathscr{U}_s(\vartheta^e_r) \, d\vartheta^e_r$$

$$= \frac{-1}{\theta_H - \theta_{H-1}} \sum_{\nu,k} \frac{K_\nu D_s}{2p\nu^2}\Big[\sin\lambda_\nu (\cos\nu\theta_H - \cos\nu\theta_{H-1}) + \cos\lambda_\nu (\sin\nu\theta_H - \sin\nu\theta_{H-1})\Big]$$

[†]For convenience of notation, let $\theta_H = \vartheta^e_{bH}$.

5.3.4 Air-gap flux density

The stator magnetic scalar potential, given in (5.7), can be expressed in the rotor reference frame. Since

$$\vartheta_s = \vartheta_r + \vartheta_m$$

it results in

$$\mathscr{U}_s(\vartheta_r, \vartheta_m) = -\sum_{\substack{\nu=6k+1\\k\in\mathbb{Z}}} \frac{\hat{K}_\nu D_s}{2p\nu} \cos[\nu p\vartheta_r + (\nu-1)p\vartheta_m - \alpha_i^e]$$

It is split using sine and cosine as

$$\mathscr{U}_s(\vartheta_r, \vartheta_m) = -\sum_{\substack{\nu=6k+1\\k\in\mathbb{Z}}} \frac{\hat{K}_\nu D_s}{2p\nu} \left[\cos(\nu p\vartheta_r)\cos((\nu-1)p\vartheta_m - \alpha_i^e) - \sin(\nu p\vartheta_r)\sin((\nu-1)p\vartheta_m - \alpha_i^e)\right] \tag{5.13}$$

$\mathscr{U}_r(\vartheta_r, \vartheta_m)$ also contains harmonics multiple of 3 and they can be grouped apart. Then, it is easy to verify that $\mathscr{U}_r$ is an even function with respect to the harmonic order ξ, so positive or negative indexes can be used indifferently:

$$\mathscr{U}_r(\vartheta_r, \vartheta_m) = \sum_{\substack{\nu=6k+1\\k\in\mathbb{Z}}} \hat{U}_\nu \sin(\nu p\vartheta_r) + \sum_{\substack{\mu=6h+3\\h\in\mathbb{Z}^+}} \hat{U}_\mu \sin(\mu p\vartheta_r) \tag{5.14}$$

The air-gap flux density can be expressed as the difference of the two magnetic scalar potentials [34,38]:

$$\mathscr{B}_g(\vartheta_r, \vartheta_m) = \mu_0 \frac{-\mathscr{U}_s(\vartheta_r, \vartheta_m) + \mathscr{U}_r(\vartheta_r, \vartheta_m)}{g} \tag{5.15}$$

Introducing (5.13) and (5.14) in (5.15), the air-gap flux density results in

$$\mathscr{B}_g = \sum_{\substack{\nu=6k+1\\k\in\mathbb{Z}}} \frac{\mu_0 D_s \hat{K}_\nu}{2gp\,\nu} \Big[\overbrace{\cos((\nu-1)\omega t - \alpha_i^e)\cos(\nu p\vartheta_r)}^{\text{stator even part}} \underbrace{- \sin((\nu-1)\omega t - \alpha_i^e)\sin(\nu p\vartheta_r)}_{\text{stator odd part}}\Big]$$
$$+ \underbrace{\sum_{\substack{\nu=6k+1\\k\in\mathbb{Z}}} \frac{\mu_0}{g\nu}\,\nu\hat{U}_\nu \sin(\nu p\vartheta_r)}_{\text{rotor without triplen harmonics}} + \underbrace{\sum_{\substack{\mu=6h+3\\h\in\mathbb{Z}^+}} \frac{\mu_0}{g\mu}\,\mu\hat{U}_\mu \sin(\mu p\vartheta_r)}_{\text{rotor triplen harmonics}}$$

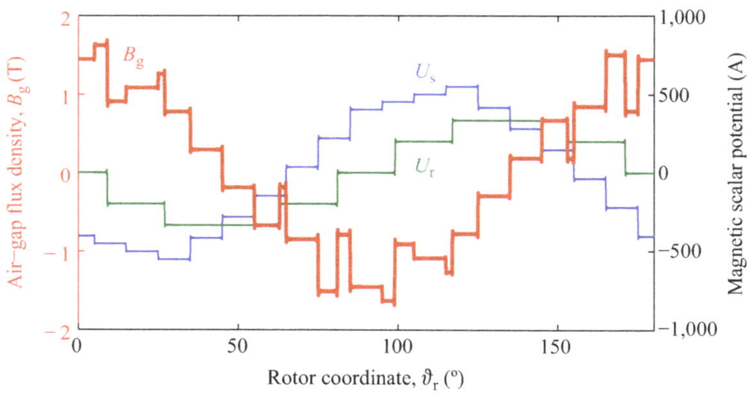

Figure 5.3 Air-gap flux density and magnetic scalar potentials of a two-barrier SyR machine without saturation at t = 0

and reordering

$$\mathscr{B}_g(\vartheta_r, \vartheta_m) = \sum_\nu \left[\frac{\alpha_\nu}{\nu} \cos(\nu p \vartheta_r) + \frac{\beta_\nu}{\nu} \sin(\nu p \vartheta_r) \right] + \sum_\mu \frac{\gamma_\mu}{\mu} \sin(\mu p \vartheta_r) \tag{5.16}$$

where α_ν, β_ν derives from the coefficients that multiply the cosine and sine functions under the summations in ν, while $\gamma_\mu = \mu_0 \mu \hat{U}_\mu / g$.

An example of the magnetic potentials and the flux density is shown in Figure 5.3.

5.3.5 Torque derivation

The torque is obtained integrating the Lorentz force density $\mathscr{K}_s \mathscr{B}_g$, both the function of ϑ_r and ϑ_m, along the air-gap surface and multiplying the result by the lever-arm $D_s/2$ [34]:

$$T_m(\vartheta_m) = -\frac{D_s}{2} \int_0^{2\pi} \mathscr{B}_g(\vartheta_r, \vartheta_m) \mathscr{K}_s(\vartheta_r, \vartheta_m) \frac{D_s L_{stk}}{2} \, d\vartheta_r$$

Remembering (5.15), it can be derived as

$$T_m = \frac{\mu_0 D_s^2 L_{stk}}{4g} \Bigg[\underbrace{\int_0^{2\pi} \mathscr{U}_s \mathscr{K}_s \, d\vartheta_r}_{A} - \underbrace{\int_0^{2\pi} \mathscr{U}_r \mathscr{K}_s \, d\vartheta_r}_{B} \Bigg]$$

The first integral, labeled as A, is zero since $\mathscr{U}_s$ and $\mathscr{K}_s$ are orthogonal functions. Therefore, the torque is only due to the interaction of electric loading $\mathscr{K}_s$ and the magnetic scalar potential of the rotor $\mathscr{U}_r$.

The final expression of the torque is

$$T_{\mathrm{m}}(\vartheta_{\mathrm{m}}) = -\frac{\mu_0 D_{\mathrm{s}}^2 L_{\mathrm{stk}}}{g} \sum_{\nu,k} \frac{\hat{K}_\nu}{\nu} \sin \nu\tfrac{\pi}{2} \cdot \cos((\nu-1)p\vartheta_{\mathrm{m}} - \alpha_i^{\mathrm{e}})[\sin(\nu\boldsymbol{\vartheta}^{\mathrm{T}})\boldsymbol{G}\boldsymbol{u}_{\mathrm{r}}]$$

5.3.6 Tooth flux density

To compute the stator tooth flux density, the air-gap flux density $\mathscr{B}_{\mathrm{g}}(\vartheta_{\mathrm{r}}, \vartheta_{\mathrm{m}})$ in (5.16) has to be referred to the stator reference frame. It is

$$\mathscr{B}_{\mathrm{g}}(\vartheta_{\mathrm{s}}, \vartheta_{\mathrm{m}}) = \sum_\nu \left[\frac{\alpha_\nu}{\nu}\cos(\nu p\vartheta_{\mathrm{s}} - \nu p\vartheta_{\mathrm{m}}) + \frac{\beta_\nu}{\nu}\sin(\nu p\vartheta_{\mathrm{s}} - \nu p\vartheta_{\mathrm{m}})\right] + \sum_\mu \frac{\gamma_\mu}{\mu}\sin(\mu p\vartheta_{\mathrm{s}} - \mu p\vartheta_{\mathrm{m}})$$

The stator tooth flux density can be obtained through the scaled average of the air-gap flux density [39–41]:

$$B_{\mathrm{t}}(\vartheta_{\mathrm{m}}) = \frac{p_{\mathrm{s}}}{\alpha_{\mathrm{s}} w_{\mathrm{t}} k_{\mathrm{pack}}} \int_{\vartheta_d}^{\vartheta_d+\alpha_{\mathrm{s}}} \mathscr{B}_{\mathrm{g}}(\vartheta_{\mathrm{s}}, \vartheta_{\mathrm{m}})\,\mathrm{d}\vartheta_{\mathrm{s}} \tag{5.17}$$

where p_{s} is the slot pitch, α_{s} the slot angle, w_{t} the tooth width, and k_{pack} is the ratio between the total effective iron and the stack length.

It can be demonstrated that the tooth flux density assumes q different behaviors, q being the number of slots per pole and per phase, defined as $q = Q/(m\,2p)$. Thus the integration extremes can be expressed as

$$\vartheta_d = (d-1)\alpha_{\mathrm{s}} - \gamma_{\mathrm{a}}\,, \qquad d = 1, \ldots, q \tag{5.18}$$

where γ_{a} identifies the angular position of the magnetic axis of phase a with respect to the first slot (see Figure 5.4).

5.3.7 Stator yoke flux density

The flux in the stator yoke—also referred to as back-iron—is obtained integrating the air-gap flux density over a pole pitch, then the flux density is

$$B_{\mathrm{y}}(\vartheta_{\mathrm{m}}) = \frac{1}{2h_{\mathrm{y}}k_{\mathrm{pack}}} \int_{-\gamma_{\mathrm{a}}}^{\pi/p-\gamma_{\mathrm{a}}} \mathscr{B}_{\mathrm{g}}(\vartheta_{\mathrm{s}}, \vartheta_{\mathrm{m}})\frac{D_{\mathrm{s}}}{2}\,\mathrm{d}\vartheta_{\mathrm{s}} \underset{q\ \text{integer}}{\overset{\uparrow}{=}} \frac{w_{\mathrm{t}}}{2h_{\mathrm{y}}}\sum_{i=1}^{Q/2p} B_{\mathrm{t},i}(\vartheta_{\mathrm{m}})$$

where $h_{\mathrm{y}} = (D_{\mathrm{se}} - D_{\mathrm{s}})/2$ is the stator yoke height.

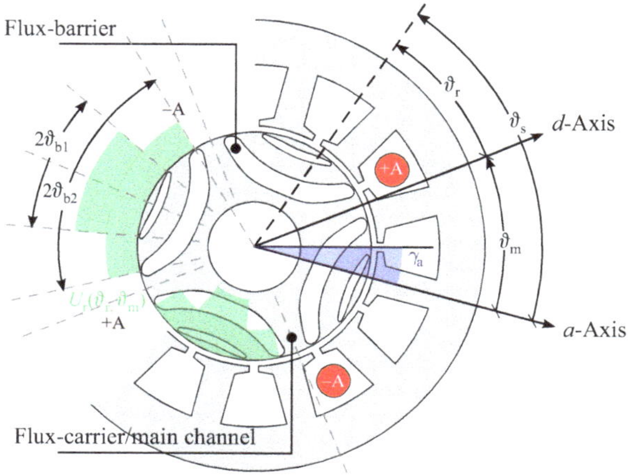

Figure 5.4 Sketch of an SyR motor with two flux-barriers per pole together with the reference axes and the barrier angle quotes. The colored area on top of the flux-barriers represents the rotor magnetic scalar potential distribution for half the machine

5.3.8 *Rotor fluxes*

In the linear case, the fluxes entering the rotor are directly obtained from the solution of the magnetic circuit in Figure 5.2. Both the island and the barrier fluxes can be computed. The iron path closer to the shaft is referred to as channel. The remaining iron paths for the d-axis flux are referred to as islands. They assume a different potential, due to the q-axis flux, and they are numbered accordingly to the number of the barrier beneath them. The barrier fluxes are obtained solving the magnetic network depicted in Figure 5.2. The d-axis channel flux comes from $\phi_{cd} = \varphi_{N_b+1} - \varphi_{2N_b+2}$ that simply sums the two fluxes to obtain the whole direct flux of the channel (Figure 5.5). Then

$$B_{cd} = \frac{\phi_{cd}}{w_{ch}\, k_{pack} L_{stk}} \tag{5.19}$$

where $w_{ch} \approx (1 - k_{air})D_r \sin((\pi/2p) - \vartheta_{bN_b})$ is an approximation of the minimum width of the channel.

For the q-axis flux of the main channel, just half of the last barrier flux can be considered, with $\phi_{cq} = \phi_{b,N_b}/2$. Thus

$$B_{cq} = \frac{\phi_{cq}}{l_{ch}\, k_{pack} L_{stk}} \tag{5.20}$$

where $l_{ch} = (D_r - D_{sh})/2$. Then, it is

$$|\boldsymbol{B}_c| = \sqrt{B_{cd}^2 + B_{cq}^2} \tag{5.21}$$

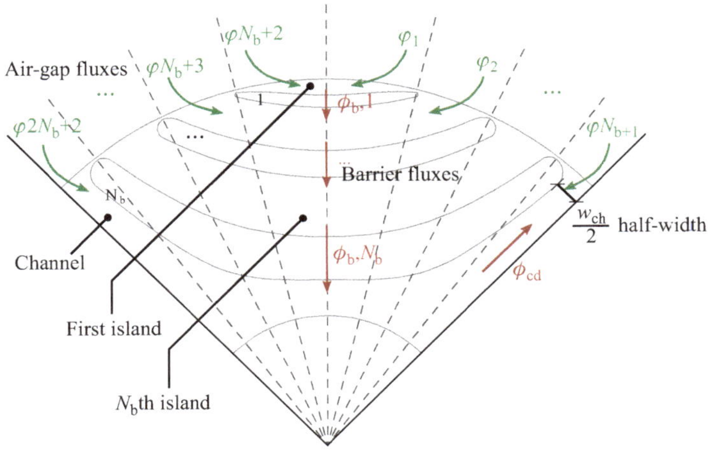

Figure 5.5 Sketch of the main parts of the rotor structure and rotor fluxes direction

The d-axis fluxes of the islands result directly from the magnetic circuit fluxes as

$$B_{\mathrm{id},j} = \frac{\varphi_j}{k_{\mathrm{pack}} L_{\mathrm{stk}} w_{\mathrm{isl},j}} \tag{5.22}$$

Regarding the q-axis fluxes of the islands, the average of the fluxes above and beneath each island is computed. An average cross-sectional area of each island is considered, so that

$$\vartheta_{\mathrm{isl},j} = \frac{\vartheta_{\mathrm{b},j} + \vartheta_{\mathrm{b},j-1}}{2}$$

$$l_{\mathrm{isl},j} = \vartheta_{\mathrm{isl},j} D_{\mathrm{s}}$$

$$\begin{pmatrix} \phi_{\mathrm{iq},1} \\ \phi_{\mathrm{iq},2} \\ \phi_{\mathrm{iq},3} \\ \vdots \\ \phi_{\mathrm{iq},N_{\mathrm{b}}} \end{pmatrix} = \frac{1}{2} \begin{bmatrix} 2 & & & & \\ 1 & 1 & & & \\ & 1 & 1 & & \\ & & \ddots & \ddots & \\ & & & 1 & 1 \end{bmatrix} \begin{pmatrix} \phi_{\mathrm{b}1} \\ \phi_{\mathrm{b}2} \\ \phi_{\mathrm{b}3} \\ \vdots \\ \phi_{\mathrm{b}N_b} \end{pmatrix}$$

$$B_{\mathrm{iq},j} = \frac{\phi_{\mathrm{iq},j}}{k_{\mathrm{pack}} L_{\mathrm{stk}} l_{\mathrm{isl},j}}$$

$$|\boldsymbol{B}_{\mathrm{i},j}| = \sqrt{B_{\mathrm{id},j}^2 + B_{\mathrm{iq},j}^2}$$

5.4 Radial iron ribs computation

The radial iron ribs are designed considering the stress due to the centrifugal force. The mechanical contribution of the tangential iron ribs is neglected and their thicknesses are set equal to the minimum manufacturing length (e.g., the lamination thickness). The main idea used to design the jth radial rib consists of three main steps. First, the jth iron cross section, $A_{\mathrm{Fe}j}$, delimited by the jth flux-barrier and rotor periphery is computed. Then, the associated centrifugal force is

$$F_{cj} = \omega_{\mathrm{m}}^2 R_{\mathrm{G},j}\, \gamma_{\mathrm{Lam}} A_{\mathrm{Fe}j} L_{\mathrm{stk}} \tag{5.23}$$

where w_{pm} is the machine maximum speed, $R_{\mathrm{G},j}$ is the iron cross-section center of gravity radius, γ_{Lam} is the lamination mass density, and L_{stk} is the machine stack length. Finally, assuming that the jth radial rib has to sustain such a centrifugal force, its thickness is given by

$$t_{\mathrm{rib},j} = \nu_\sigma \frac{F_{cj}}{\sigma_{\mathrm{r}} L_{\mathrm{stk}}} \tag{5.24}$$

where σ_{r} is the tensile strength of the lamination while ν_σ is a safety factor, typically set in the range between 2 and 3 [42], which takes into account the fatigue effect.

The two methods proposed in the following differs from each other only for the computation of the jth iron cross-sectional area.

5.4.1 Simplified method

The method proposed in this section allows us to compute analytically both the rotor iron channel masses and radial ribs size. Such a method refers to the simplified rotor geometry reported in Figure 5.6 in which the jth iron channel cross section can be easily computed from the circular segment delimited by the jth flux-barrier [43].

Figure 5.6(a) shows that the first iron channel cross section is twice the circular segment, A_{cs1}, defined by the first flux-barrier, $2\vartheta_{\mathrm{b1}}$, so that it can be computed as follows:

$$A_{\mathrm{Fe1}} = 2A_{\mathrm{cs1}} = \frac{D_{\mathrm{r}}^2}{4}\left(2\vartheta_{\mathrm{b1}} - \sin 2\vartheta_{\mathrm{b1}}\right) \tag{5.25}$$

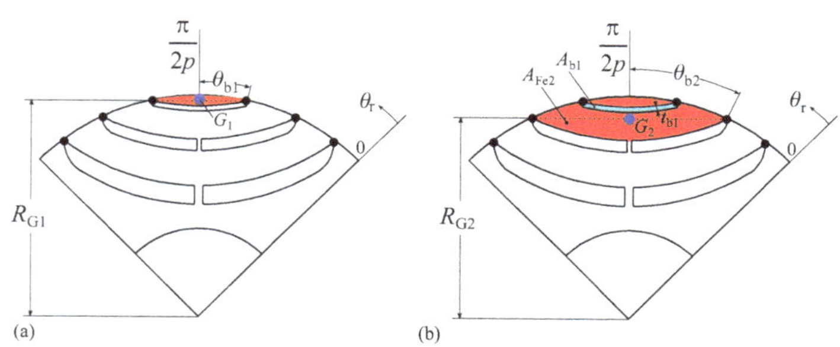

Figure 5.6 Computation of iron channel cross sections in simplified method. (a) First iron area and its barycenter definitions and (b) Second iron area and its barycenter definitions

where D_r is the rotor diameter. The radius of the center of gravity, G_1, is

$$R_{G,1} = \frac{D_r}{2} \cos \vartheta_{b1} \tag{5.26}$$

and the centrifugal force is given by

$$F_{c1} = \gamma_{Lam} \omega_m^2 R_{G,1} A_{Fe,1} L_{stk} = \gamma_{Lam} \omega_m^2 \frac{D_r}{2} A_{Fe,1} L_{stk} \cos \vartheta_{b1} \tag{5.27}$$

with (5.24), the iron rib thickness results in

$$t_{rib,1} = \nu_\sigma \frac{F_{c1}}{\sigma_r L_{stk}} = \nu_\sigma \frac{\gamma_{Lam} \omega_m^2 D_r^3 (2\vartheta_{b1} - \sin 2\vartheta_{b1}) \cos(\vartheta_{b1})}{8\sigma_r} \tag{5.28}$$

The red shaded area highlighted in Figure 5.6(b) is the total iron area used to design the second radial rib. In this case, the cross-sectional area of the first flux-barrier, A_{fb1}, has to be subtracted by the total area delimited by the rotor periphery and the second flux-barrier (i.e. $2A_{cs2}$). To do that, it is convenient to define the following ratio:

$$k_2 = \frac{A_{fb1}}{2A_{cs2}} = \frac{4 t_{b1} \vartheta_{b1}}{D_r^2 [2\vartheta_{b2} - \sin(2\vartheta_{b2})]} \tag{5.29}$$

from which it is possible to compute the actual iron cross-sectional area:

$$A_{Fe2} = (1 - k_2) \frac{D_r^2}{4} [2\vartheta_{b2} - \sin(2\vartheta_{b2})] \tag{5.30}$$

the relative centrifugal force and radial rib thickness can be computed, with the proper substitutions, by means of (5.23) and (5.24), respectively. In general, the computation of the jth iron cross-sectional area can be extended as follows:

$$A_{Fej} = (1 - k_j) \frac{D_r^2}{4} [2\vartheta_{bj} - \sin(2\vartheta_{bj})] \tag{5.31}$$

where k_j is

$$k_j = \frac{4 \sum_{i=1}^{j-1} t_{bi} \vartheta_{bi}}{D_r^2 [2\vartheta_{bj} - \sin(2\vartheta_{bj})]} \tag{5.32}$$

Assuming a uniform distribution of the magnetic insulation, for a preliminary design, it is possible to substitute k_j with the magnetic insulation ratio k_{air}, defined in section 3.3.

5.4.2 Accurate method

The method proposed in this section is more general than the previous one since it does not refer to a specific rotor geometry. Its accuracy relies on the fact that the iron cross-sectional areas, referred to as iron channels, and centers of gravity are computed using the FE mesh. In particular, the jth iron channel cross-sectional area is given by

$$A_{Fej} = \sum_{i=1}^{n} a_{ij} \tag{5.33}$$

where a_{ij} is the ith mesh element of the jth iron channel and n is its mesh element number. The relative center of gravity, $R_{G,j}$, is

$$R_{G,j} = \frac{\sum_{i=1}^{n} a_{ij} R_{G,ij}}{\sum_{i=1}^{n} a_{ij}} \tag{5.34}$$

where $R_{G,ij}$ is the center of radius of the element a_{ij}. Finally, the centrifugal force and radial rib thickness are computed by means of (5.23) and (5.24), respectively.

5.5 Computation of the sleeve thickness

A retaining sleeve can be adopted to counteract the centrifugal force through a uniform radial pressure. The formulation to design a retaining sleeve comes from the membrane theory since its thickness, t_{sleeve}, is much smaller than the rotor radius. Hence, the relationship between the centrifugal force and the radial pressure is the following:

$$F_{c,sleeve} = \int_{-\pi/2}^{\pi/2} P \cos\vartheta \, \frac{DL_{stk}}{2} \, d\vartheta = PDL_{stk} \tag{5.35}$$

where P is the pressure at the air gap (i.e., contact pressure) and ϑ is the angular coordinate. As reported in Figure 5.7, the pressure is assumed to be uniformly distributed along the rotor periphery so as to link the centrifugal force and the sleeve thickness through

$$t_{sleeve} = \nu_{sleeve} \frac{F_{c,sleeve}}{2\sigma_{sleeve} L_{stk}} \tag{5.36}$$

where ν_{sleeve} and σ_{sleeve} are the sleeve material safety factor and tensile strength, respectively.

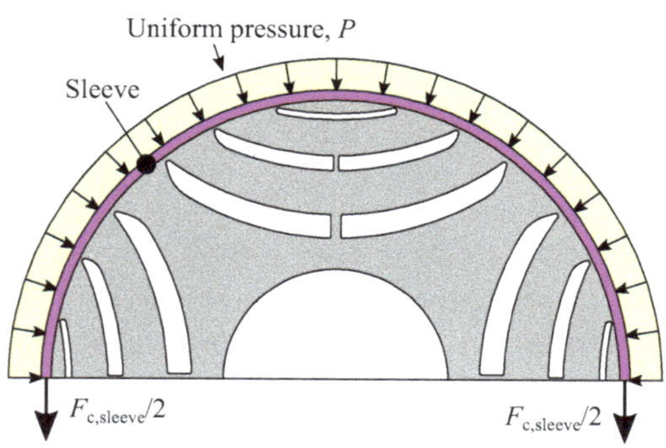

Figure 5.7 Computation of the sleeve thickness

5.6 Analytical model of the SyR motor

In this section, the analytical model of the SyR machine is developed. Only the fundamental MMF is considered. This leads to a simple computation of the d- and q-axis inductances, through which the average electromagnetic torque is derived.

As per usual, only the radial ribs are taken into account while the tangential ones are neglected.

5.6.1 d-Axis inductance computation

The fundamental harmonic amplitude of the electric loading along the d-axis is

$$\hat{K}_{sd} = \frac{3k_w N_s \hat{I}_d}{\pi D_s} \tag{5.37}$$

where k_w is the winding factor, N_s is the number of series conductors per phase, D_s is the stator inner diameter, and I_d is the d-axis current.

Figure 5.8(a) sketches the flux lines through the rotor with d-axis electric loading. Usually, in SyR machines the magnetic reluctance along the d-axis is mainly due to the air-gap reluctance. However, in High-Speed (HS) applications, the rotor saturation is higher and greatly affected by the radial rib sizes so that it has to be considered to obtain reliable results from the analytical model. In the present model, the rotor saturation is taken into account multiplying the actual air-gap length by the saturation factor, k_{sat}, that is greater than 1.

The fundamental of the scalar magnetic potential, $\hat{U}_{sd}$, along the stator inner diameter can be computed integrating the electric loading yielding

$$\hat{U}_{sd} = \frac{\hat{K}_{sd} D_s}{2p} \tag{5.38}$$

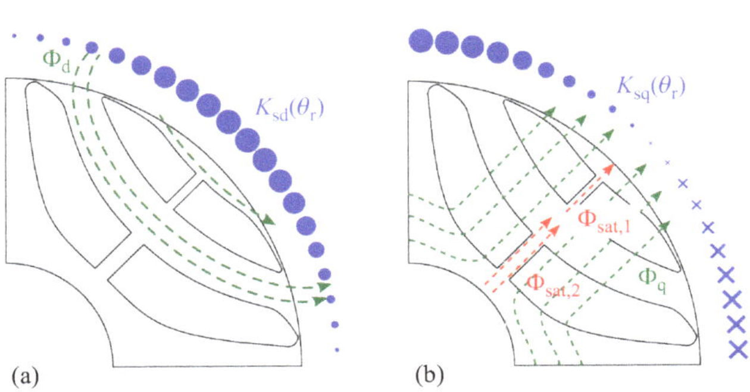

Figure 5.8 Qualitative flux lines for a rotor with two flux-barriers per pole: (a) d-axis excitation and (b) q-axis excitation

and the fundamental harmonic amplitude of the d-axis air-gap flux density, $\hat{B}_{gd}$, can be computed by means of Ampère's law as follows:

$$\hat{B}_{gd} = \mu_0 \frac{\hat{U}_{sd}}{g\, k_C\, k_{sat}} \tag{5.39}$$

where p is the number of pole pairs and k_C is the Carter coefficient that considers the slot opening effect. The d-axis flux linkage is

$$\hat{\Lambda}_d = \mu_0 \frac{k_w N_s}{(2p)^2} \frac{D_s^2 L_{stk} \hat{K}_{sd}}{g\, k_C\, k_{sat}} \tag{5.40}$$

the iron saturation and the slot openings reduce the value of d-axis flux, since these factors decrease the maximum anisotropy of the rotor and have negative effect on the average torque.

5.6.2 q-Axis inductance computation

Along the q-axis the flux flows through the rotor flux-barriers and the air gap as reported qualitatively in Figure 5.8(b).

The q-axis magnetic circuit of a three flux-barriers per pole rotor and its schematic representation are shown in Figure 5.9.

The fundamental harmonic amplitude of the electric loading along the q-axis is

$$\hat{K}_{sq} = \frac{3 k_w N_s \hat{I}_q}{\pi D_s} \tag{5.41}$$

where I_q is the q-axis current. The relative q-axis scalar magnetic potential amplitude is

$$\hat{U}_{sq} = \frac{\hat{K}_{sq} D_s}{2p} \tag{5.42}$$

Its waveform is split into parts according to the rotor geometry. The jth generator, $\mathscr{U}_s$ (with $j = 1, 2, 3$ in case of three flux-barriers per pole), represents the average value

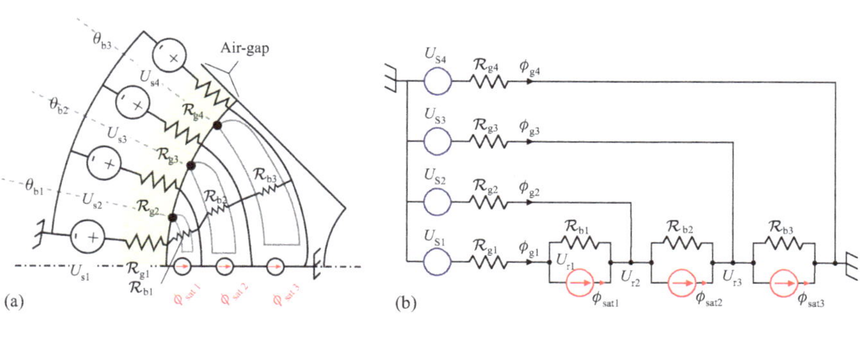

Figure 5.9 Magnetic circuit of rotor with three flux-barriers per pole: (a) schematic representation of the magnetic circuit and (b) magnetic circuit

of the stator scalar magnetic potential computed between two flux-barrier angles. It is given by integrating the stator scalar magnetic potential as follows:

$$\mathscr{U}_{s} = \frac{1}{\vartheta_{bj} - \vartheta_{b(j-1)}} \cdot \int_{\vartheta_{b(j-1)}}^{\vartheta_{bj}} \hat{U}_{sq} \sin(p\vartheta_{r})\, d\vartheta_{r} \tag{5.43}$$

The equivalent air-gap magnetic reluctance in front of the jth flux-barrier, already considering the parallel of the two reluctances in symmetrical position with respect to the q-axis, is proportional to the air-gap region delimited by the flux-barrier angles ϑ_{bj} and $\vartheta_{b(j-1)}$, computed as

$$\mathfrak{R}_{gj} = \frac{g''}{\mu_0 L_{stk} D_s \left[\vartheta_{bj} - \vartheta_{b(j-1)}\right]} \tag{5.44}$$

where $\vartheta_{b0} = 0$ and $\vartheta_{b(N_b+1)} = (\pi/2)$. Each flux-barrier is modeled by a magnetic reluctance in parallel to a flux generator. The magnetic reluctance of jth flux-barrier is a function of the geometrical parameters as

$$\mathfrak{R}_{bj} = \frac{t_{bj}}{\mu_0 L_{stk} l_{bj}} \tag{5.45}$$

where t_{bj} and l_{bj} are the equivalent thickness and length of the jth flux-barrier, respectively. The magnetic flux generator, ϕ_{satj}, represents the flux flowing through the iron ribs, computed as

$$\phi_{satj} = B_{sat} t_{rib,j} L_{stk} \tag{5.46}$$

where $t_{rib,j}$ is the thickness of the jth iron rib and B_{sat} is the iron saturation flux density (usually set in the range between 1.8 T and 2 T). Let us note that the $t_{rib,}$ thickness affects the rotor anisotropy and it must be kept as thin as possible, according to the centrifugal force stress.

The scalar magnetic potentials of the iron ribs are computed by means of Thévenin's theorem multiplying the flux flowing through the iron rib by the magnetic reluctances:

$$U_{satj} = \mathfrak{R}_{bj}\phi_{satj} = \mathfrak{R}_{bj} B_{sat} t_{rib,j} L_{stk} \tag{5.47}$$

the transformation is explained in Figure 5.10(a) and (b). The first one shows the initial one-port component with the jth flux generator in parallel with the jth flux-barrier reluctance. The second one shows the equivalent circuit composed by a potential generator U_{satj} connected in series with the flux-barrier reluctance $\mathfrak{R}_{bj}$.

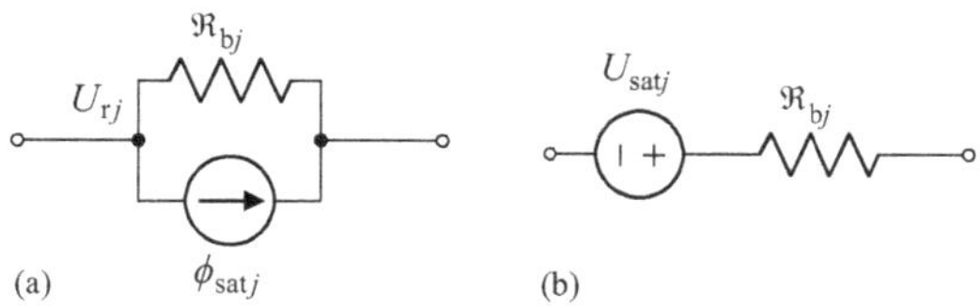

Figure 5.10 (a) One-port component of a flux-barrier with a radial rib and (b) equivalent circuit with Thévenin's theorem

Solving the magnetic circuit, the three islands potentials U_{r1}, U_{r2} and U_{r3} are found. They are, respectively,

$$\begin{aligned} U_{r1} &= \Re_{b1}\phi_{g1} + U_{r2} - U_{sat1} \\ &= -\frac{t_{b1}}{\mu_0 L_{stk} l_{b1}} \int\limits_{(\pi/2p)-\vartheta_{b1}}^{(\pi/2p)+\vartheta_{b1}} B_g(\vartheta_r)\frac{D_s}{2}L_{stk}d\vartheta_r + U_{r2} - U_{sat,1} \end{aligned}$$

$$\begin{aligned} U_{r2} &= \Re_{b2}(\phi_{g1} + \phi_{g2}) + U_{r3} - U_{sat2} \\ &= \Re_{b2}\left[\phi_{g1} - 2\int\limits_{(\pi/2p)+\theta_{b1}}^{(\pi/2p)+\theta_{b2}} B_g(\theta_r)\frac{L_{stk}D_s}{2}d\theta_r\right] + U_{r3} - U_{sat2} \end{aligned}$$

$$\begin{aligned} U_{r3} &= \Re_{b3}(\phi_{g1} + \phi_{g2} + \phi_{g3}) - U_{sat3} \\ &= \Re_{b3}\left[(\phi_{g1} + \phi_{g2}) - 2\int\limits_{(\pi/2p)+\vartheta_{b3}}^{(\pi/2p)+\vartheta_{b2}} B_g(\theta_r)\frac{L_{stk}D_s}{2}d\theta_r\right] - U_{sat3} \end{aligned}$$

The rotor island potentials can be finally written as

$$\begin{aligned} U_{r1} &= -\frac{\hat{K}_{sq}}{p^2}D_s\rho_1 + f_{r1} \\ U_{r2} &= -\frac{\hat{K}_{sq}}{p^2}D_s\rho_2 + f_{r2} \\ U_{r3} &= -\frac{\hat{K}_{sq}}{p^2}D_s\rho_3 + f_{r3} \end{aligned}$$

where the three terms f_{rj} (with $j = 1, 2, 3$) are functions of the scalar magnetic potentials U_{sat1}, U_{sat2} and U_{sat3} of the rotor ribs, that is $f_{rj} = f_{rj}(U_{sat1}, U_{sat2}, U_{sat3})$. They depend on the rotor geometry and can be obtained solving the magnetic model presented earlier.

Equation (6.2) highlights the impact of the rotor ribs. In SyR machines operating at low speed, the impact of the rotor ribs is negligible [44,45], so that the terms f_{rj} are often omitted in (5.14).

In HS applications, the radial rib thicknesses are so big that their impact on the machine performance cannot be neglected. Therefore, it is mandatory to consider the scalar magnetic potentials U_{sat1}, U_{sat2} and U_{sat3} in the model.

As for the d-axis, the flux along the quadrature axis is computed integrating the air-gap flux density in a pole pitch:

$$\phi_c = \int_0^{\pi/p} \mu_0 \frac{-U_s(\theta_r) + U_r(\theta_r)}{g''} \frac{D_s}{2} L_{stk}\, d\vartheta_r \tag{5.48}$$

finally, the flux linkage, Λ_q, is

$$\Lambda_c = \mu_0 \frac{k_w N_s}{(2p)^2} \frac{D_s^2 L_{stk} \hat{K}_{sq}}{g''} \left[1 - 4\sum_{j=1}^{3} (\rho_j - \rho_{j-1})(\vartheta_{bj} - \vartheta_{b(j-1)})\right] \tag{5.49}$$

where the terms ρ_j are detailed reported in [36]. The first term of (5.49) does not consider neither the flux-barriers nor the iron ribs. For this reason, its expression is the same as for the d-axis flux linkage (5.40) and can be considered constant. The terms with ρ_j (with $j = 1, 2, 3$) depend on the geometry of the rotor flux-barriers. They are negative, in the sense, that they tend to decrease the flux along the q-axis for given current, increasing the rotor anisotropy. Therefore, their effect is beneficial as far as the electromagnetic torque is concerned.

5.6.3 Torque and power equations

Substituting (5.47) in (5.49), and performing some computations, the electromagnetic torque results in a function of the rotor geometry, the two axis currents and the rotor speed. It can be expressed as

$$T = X - Y\omega_m^2 \tag{5.50}$$

where the variables $X = X(I_d, I_q)$ and $Y = Y(I_d)$, reported in Appendix C, contain geometrical and winding parameters.

Let us remark that (5.50) indicates the torque exhibited by a motor designed for reaching the maximum speed ω_{pm}. It is worth noticing that the first term is independent of the rotor speed, i.e., it is constant, while the second one is a square function of the speed. Multiplying (5.50) by the mechanical speed ω_m, the output power is

$$P = X\omega_{pm} - Y\omega_{pm}^3 \tag{5.51}$$

The critical speed at which the output power (5.51) is maximum is obtained setting to zero the derivative of the power with respect to the speed. The critical speed then results in

$$\omega_{m,cr} = \sqrt{\frac{X}{3Y}} \tag{5.52}$$

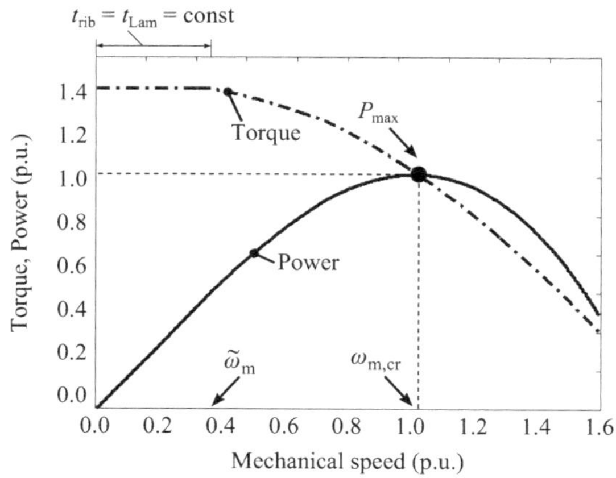

Figure 5.11 Torque and power versus mechanical speed. Each point of the curves refers to a different rotor geometry designed for a specific speed

Figure 5.11 shows the qualitative behavior of the torque and the power versus the mechanical speed. In the shaded region, the torque is constant since all the rib thicknesses, equal to the lamination width (t_{Lam}), do not vary because they are sufficient to counteract the low centrifugal stress. Thus, the power increases linearly with the speed. The speed limit, $\tilde{\omega}_m$, after which the radial ribs have to be modified, can be computed setting $t_{rib} = t_{Lam}$ in (5.24) for each flux-barrier.

For $\omega_{pm} > \tilde{\omega}_m$ the thickness of the iron ribs increases with the square of the rotor speed as shown in (5.24). The torque decreases monotonically because of (5.50), while the power increases until the critical speed, $\omega_{m,cr}$, computed by (5.52). After this point, the power decreases: this means than the anisotropy of the machine is compromised by the large iron ribs.

5.7 Analytical model with saturation

The previous analytical model works properly in the case of linear behavior of ferromagnetic material, that is in the first part of the iron *B–H* curve (the curve is reported in Figure 5.12). However, this condition is rarely met. In SyR motors the rotor iron paths are saturated to achieve a quite high torque density and power factor. Thus, in practice, the saturation of the machine has to be taken into account.

Hereafter, the saturation is taken into account through a discretization of the flux line paths into sections. In particular, a section can be a stator tooth, a stator yoke sector or a rotor half iron path. The path of a flux line results in a succession of these sections. Therefore, a magnetic voltage drop and a saturation factor are associated with each

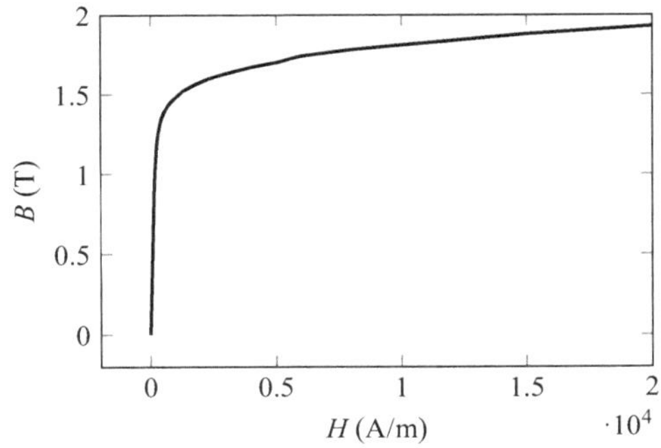

Figure 5.12 Magnetization curve of the ferromagnetic material used

flux line and referred to the air-gap. At first the rotor iron paths are considered, and then the stator teeth and yoke.

5.7.1 Rotor channel and islands magnetic voltage drop

The magnetic voltage drop of the rotor channel is estimated from (5.21). Since the flux density is computed in the smallest cross section, it is the maximum value of the channel. The average is estimated as $B_{\rm c}^{\star} = 0.85\,|\boldsymbol{B}_{\rm c}|$ and, from the iron B–H curve, the magnetic field $H_{\rm c}$ is obtained (in the following, it is indicated as $B \rightarrow H$). Then the magnetic voltage drop results in $\psi_{\rm c} = H_{\rm c}\, l_{\rm ch}$. Similarly, the islands have a complex geometry and a different flux density in every point. The middle section of the island is considered, where the width is precisely known. It should be noted that the two half-island paths have different voltage drops due to part of the entering elementary flux that crosses the barrier and does not reclose on the other side. To recover this difference, a magnetic voltage drop is computed for every half island using the elementary flux.The procedure is sketched as follows:

$$|\boldsymbol{B}_{{\rm i},j}| \quad \longrightarrow \quad H_{{\rm i},j} \quad \text{then} \quad \psi_{{\rm i},j} = H_{{\rm i},j}\,\frac{l_{{\rm isl},j}}{2}$$

The same procedure is carried out for the second (south) half pole.

The vector of rotor magnetic voltage drops, corresponding to the sequence of channel and islands starting from the origin of $\vartheta_{\rm r}$, is

$$\boldsymbol{\psi}_{\rm rp} = \{\psi_{\rm c},\, \underbrace{\psi_{{\rm i},N_{\rm b}},\ldots,\psi_{{\rm i},1}}_{\boldsymbol{\psi}_{\rm i}^{\rm N}}\ \underbrace{\psi_{{\rm i},N_{\rm b}+2},\ldots,\psi_{{\rm i},2N_{\rm b}+2}}_{\boldsymbol{\psi}_{\rm i}^{\rm S}},\, \psi_{\rm c}\}^{\rm T}$$

where $\boldsymbol{\psi}_{\rm i}^{\rm N}$ and $\boldsymbol{\psi}_{\rm i}^{\rm S}$ are the vectors of the first half pole (identified by the superscript N meaning north) and the second half pole (identified by the superscript S meaning south) voltage drops, respectively.

Numerically, this vector is distributed into a number of points according to the angle spanned. In the end, the distribution of the rotor magnetic voltage drops is equivalent to the distribution of the scalar magnetic potentials along the air-gap:

$$\boldsymbol{\psi}_{\mathrm{r}} = \{\underbrace{\underbrace{\psi_{\mathrm{c}}, \ldots, \psi_{\mathrm{c}}}_{N_{N_{\mathrm{b}}}+1}, \underbrace{\psi_{\mathrm{i},N_{\mathrm{b}}}, \ldots, \psi_{\mathrm{i},N_{\mathrm{b}}}}_{N_{N_{\mathrm{b}}}}, \ldots, \underbrace{\psi_{\mathrm{i},1}, \ldots, \psi_{\mathrm{i},1}}_{N_1}, \ldots}_{Res/2p}\}^{\mathrm{T}}$$

where Res is the number of points used for discretizing the space $[0, 2\pi]$ along the air-gap, and the subscript of N refers to the corresponding island.

5.7.2 Stator teeth and yoke magnetic voltage drops

The instantaneous tooth flux density in the saturation case is obtained through the numerical integration of (5.17). Once the flux density for every tooth of interest (typically Q/p teeth for an integer-slot winding) is computed, the magnetic voltage drops are derived, as follows:

$$|B_{\mathrm{t},j}| \longrightarrow H_{\mathrm{t},j} \quad \text{then} \quad \psi_{\mathrm{t},j} = H_{\mathrm{t},j}\, h_{\mathrm{s}} \tag{5.53}$$

where h_{s} is the tooth height.

The back-iron corresponding to a pole pair is split into Q/p parts, each one covering a slot angle, as illustrated in Figure 5.13 (dotted lines). The flux density can be derived for each part as

$$\tilde{\phi}_{\mathrm{y}}(k) = w_{\mathrm{t}} L_{\mathrm{stk}} \sum_{j=1}^{k} B_{\mathrm{t},j} \qquad k = 1, \ldots, Q/p$$

$$\phi_{\mathrm{y}}(k) = \tilde{\phi}_{\mathrm{y}}(k) - \frac{1}{Q/p} \sum_{j=1}^{Q/p} \tilde{\phi}_{\mathrm{y}}(j)$$

$$B_{\mathrm{y}}(k) = \frac{|\phi_{\mathrm{y}}(k)|}{h_{\mathrm{y}} L_{\mathrm{stk}}} \longrightarrow H_{\mathrm{y}}(k)\,, \quad \psi_{\mathrm{y}}(k) = H_{\mathrm{y}}(k)\, \Delta l$$

where $\Delta l = \pi(D_{\mathrm{se}} - h_{\mathrm{y}})/Q$ is the average length of one back-iron part. Once the stator yoke voltage drops are computed in all sections, the flux path for all flux lines of interest have to be correctly identified. Two of these lines are shown in Figure 5.13. Summing the voltage drops along these lines is equivalent to consider that each flux line flows for a full pole and it encounters $Q/2p$ voltage drops along its path. The voltage drops are summed with the sign of the corresponding flux in the same

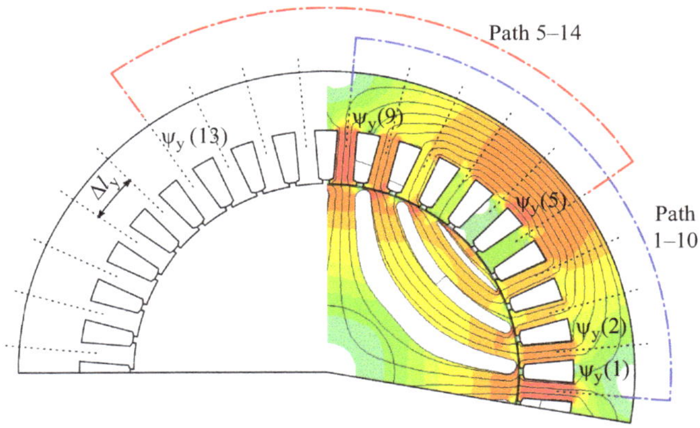

Figure 5.13 Real flux lines and main integration paths ($Q = 36$, $2p = 4$)

part. Finally, every flux line has gathered the same overall magnetic voltage drop of the actual flux line of that particular tooth. Therefore, it is

$$\psi_{\mathrm{yt}}(n) = \left| \frac{1}{2} \sum_{k=n}^{n+\frac{Q}{2p}-1} \psi_{\mathrm{y}}(k) \,\mathrm{sign}\, \phi_{\mathrm{y}}(k) \right|, \quad n = 1, \ldots, Q/2p \tag{5.54}$$

where the double subscript "yt" indicates that the yoke drop is referred to the tooth. The $1/2$ occurs since half the path is counted. For example, let us consider the path 1–10 shown in Figure 5.13, which covers $\psi_{\mathrm{y}}(1)$ to $\psi_{\mathrm{y}}(9)$: it starts from tooth number one (below the first slot) and it ends in tooth ten. The total magnetic voltage drop associated with this line is

$$2\psi_{\mathrm{yt}}(1) = \psi_{\mathrm{y}}(1) + \psi_{\mathrm{y}}(2) + \cdots + \psi_{\mathrm{y}}(9) \tag{5.55}$$

On the other hand, along the path 5–14 it is

$$\begin{aligned} 2\psi_{\mathrm{yt}}(5) &= \psi_{\mathrm{y}}(5) + \cdots + \psi_{\mathrm{y}}(9) - \psi_{\mathrm{y}}(10) - \cdots - \psi_{\mathrm{y}}(13) \\ &\cong \psi_{\mathrm{y}}(5) \end{aligned}$$

The last equivalence is the main assumption of this computation. In fact, it implies that $\psi_{\mathrm{y}}(6)$ is equal to $\psi_{\mathrm{y}}(13)$ and so on, and the remaining term is just $\psi_{\mathrm{y}}(5)$. Even if the machine is not symmetrically magnetized, this is still a fairly good approximation thanks to the lower value of the q-axis flux with respect to the d-axis flux. Despite the high values of α_i^{e}, it can be shown that this approximation is good enough. Similarly to the rotor, the stator magnetic voltage drops can be collected and distributed along the discretization of the angular coordinate at the air-gap.

5.7.3 Total magnetic voltage drop and saturation factor

In order to combine the rotor and stator magnetic voltage drop distributions, the rotor position and the first tooth displacement have to be taken into account through the shift of one of the two distributions, according to the adopted reference frame. The air-gap magnetic voltage drop can be easily computed from the flux density obtained in the previous iteration:

$$\boldsymbol{\Psi}_{\mathrm{g}} = \boldsymbol{H}_{\mathrm{g}}\, g = \frac{\boldsymbol{B}_{\mathrm{g}}}{\mu_0}\, g \tag{5.56}$$

All the magnetic voltage drops are summed to obtain the total voltage drop, which is again referred to the air-gap:

$$\boldsymbol{\Psi}_{\mathrm{tot}} = \boldsymbol{\Psi}_{\mathrm{g}} + \boldsymbol{\Psi}_{\mathrm{s}}^{\mathrm{r}} + \boldsymbol{\Psi}_{\mathrm{r}} \tag{5.57}$$

where $\boldsymbol{\Psi}_{\mathrm{s}}^{\mathrm{r}}$ is the vector of stator voltage drops in the rotor reference frame. Then

$$k_{\mathrm{sat}} = \frac{\boldsymbol{\Psi}_{\mathrm{tot}}}{\boldsymbol{\Psi}_{\mathrm{g}}} \tag{5.58}$$

This saturation factor is different from the usually adopted factor. In fact (5.58) is a distribution of saturation factors along the air-gap that better represents the saturation of the machine.

The adopted iteration scheme is fixed-point like, with a random relaxation to improve the stability of the convergence:

$$\boldsymbol{k}_{\mathrm{sat}}^{(m+1)} \leftarrow \boldsymbol{k}_{\mathrm{sat}}^{(m)} + 0.5\,\mathrm{r\ and}\left(\boldsymbol{k}_{\mathrm{sat}}^{(m+1)} - \boldsymbol{k}_{\mathrm{sat}}^{(m)}\right) \tag{5.59}$$

For the next iteration the updated air-gap flux density is simply

$$\boldsymbol{B}_{\mathrm{g}}^{(m+1)} = \mu_0 \frac{-\boldsymbol{U}_{\mathrm{s}} + \boldsymbol{U}_{\mathrm{r}}}{\boldsymbol{k}_{\mathrm{sat}}^{(m+1)}\, g} \tag{5.60}$$

and the iteration cycle restarts. The error of the method was evaluated through

$$\varepsilon_{\mathrm{sat}} = \left\| \boldsymbol{k}_{\mathrm{sat}}^{(\mathrm{m}+1)} - \boldsymbol{k}_{\mathrm{sat}}^{(m)} \right\| \tag{5.61}$$

5.8 Torque maps

The nonlinear model is used to compute the impact of the rotor geometry on both the average torque and torque ripple. In particular, the impact of the flux-barrier-end angles is analyzed, since they heavily affect the torque ripple [34]. The average torque and some torque ripple harmonics are computed as a function of the flux-barrier angles, $(\vartheta_{\mathrm{b1}}, \vartheta_{\mathrm{b2}})$, and their amplitude is shown graphically by using maps. Figure 5.14 reports such torque maps, computed for a reference motor with two flux barriers whose data is reported in Table 5.1.

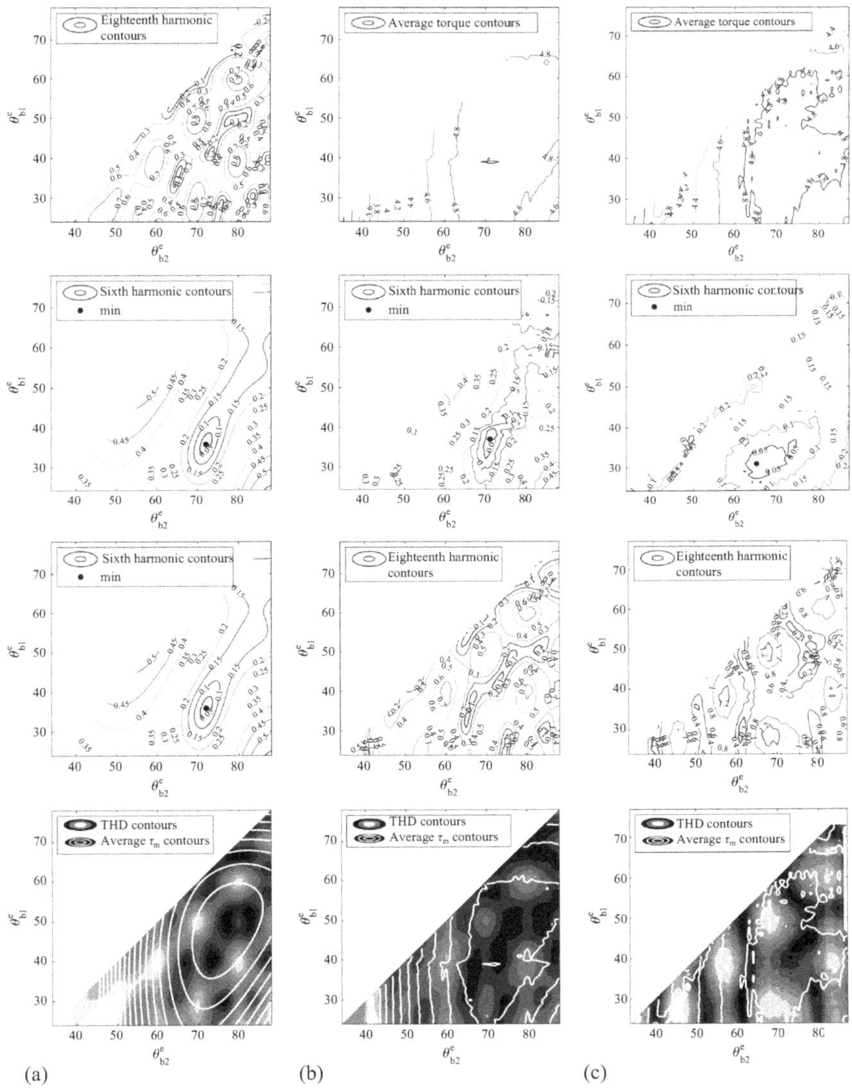

Figure 5.14 Main torque maps of the two-barrier rotor. The darker and cooler colors represent the valleys of the corresponding quantity. (a) Analytic linear, (b) analytic saturation and (c) FE

For the sake of an easy comparison, Figure 5.14 reports three *columns* which refer to the results obtained by the means of

1. the analytical linear model [34],
2. the nonlinear model, described earlier,
3. the FE method applied on the same motor geometries [46–48].

Table 5.1 Parameters of the reference motor

$Q = 36$	number of slots
$2p = 4$	number of poles
$y_q = 9$	coil pitch
$D_e = 200$ mm	stator outer diameter
$D_s = 125$ mm	stator inner diameter
$L_{stk} = 40$ mm	lamination stack length
$g = 0.35$ mm	air-gap thickness
$S_{slot} = 100$ mm^2	slot section area
$J = 3$ A/mm^2	conductor current density
$k_{fill} = 0.45$	slot fill factor
$k_{pack} = 0.95$	lamination pack factor
$k_{air} = 0.35$	ratio of rotor magnetic insulation

The results on SyR motor achieved through FE analysis have been compared with experimental test several times by the authors, obtaining satisfactory agreements [34, 49–51] but also in other works [52–54]. For this reason, in the following comparison, the results computed analytically are considered to be valid if they agree with the FE results.

The *first row* reports the comparison among average torque maps. The map computed with the linear model shows higher average torques with respect to the other two maps. This is the obvious consequence of the fact that since the model does not take into account the iron saturation. However, it is worth noticing that the behavior of the torque curves is almost identical.

Comparing the maps of the second and third column, it is possible to note that the nonlinear analytical model predicts correctly not only the behavior of the torque maps as a function of ϑ_{b1} and ϑ_{b2}, but also its amplitude. Therefore, the nonlinear model can be used as an alternative to FE simulation to derive such a map.

For this machine configuration, the average torque is almost independent of the first barrier angle. It can be noted that the average torque reaches its maximum in a wide region. For instance, for $\vartheta_{b2}^{e} > 60°$ the maps are quite flat. Thus, the designer is free to move within this space looking for torque ripple minima. This behavior is reflected also by the FE map.

The *second row* shows the maps of the torque harmonic of order six, which is the lowest order one. Independently from the model used, it appears that there is an evident minimum—highlighted by the black dot—corresponding to the angle combination $(\vartheta_{b1}^{e}, \vartheta_{b2}^{e}) = (36°, 72°)$. Such a point is virtually coincident in the maps obtained from the linear and nonlinear analytical models, while it is a bit shifted when FE is used. This is caused by the local saturation of tooth tips and iron parts, which is not considered in the analytical models. In addition, it can be noted that the torque ripple contours obtained by the three models are in a satisfactory agreement in the whole region.

The same considerations can be made comparing the maps for the *third row*, which report the torque harmonic of order 18. Such a torque harmonic corresponds to the first magnetic scalar potential (also referred to as MMF) slot harmonic. They are the MMF harmonics produced by the winding discretized inside the slots (18 = 36 slots/2 pole pairs), which are characterized by a winding factor equal to the fundamental one. They typically cause the highest ripple. It can be noted that the number of peaks and valleys is increased with respect to the sixth torque harmonic. This trend is general: the higher the harmonic order, the higher the number of maxima and minima. Once more, peaks and valleys predicted by the linear analytical model and nonlinear analytical models are in a good agreement with those found by FE.

Finally, in the *last row* the maps compare the torque Total Harmonic Distortion (*THD*), defined as

$$THD = \frac{\sqrt{\sum_{h\neq 0} \hat{\tau}_h^2}}{\tau_0} \tag{5.62}$$

where τ_0 is the average torque. The filled map shows *THD* contours, while the white superimposed contours refer to the average torque map. The brighter the color, the higher the *THD*. Thus, as far as the torque ripple is concerned, the better combinations of $(\vartheta_{\mathrm{b1}}^{\mathrm{e}}, \vartheta_{\mathrm{b2}}^{\mathrm{e}})$ are those corresponding to the darker areas.

Furthermore, by comparing the results of the third and fourth rows, it can be observed that bright colors correspond to the peaks of torque harmonic of 18th order, which is due to the MMF slot harmonics. This highlights the heavy impact of the MMF slot harmonics on the overall torque ripple.

Finally, as an overall conclusion, the linear and nonlinear analytical models produce a behavior of average torque map very similar to the one obtained through FE, with the difference that the linear map has higher values than the other two models. Therefore, the nonlinear analytical model has to be adopted to predict the average torque. On the other hand, both the analytical models are able to find the position of maxima and minima of the torque *THD*. Thus, it can be stated that even the linear model can be used in spite of the nonlinear one when searching for the flux-barrier angle combinations exhibiting a minimum torque ripple, with the advantage of a higher speed and similar accuracy. The slight shifts of maxima and minima found by the FE maps are mainly due to the local saturation of iron, which is not taken into account in the linear analytical model.

5.9 Discussion

This chapter has show an accurate nonlinear analytical model for the SyR machine. A good agreement between analytical and FE simulations has been achieved, even in highly saturated machines.

Thanks to the speed of the analytical model, it is possible to quickly obtain some maps of torque harmonics as a function of the flux-barrier-end angles. These maps can be a useful design tool for the design of an SyR machine. Overall, the analytical maps are able to properly approximate the FE maps.

The average torque behavior is correctly predicted by the nonlinear analytical model, while the linear model overestimates it, obviously. However, even if there are some differences in the amplitude prediction, the flux-barrier angle combinations corresponding to the minima and maxima are estimated well, by means of both the linear and nonlinear models. This fact is quite significant because it suggests that the analytical linear model can be employed to get good design points in the flux-barrier-angle plane for a specific motor in a small amount of time (some minutes). On the other hand, the nonlinear analytical model can be used to properly predict the average torque.

Chapter 6

Design criteria of flux-barriers in synchronous reluctance machines

In this chapter, a design criterion for multi flux-barrier Synchronous Reluctance (SyR) machines is proposed. In particular, the focus is on the proper flux-barrier design [55]. An analytical model is adopted to compute the impact of the rotor flux-barriers on the torque, focusing on its ripple.

Contrary to the model proposed in [56] and further extended in Chapter 5, not all the magnetomotive force harmonics are considered, but only those which cause the highest torque ripple. As a consequence, the analytical model becomes lighter and it is possible to derive some simple equations to predict the suitable geometry of the rotor flux-barriers. The previous simplification can be verified in many publications, such as [34] where the 12th torque harmonic is dominant, and in [52] where the main source of torque ripple is the 18th harmonic.

Rotors with one, two and three flux-barriers per rotor pole are taken into account. Some examples referring to four-pole SyR machines with different numbers of stator slots are investigated and analyzed. The results are compared with a full-featured analytical model and validated using Finite-Element Analysis (FEA).

6.1 The simplified analytical model

The analytical model presented hereafter is derived from the one proposed in [34] and further developed in Chapter 5, and it is used to determine the geometry of an SyR motor with one, two or three flux-barriers per pole (Table 6.1). The assumptions that are considered in this model are

1. to neglect the stator and rotor magnetic voltage drops,
2. to neglect the stator slotting effect,
3. to neglect the rotor iron ribs.

For a given stator winding and rotor geometry, the aim of the model is to correctly predict the combinations of flux-barrier angles that lead to the lower (and higher too) torque ripple values.

Table 6.1 Geometrical data of the Synchronous Reluctance (SyR) machine

Quantity	Symbol	Value	u.m.
Stack length	L_{stk}	40	mm
Outer stator diameter	D_s	200	mm
Inner stator diameter	D_e	125	mm
Air-gap length	g	0.35	mm
Number of pole pairs	p	2	

6.1.1 Application of a nonlinear model

A critique to the model mentioned previously might be that it is basically a simple linear model that does not consider the actual iron B–H characteristic while usually the saturation is quite high in SyR machines. Consequently, the model should not be able to predict the proper torque waveform (and consequently design the proper rotor geometry) when the SyR machine is strongly saturated. This remark is only partially true. Experimental results shown in [56–58] highlight that the optimal solutions of the rotor flux-barrier angles found through the linear model are effective to achieve smooth torque even at high load. The torque ripple, due to the interaction between electromagnetic fields, is properly predicted [34,52,59].

6.1.2 Simplification of the model

It can be proven that electromagnetic torque is produced by the interaction between the electric loading, $\mathscr{K}_s(\vartheta_r, \vartheta_m)$, and the rotor magnetic potential, $\mathscr{U}_r(\vartheta_r, \vartheta_m)$.* The electric loading contains many space harmonics, produced by the arrangement of the winding within the slots. The electric loading creates the stator magnetic scalar potential, whose fundamental moves synchronously with the rotor, while the other harmonics move asynchronously (Figure 6.1).

The rotor magnetic potential, U_{r1}, is caused by the flux crossing the flux-barrier. It is mainly produced by the fundamental of the electric loading, while the harmonics cause slight oscillations of it.

The torque expression can be split into two terms:

$$\begin{aligned} T_m(\vartheta_m) &= k_T \int_0^{2\pi} \mathscr{K}_s(\vartheta_r, \vartheta_m)\, \mathscr{U}_r(\vartheta_r, \vartheta_m)\, d\vartheta_r \\ &= \langle T_m \rangle + \Delta T_m(\vartheta_m) \end{aligned} \tag{6.1}$$

*See Chapter 5.

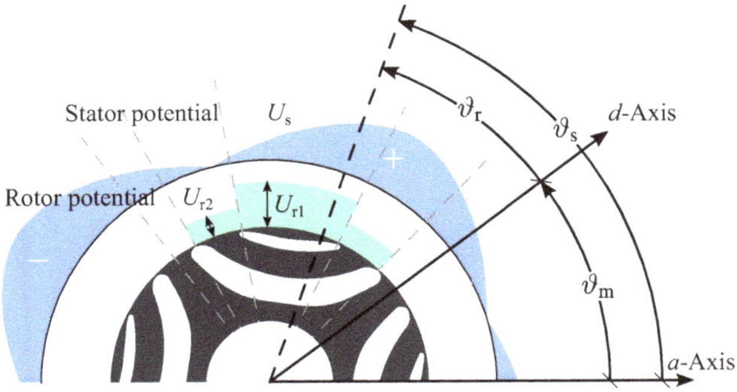

Figure 6.1 Sketch of the fundamental stator magnetic scalar potential and rotor reaction magnetic scalar potential

where k_T is a factor depending on the motor geometry. The first term, $\langle T_m \rangle$, is constant and corresponds to the average torque while the second term, ΔT_m, represents the torque ripple, which is a function of the rotor position, ϑ_m. All the harmonics of electric loading cause torque oscillations, but the highest ripple is due to the slot harmonics (i.e. those characterized by a winding factor equal to the fundamental one). In the following, the focus is on the first two slot harmonics, with order $\nu = 1 \pm Q/p$, which are responsible for the highest torque ripple. As an example, a machine with a combination of 24 slots and 4 poles has the highest ripple component that oscillates 12 times the frequency of the fundamental, caused by the electric loading harmonics of order 11 and 13.

6.2 One flux-barrier rotor

At first, a rotor with one flux-barrier has been studied (Figure 6.2). The complete expression of the rotor magnetic scalar potential is

$$\begin{aligned} U_{r1} &= a \int_{-\vartheta_b}^{\vartheta_b} \mathscr{U}_s(\vartheta_r, \vartheta_m)\, d\vartheta_r \\ &= -aD_s \sum_{\nu} \frac{\hat{K}_\nu}{(\nu p)^2} \cos(\lambda_\nu) \sin(\nu p \vartheta_b) \end{aligned} \tag{6.2}$$

where

$$a = \frac{\frac{D_s}{2g}\frac{t_b}{l_b}}{1 + \frac{D_s}{2g}\frac{t_b}{l_b} 2\vartheta_b}, \quad \lambda_\nu = \frac{\nu\pi}{2} + (\nu - 1)p\vartheta_m - \alpha_i^e$$

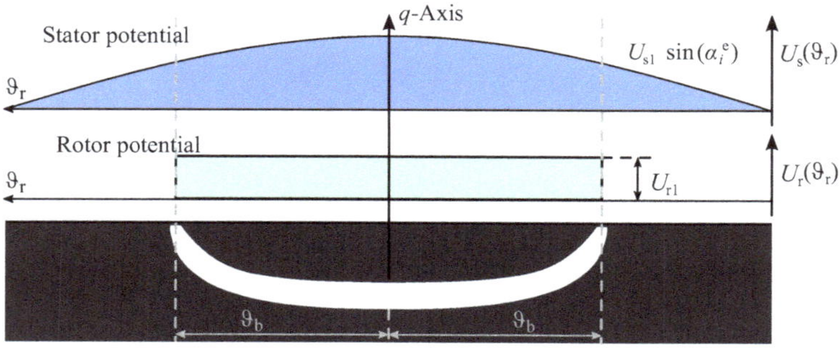

Figure 6.2 Sketch of a linearized one flux-barrier SyR rotor, with the stator and rotor scalar magnetic potentials shown

By assumption, any variation of the rotor magnetic potential is neglected. U_{r1} is assumed to be independent of the rotor position and due to the fundamental of $\mathscr{K}_s$ only. Thanks to such hypothesis, its expression becomes

$$U_{r1} = -\frac{aD_s\hat{K}_1}{p^2}\cos(\tfrac{\pi}{2} - \alpha_i^e)\sin(p\vartheta_b)$$

The torque is computed integrating the product between electric and magnetic loadings over the flux-barrier:

$$T_m(\vartheta_m) = k_T 2p \int_{-\vartheta_b}^{\vartheta_b} U_{r1}\mathscr{K}_s(\vartheta_r, \vartheta_m)\,d\vartheta_r \tag{6.3}$$

Let us consider the electric loading harmonic of order ν that travels asynchronously with respect to the rotor. It is

$$\mathscr{K}_s^{(\nu)}(\vartheta_r, \vartheta_m) = \hat{K}_\nu \sin[\nu p\vartheta_r + (\nu - 1)p\vartheta_m - \alpha_i^e]$$

Such electric loading harmonic produces a torque contribution expressed as

$$T_\nu = k_T\,2p \int_{-\vartheta_b}^{\vartheta_b} U_{r1}\,\mathscr{K}_s^{(\nu)}(\vartheta_r, \vartheta_m)\,d\vartheta_r \tag{6.4}$$

Regardless of the rotor position, the torque ripple produced by this harmonic can be simply evaluated considering two opposite situations, according to the waveform of the $\mathscr{K}_s^{(\nu)}$ harmonic with respect to the q-axis: (a) when it is odd and (b) when it is even. The two situations are sketched in Figure 6.3.

In the time instant when the waveform is odd (sine waveform in Figure 6.3(a)), the torque contribution resulting from (6.4) is zero. On the contrary, in the time instant when the waveform is even (cosine waveform in Figure 6.3(b)), there is either a positive or negative instantaneous torque contribution. The resulting torque ripple caused

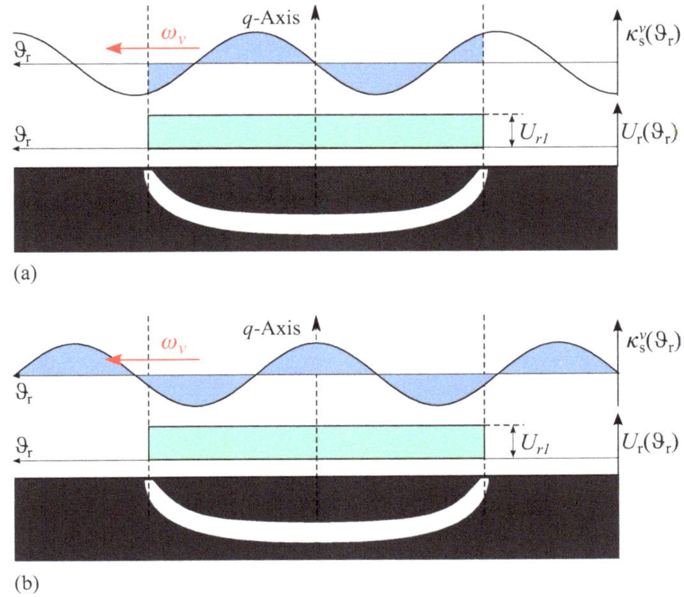

Figure 6.3 Electric loading harmonic waveform (moving at ω_ν) with respect to the rotor flux-barrier (fixed) in two different rotor positions: (a) sine and (b) cosine electric loading harmonic waveform

by the considered harmonic is estimated by the amplitude found when the waveform is even with respect to the barrier. Therefore, the torque contribution becomes

$$
\begin{aligned}
T_\nu &\propto \hat{K}_\nu \int_{-\vartheta_b}^{\vartheta_b} \cos(\nu p \vartheta_r)\, d\vartheta_r \\
&\propto 2\hat{K}_\nu \frac{\sin(\nu p \vartheta_b)}{p\nu}
\end{aligned}
$$

However, each torque harmonic is produced by a pair of electric loading harmonics. For a better understanding, let us consider the complete torque expression given in [36] by

$$
\begin{aligned}
T_m(\vartheta_m) = &-\frac{\mu_0}{g}\frac{D_s^2 L_{stk}\, U_{r1}}{2}\cdot \\
&\cdot \sum_\nu \frac{\hat{K}_\nu}{\nu} \sin(\nu p \vartheta_b) \sin \nu\tfrac{\pi}{2}\, \cos[(\nu - 1)p\vartheta_m - \alpha_i^e]
\end{aligned}
\tag{6.5}
$$

The torque ripple of order h is due to two electric loading harmonics, namely $\nu_1 = (1 - h)$ and $\nu_2 = (1 + h)$, where $h = 6n$, $n \in \mathbb{N}$. The torque contributions of these two harmonics are

$$T_{\mathrm{m},\nu_1}(\vartheta_{\mathrm{m}}) = T_{\nu_1} \cos[(\nu_1 - 1)p\vartheta_{\mathrm{m}} - \alpha_i^{\mathrm{e}}]$$
$$T_{\mathrm{m},\nu_2}(\vartheta_{\mathrm{m}}) = T_{\nu_2} \cos[(\nu_2 - 1)p\vartheta_{\mathrm{m}} - \alpha_i^{\mathrm{e}}]$$

After some computations, the torque amplitude due to the considered harmonic results in

$$T_\nu = k_T^* U_{\mathrm{r}1} \frac{\hat{K}_\nu}{\nu} \sin(\nu \tfrac{\pi}{2}) \sin(\nu p \vartheta_{\mathrm{b}})$$
$$k_T^* = -\frac{\mu_0 D_{\mathrm{s}}^2 L_{\mathrm{stk}}}{g}$$

and it can be either positive or negative. It is then possible to compute amplitude and phase of the torque harmonic of order h. Referring to the current vector angle $\alpha_i^{\mathrm{e}} = 45°$, they are

$$T_h = \sqrt{T_{\nu_1}^2 + T_{\nu_2}^2}$$
$$\phi_h = \arctan\left(\frac{T_{\nu_2} - T_{\nu_1}}{T_{\nu_2} + T_{\nu_1}}\right)$$

From (6.2), the torque ripple expression for the two harmonics becomes

$$T_h^2 \propto \left[\frac{\sin \nu_1 \vartheta_{\mathrm{b}}^{\mathrm{e}}}{\nu_1}\right]^2 + \left[\frac{\sin \nu_2 \vartheta_{\mathrm{b}}^{\mathrm{e}}}{\nu_2}\right]^2$$

employing electrical angles, $\vartheta_{\mathrm{b}}^{\mathrm{e}} = p\vartheta_{\mathrm{b}}$. In general,

$$\frac{\mathrm{d}}{\mathrm{d}\vartheta_{\mathrm{b}}^{\mathrm{e}}} \left\{\left[\frac{\sin \nu_1 \vartheta_{\mathrm{b}}^{\mathrm{e}}}{\nu_1}\right]^2 + \left[\frac{\sin \nu_2 \vartheta_{\mathrm{b}}^{\mathrm{e}}}{\nu_2}\right]^2\right\} = 0$$
$$\frac{2}{\nu_1} \sin \nu_1 \vartheta_{\mathrm{b}}^{\mathrm{e}} \cos \nu_1 \vartheta_{\mathrm{b}}^{\mathrm{e}} + \frac{2}{\nu_2} \sin \nu_2 \vartheta_{\mathrm{b}}^{\mathrm{e}} \cos \nu_2 \vartheta_{\mathrm{b}}^{\mathrm{e}} = 0$$
$$\nu_2 \sin(2\nu_1 \vartheta_{\mathrm{b}}^{\mathrm{e}}) + \nu_1 \sin(2\nu_2 \vartheta_{\mathrm{b}}^{\mathrm{e}}) = 0$$

Imposing to zero the derivative of the torque ripple with respect to $\vartheta_{\mathrm{b}}^{\mathrm{e}}$, it results in

$$\nu_2 \sin(2\nu_1 \vartheta_{\mathrm{b}}^{\mathrm{e}}) + \nu_1 \sin(2\nu_2 \vartheta_{\mathrm{b}}^{\mathrm{e}}) = 0 \tag{6.6}$$

This transcendental equation can be solved numerically for the flux-barrier angle $\vartheta_{\mathrm{b}}^{\mathrm{e}}$. The solutions found through (6.6) are the minima and maxima of the torque ripple.

6.2.1 Practical results

To validate the model previously derived, Figure 6.4 reports the torque behaviors due to the first two slot harmonics ($\nu = 1 \pm Q/p$) versus the electrical flux-barrier angle, $\vartheta_{\mathrm{b}}^{\mathrm{e}}$. For a four-pole machine ($2p = 4$), they are the 5th- and 7th-order harmonics for the machine with $Q = 12$, the 11th and 13th for $Q = 24$, and the 17th and 19th

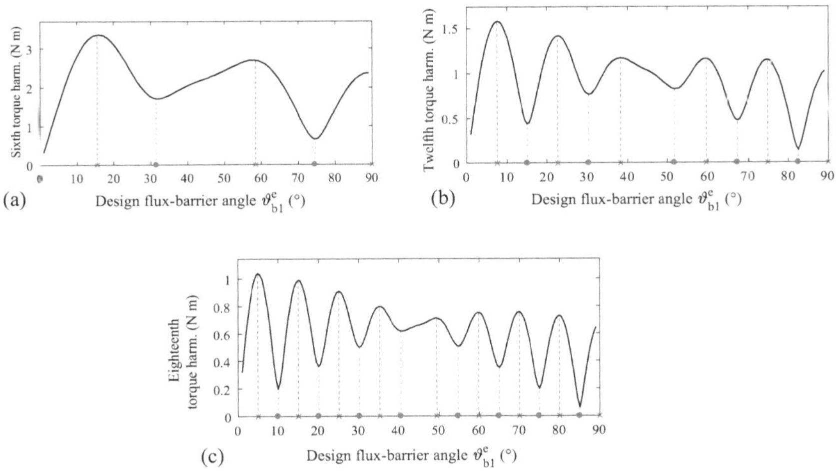

Figure 6.4 Analytical torque maps of the first slot harmonics for different number of slots Q and number of poles 2p combinations in one-barrier machines: (a) $Q = 12, 2p = 4$; (b) $Q = 24$, $2p = 4$; (c) $Q = 36$, $2p = 4$. The vertical lines represent the positions of the torque ripple minima found by means of (6.6)

for $Q = 36$. These waveforms are obtained through the complete model of the SyR machine expressed by (6.5) as in [34,36]. In addition, the vertical lines identify the flux-barrier angles found by the simple formulation derived in (6.6).

The perfect prediction of the torque minima by (6.6) can be noted: they correspond to the minima computed by means of the complete analytical magnetic model of the machine. Furthermore, this occurs for different combinations of slots per pole.

6.2.2 FE validation

To support the previous results, two rotors with different flux-barrier angles are selected from Figure 14.6(a) and the corresponding motor features are compared. A low- and a high-torque ripple points have been chosen for some Finite Element (FE) simulations. The considered motor has $Q = 36$ and $2p = 4$, so that it refers to Figure 14.6(a). An RMS current density of about 6.3 A/mm^2 is imposed. The magnetic insulation ratio, k_{air} (which accounts for the air length along the q-axis with respect to the total available length), is set to 0.3. The flux-barrier electrical angle, ϑ^{e}_{b}, is set as follows:

1. $\vartheta^{e}_{b} = 74.95°$ for the low-torque ripple solution;
2. $\vartheta^{e}_{b} = 69.93°$ for the high-torque ripple solution.

The corresponding mechanical angles result in 37.48° and 34.97°, respectively, so that the difference is 2.5° only. This should not represent a problem since common manufacturing tolerances are well below these values. With one flux-barrier per pole, the chosen flux-barrier angle is quite large because it generally leads to higher average torque values. In fact, it can be proven that the average torque of one flux-barrier SyR motor is proportional to $\sin^2 \vartheta_b^e / \vartheta_b^e$. Such a function exhibits a maximum at $\vartheta_b^e = 66.7°$ (i.e. $\vartheta_b = 33.35°$). The geometries and the results are reported in Figures 6.5 and 6.6. It can be seen that the two geometries are virtually identical in the middle of the rotor, and their only difference lies in the flux-barrier-end position. At first, the

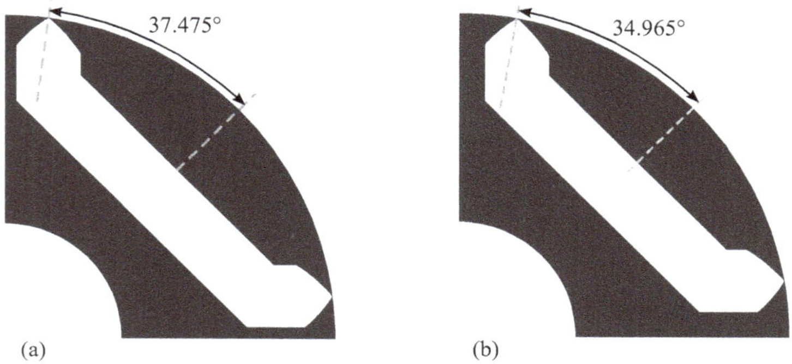

Figure 6.5 Geometries of two one flux-barrier solutions with different flux-barrier angles: (a) low- and (b) high-torque ripple geometry

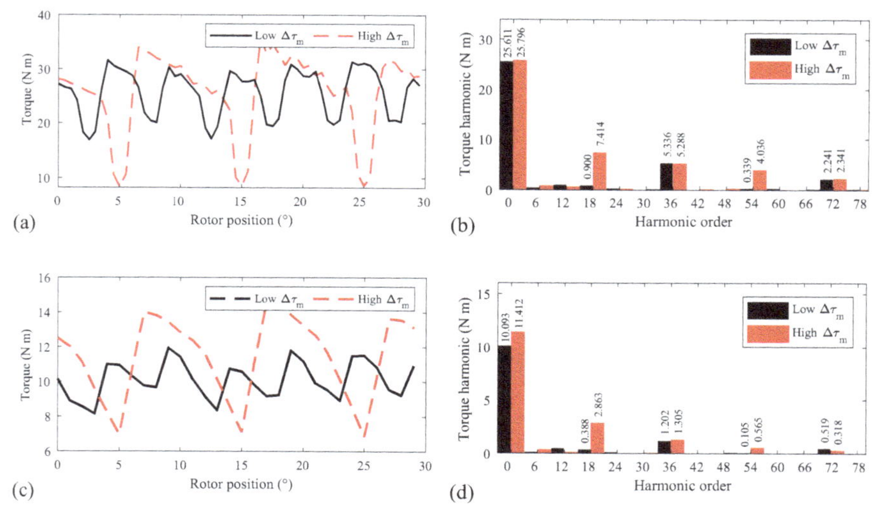

Figure 6.6 Torque waveforms and spectra of one flux-barrier solutions with different flux-barrier angles: comparison of (a) the torque waveforms with linear iron, (b) the torque spectra with linear iron, (c) the waveforms spectra with nonlinear iron and (d) the torque spectra with nonlinear iron

simulations are carried out with linear ferromagnetic material in order to obtain a fair comparison with the analytical model.

Figure 6.6 reports the harmonic content of the torque behavior achieved from FE simulations. Black bars correspond to the low ΔT_m solution, and white bars correspond to the high ΔT_m solution. To achieve accurate torque results with FEA, it is mandatory to design a proper mesh at the air gap of the machine. Typically, three to four elements along the radial direction are sufficient.

Figure 6.6(b) highlights that the correct choice of the flux-barrier-end position heavily affects the 18th harmonic amplitude and its odd multiple harmonics. In this particular case, the overall torque ripple results 53.5% for the low-ripple geometry and 106.7% for the high-ripple one. This results in remarks that such a notable improvement is due to the proper choice of flux-barrier angle ϑ_b so as to minimize the torque ripple contribution due to the first two slot harmonics.

Figure 6.6(d) reports the torque harmonic content achieved considering nonlinear ferromagnetic material. Obviously, the resulting average torque and torque harmonic amplitudes are quite lower with respect to the linear iron case. Also the overall torque ripple results to be lower, being 37.6% and 67.9% for the low ΔT_m and high ΔT_m solutions, respectively. However, the choice of the flux-barrier-end angle, ϑ_b, still affects the torque ripple caused by the slot harmonics.

Other simulations have been performed changing the magnetic insulation coefficient (i.e. the coefficient k_{air}) of the flux-barriers to show the effect of a higher ratio on the low-ripple geometry. While the average torque results lower (due to increased saturation), the torque ripple remains the same.

At last, it was also observed that, in the case of nonlinear iron, the shape of the flux-barrier-end affects the torque harmonic content. In fact, they affect the distribution of the flux entering the rotor. This effect, particularly evident in one flux-barrier rotors, may change the effective position of the flux-barrier-end, leading to a different torque spectrum. This is why the analyzed motors shown in Figure 6.5(a) and (b) exhibit a quite sharp flux-barrier-end.

6.3 Two flux-barrier rotor

Similarly as before, the two magnetic scalar potentials of the rotor islands, U_{r1} and U_{r2}, are assumed to be constant (as depicted in Figure 6.1). As said earlier, the machine is considered to be magnetized only by the fundamental harmonic of the stator magnetic scalar potential, U_{s1}. The resulting complete torque expression is given by [34,36]

$$T_m(\vartheta_m) = -\frac{\mu_0 D_s^2 L_{stk}}{4g}\left[U_{r2}\sum_{\nu}\frac{K_\nu}{\nu}\sin\nu\tfrac{\pi}{2}\sin\nu\vartheta_{b2}^{e} + (U_{r1}-U_{r2})\sum_{\nu}\frac{K_\nu}{\nu}\sin\nu\tfrac{\pi}{2}\sin\nu\vartheta_{b1}^{e}\right] \cdot\cos[(\nu-1)p\vartheta_m-\alpha_i^{e}] \tag{6.7}$$

Therefore, the torque produced by a single harmonic is proportional to

$$T_\nu \propto (U_{r1} - U_{r2})\frac{\sin \nu\vartheta^e_{b1}}{\nu} + U_{r2}\frac{\sin \nu\vartheta^e_{b2}}{\nu} \tag{6.8}$$

A further assumption can be made as follows: the rotor magnetic voltage drops are considered to be equal (as depicted in Figure 6.1), such that

$$\Delta U_r = U_{r1} - U_{r2} = U_{r2}$$

This assumption is often verified, thanks to the choice of the magnetic insulation (i.e. the flux-barrier thicknesses) distribution in the rotor. Thus, (6.8) becomes

$$T_\nu \propto \frac{\sin \nu\vartheta^e_{b1}}{\nu} + \frac{\sin \nu\vartheta^e_{b2}}{\nu}$$

As mentioned earlier, the actual torque ripple harmonic amplitude is obtained by considering a couple of harmonics. As usual, the first two slot harmonics are considered, which means $h = Q/p$ and $\nu_1 = 1 - h$, $\nu_2 = 1 + h$, so that

$$T_h^2 \propto \left[\frac{\sin \nu_1\vartheta^e_{b1}}{\nu_1} + \frac{\sin \nu_1\vartheta^e_{b2}}{\nu_1}\right]^2 + \left[\frac{\sin \nu_2\vartheta^e_{b1}}{\nu_2} + \frac{\sin \nu_2\vartheta^e_{b2}}{\nu_2}\right]^2 \tag{6.9}$$

In order to get the minima, (6.9) is derived by the first barrier angle, ϑ^e_{b1}, and then by the second one, ϑ^e_{b2}:

$$\begin{cases} 2\left[\frac{\sin \nu_1\vartheta^e_{b1}}{\nu_1} + \frac{\sin \nu_1\vartheta^e_{b2}}{\nu_1}\right]\cos \nu_1\vartheta^e_{b1} + 2\left[\frac{\sin \nu_2\vartheta^e_{b1}}{\nu_2} + \frac{\sin \nu_2\vartheta^e_{b2}}{\nu_2}\right]\cos \nu_2\vartheta^e_{b1} = 0 \\ 2\left[\frac{\sin \nu_1\vartheta^e_{b1}}{\nu_1} + \frac{\sin \nu_1\vartheta^e_{b2}}{\nu_1}\right]\cos \nu_1\vartheta^e_{b2} + 2\left[\frac{\sin \nu_2\vartheta^e_{b1}}{\nu_2} + \frac{\sin \nu_2\vartheta^e_{b2}}{\nu_2}\right]\cos \nu_2\vartheta^e_{b2} = 0 \end{cases} \tag{6.10}$$

It results in (6.10). This represents a transcendental system of two equations in two variables, $(\vartheta^e_{b1}, \vartheta^e_{b2})$, and it is solved numerically, both searching for minima and maxima.

6.3.1 Practical results

Again, to validate the model, Figure 6.7 reports the torque behaviors due to the first two slot harmonics in the plane of the two electrical flux-barrier angles ϑ^e_{b1} and ϑ^e_{b2}. These harmonics are of the same order of the previous ones. The 2D torque harmonic maps are again obtained through a complete model of the two flux-barrier SyR machine, from (6.7).

Then the points marked by black dots and blue crosses are found by means of the model proposed in (6.10). They identify the flux-barrier angles that correspond to the torque ripple minima and maxima, respectively. Once again, this model is able to predict these points with satisfactory precision, despite the additional assumption of constant magnetic voltage drop across each flux-barrier.

6.3.2 FE validation

Also in this case a low-torque ripple and two high-torque ripple points have been chosen for FE simulations (Figure 6.8). The main data of the considered motor are

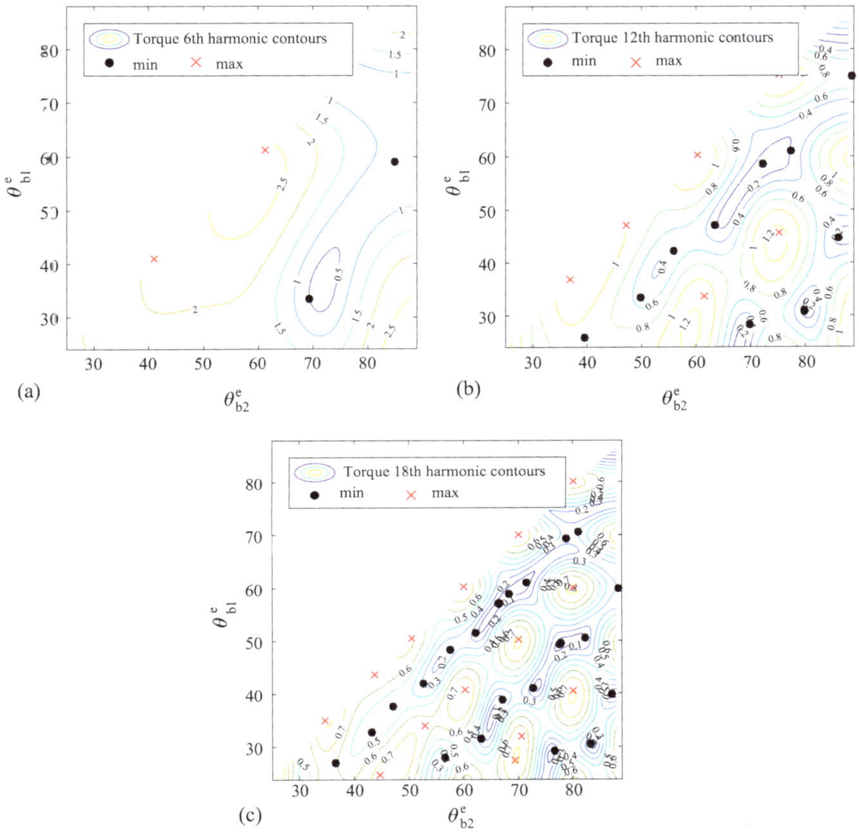

Figure 6.7 Analytical torque maps of the first slot harmonics for different slots and poles combinations in a two flux-barrier machine: (a) $Q = 12$, $2p = 4$; (b) $Q = 24$, $2p = 4$; (c) $Q = 36$, $2p = 4$. The black dots show the minima found by means of the model proposed here. Contour lines are obtained by means of the complete model

the same, so $Q = 36$ and $2p = 4$, with an RMS current density equal to 6.3 A/mm^2, and it refers to Figure 14.6(b). The flux-barrier electrical angles, $(\vartheta^e_{b1}, \vartheta^e_{b2})$, are set to

1. $(49.5^\circ, 78^\circ)$ for the low-torque ripple solution;
2. $(49.5^\circ, 70^\circ)$ for the first high-torque ripple solution;
3. $(40.5^\circ, 80^\circ)$ for the second high-torque ripple solution.

Figure 6.9(c) reports the comparison between the torque versus rotor position behaviors of the low- and high-torque ripple solutions with nonlinear ferromagnetic materials. It can be observed how the different choice of just one flux-barrier angle (between the first two solutions) greatly affects the resulting torque waveform and the

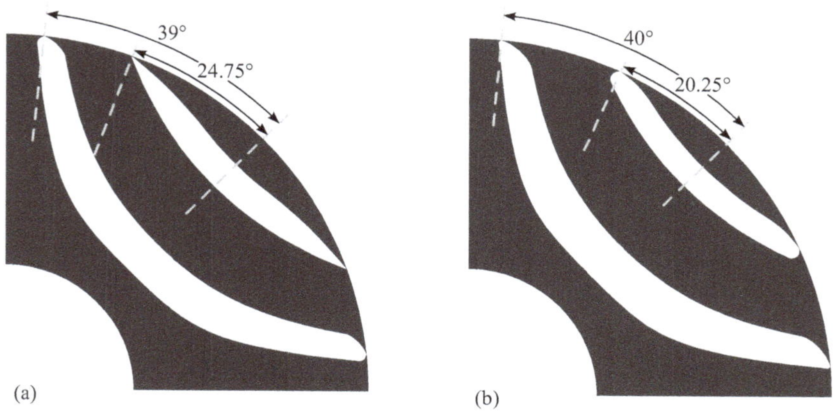

Figure 6.8 (a) Low- and (b) high-torque ripple geometries with two flux-barriers

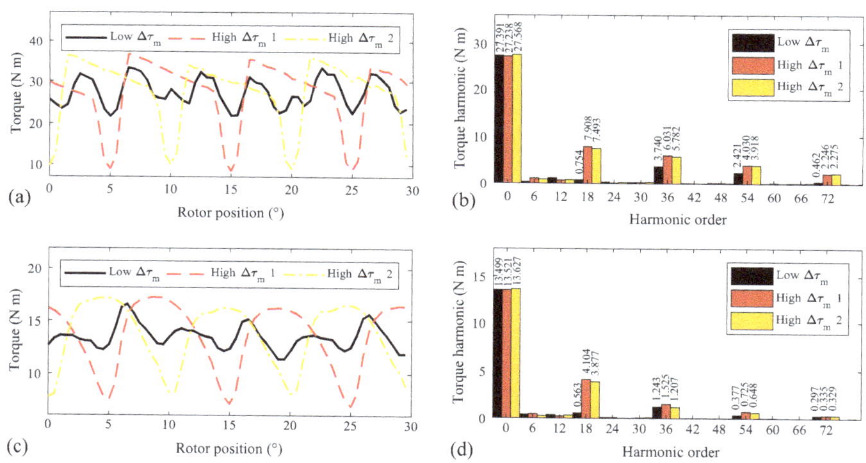

Figure 6.9 Torque waveforms and spectra of two flux-barriers solutions with different flux-barrier angles: comparison of the torque spectra with (a) linear (rotor position) and (b) linear iron (harmonic order), and (c) nonlinear (rotor position) and (d) nonlinear iron (harmonic order)

relative harmonic content. This is highlighted in Figure 6.9(d) which reports the low ΔT_m solution torque spectrum with black bars, and the high ΔT_m solutions torque spectra in gray and white. The saturation of the ferromagnetic material affects the shape of all the torque waveforms, which appear smoother, so all the ripples result to be lower.

Table 6.2 Comparison of average torque and torque ripple for low- and high-ripple solutions

	$\langle T_m \rangle$ [Nm]	ΔT_m [%]
Linear case		
Low-ripple solution	27.39	42.66
First high-ripple solution	27.24	103.64
Second high-ripple solution	27.57	99.06
Nonlinear case		
Low-ripple solution	13.50	38.41
First high-ripple solution	13.52	76.23
Second high-ripple solution	13.63	68.34

A final comparison between the average torques and ripples is reported in Table 6.2 in both linear and nonlinear cases. As expected, the average torque and ripple are higher with linear ferromagnetic materials, but the gap between low- and high-ripple solutions remains.

6.4 Three flux-barriers rotor

At this point, the model can easily be extended to three flux-barriers rotors. In fact, using the same assumptions as before

$$\Delta U_r = U_{r1} - U_{r2} = U_{r2} - U_{r3} = U_{r3}$$

the torque harmonic expression due the electric loading harmonic becomes

$$T_\nu \propto \frac{\sin \nu\vartheta_{b1}^e}{\nu} + \frac{\sin \nu\vartheta_{b2}^e}{\nu} + \frac{\sin \nu\vartheta_{b3}^e}{\nu}$$

and, therefore, the actual torque ripple harmonic results in

$$T_h^2 \propto \left[\frac{\sin \nu_1\vartheta_{b1}^e}{\nu_1} + \frac{\sin \nu_1\vartheta_{b2}^e}{\nu_1} + \frac{\sin \nu_1\vartheta_{b3}^e}{\nu_1}\right]^2 + \left[\frac{\sin \nu_2\vartheta_{b1}^e}{\nu_2} + \frac{\sin \nu_2\vartheta_{b2}^e}{\nu_2} + \frac{\sin \nu_2\vartheta_{b3}^e}{\nu_2}\right]^2 \tag{6.11}$$

Once again, this equation can be minimized numerically to find suitable flux-barrier angles. In this case, the number of minima far exceeds the number of maxima. In particular, the maxima are either degenerated three flux-barriers rotors, or the flux-barrier angles are displaced by a stator slot angle.

On the other hand, the minima can be found on some lines in the 3D space created by the three flux-barrier angles and so they are very numerous.

6.4.1 FE validation

One low- and one high-torque ripple geometries have been chosen for FE validation. The respecting flux-barrier angles ($\vartheta^e_{b1}, \vartheta^e_{b2}, \vartheta^e_{b3}$) are

- (21.67°, 51.31°, 76.34°) for low-torque ripple solution;
- (19.89°, 59.98°, 79.99°) for high-torque ripple solution.

It can be noted that the high ripple solution is characterized by flux-barrier angles which are multiples of the slot angle, while the low ripple solution does not have evenly spaced angles (Figure 6.10).

Figure 6.11 shows the comparison between the low and the high-torque ripple geometries. As expected, the high-torque ripple geometry exhibits high-torque harmonics, in particular the first slot harmonic (the 18th). Overall, the torque ripple results to be 17.9% for the low ripple geometry while it is 72.9%.

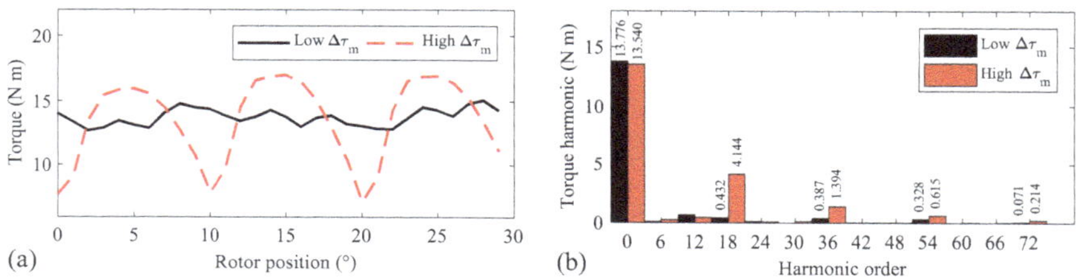

Figure 6.10 Geometries and spectra of three flux-barriers solutions with different flux-barrier angles: comparison of the torque spectra with (a) nonlinear iron (rotor position) and (b) nonlinear iron (harmonic order)

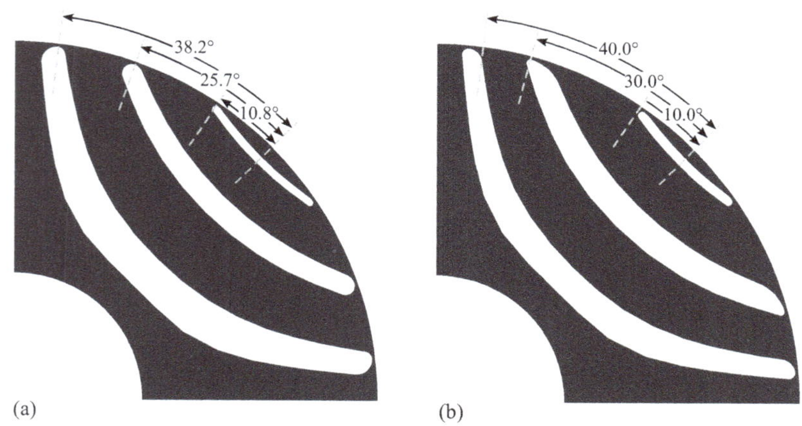

Figure 6.11 (a) Low- and (b) high-torque ripple geometries with three flux-barriers

6.5 Design of asymmetric flux-barriers in SyRM

In this section, the previous simple model is extended to asymmetric SyR machines, so machines with different flux-barriers between adjacent rotor poles, in order to obtain a sort of compensation of the torque harmonics. This time, a different machine is studied, and its data are reported in Table 6.3.

6.5.1 Determination of solutions getting minimum torque harmonic

Equation (6.11) can be minimized numerically for the barrier angle $\vartheta_b^e = p\vartheta_b$. It allows one to determine the flux-barrier angles ϑ_b^e that minimize the torque oscillations due to the harmonics of electric loading of order ν_1 and ν_2.

Figure 6.12 reports the torque behaviors of the first two slot harmonics versus the electrical flux-barrier angle, ϑ_b^e. Referring to a machine with $Q = 24$ slots and $2p = 4$ poles, they are the harmonics of 11th- and 13th-order and 23rd- and 25th-order (which are the first two and the second two electric loading slot harmonics). The black dots identify the flux-barrier angles found by the minimization of (6.2). Comparing with the waveforms obtained through a complete model of the SyR machine, the vertical lines starting from the dots show that the positions of the minima are properly predicted.

Table 6.3 Geometrical data of the SyR machine used as an example in the analysis

Quantity	**Symbol**	**Value**	**u.m.**
Stack length	L_{stk}	40	mm
Outer stator diameter	D_e	120	mm
Inner stator diameter	D_s	70	mm
Air-gap length	g	0.4	mm
Number of stator poles	$2p$	4	
Number of stator slots	Q	24	

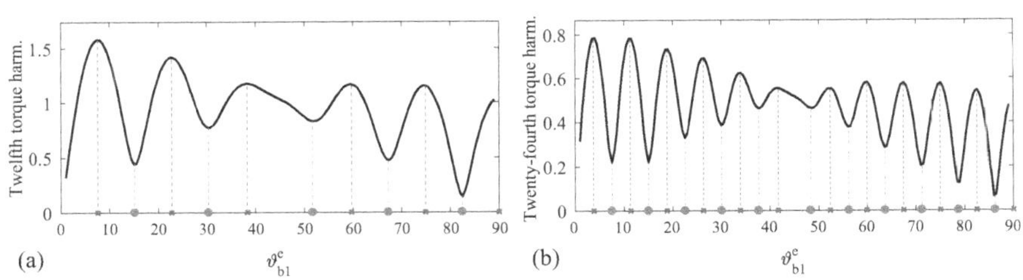

Figure 6.12 Analytical torque maps of the first four slot harmonics for a motor with $Q = 24$, $2p = 4$ with one-barrier per pole: (a) $\nu_1 = 11$, $\nu_2 = 13$; (b) $\nu_1 = 23$, $\nu_2 = 25$. Green dots correspond to angles of minimum torque harmonic

Both the simple and the complete analytical model [34,55] considered the current vector angle to be $\alpha_i^e = 45°$. On the other hand, rarely an SyR machine has the maximum torque per ampere trajectory at 45° due to saturation. It works at higher angles and even up to 80° or more when flux-weakening. More generally, instead of (6.11), the actual *h*th torque harmonic amplitude and phase are

$$\begin{aligned} T_h &= \sqrt{T_{\nu_1}^2 + T_{\nu_2}^2 + 2T_{\nu_1}T_{\nu_2}\cos 2\alpha_i^e} \\ \phi_h &= \arctan\left(\frac{T_{\nu_2} - T_{\nu_1}}{T_{\nu_1} + T_{\nu_2}}\tan\alpha_i^e\right) \end{aligned} \tag{6.12}$$

It can be seen that the final amplitude of the torque harmonic depends on the amplitudes of the two components of torque harmonic and also on the current vector angle. The choice of such an angle could lead to quite different results, especially for small flux-barrier angles, as shown in Figure 6.13. However, high flux-barrier angles (typically greater than $\vartheta_b^e = 60°$) are of interest, since they generally lead to a higher average torque and, hence, better performance.

In order to consider a general case, the torque function (6.12) can be tested at 67.5°, which is halfway from 45° through 90°. In this way, (6.12) are simplified to

$$\begin{aligned} T_h &= \sqrt{T_{\nu_1}^2 + T_{\nu_2}^2 - \sqrt{2}T_{\nu_1}T_{\nu_2}} \\ \phi_h &= \arctan\left(\frac{T_{\nu_2} - T_{\nu_1}}{T_{\nu_1} + T_{\nu_2}}(1+\sqrt{2})\right) \end{aligned} \tag{6.13}$$

Depending on the sign of the two torque functions, using (6.13) or (6.2) could lead to different results varying the current angle. The results are reported in Figure 6.13 with a red line. The blue dash-dotted line is the function given by (6.2), whose minima and maxima perfectly correspond to the one of Figure 6.12. It is evident that for small flux-barrier angles, the two functions minima match perfectly for flux-barrier angles greater than 45°. The list of the minima found for the 24th harmonic is reported in Table 6.4.

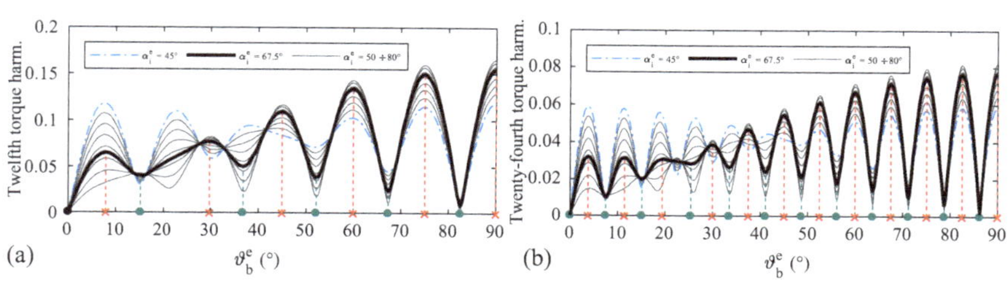

Figure 6.13 Torque maps of the first four slot harmonics for a motor with $Q = 24$, $2p = 4$ with one-barrier per pole for different current angles: (a) $\nu_1 = 11$, $\nu_2 = 13$; (b) $\nu_1 = 23$, $\nu_2 = 25$. Black dots correspond to angles of minimum torque harmonic at $\alpha_i^e = 67.5°$

Table 6.4 Set $S_h = \{\vartheta^e_{bi}\}^n_{i=1}$ of flux-barrier angles corresponding to minima of the 24th torque harmonic (Figure 6.13(b))

#	1	2	3	4	5	6
ϑ^e_b [°]	7.5	15.0	25.4	33.5	41.1	48.6
#	7	8	9	10	11	
ϑ^e_b [°]	56.2	63.7	71.2	78.7	86.2	

6.5.2 Combination of minima solutions

The previous analysis allowed to find those flux-barrier angles, ϑ^e_b, yielding a minimum of a fixed harmonic of the torque. A pair of such solutions can be adopted and combined together in the same rotor obtaining an asymmetric structure [34]. It exhibits different flux-barrier geometries that are alternated between the poles. In this way,

1. the torque harmonic that has been minimized as described earlier remains of low value,
2. a sort of compensation of the other torque harmonics is achieved.

This means that two torque harmonics can be minimized: one through the choice of flux-barrier angles and the other one through an asymmetric rotor. The two harmonics are referred to as h and k in the following.

The easiest way to apply torque harmonics compensation is to use phasors. Let T'_k be the kth torque harmonic produced by the first pole, with phase ϕ'_k, and T''_k be the kth torque harmonic produced by the second pole (different from the first), with phase ϕ''_k. Both poles have flux-barrier angles that minimize the hth torque harmonic. Then their vector sum results in

$$T'_k e^{j\phi'_k} + T''_k e^{j\phi''_k}$$

and so the problem to solve is as follows:

$$\min_{i \neq j} \left| T^{(i)}_k e^{j\phi^{(i)}_k} + T^{(j)}_k e^{j\phi^{(j)}_k} \right| \tag{6.14}$$

where i and j are the indexes of the set $S_h = \{\vartheta^e_{bi}\}^n_{i=1}$ of flux-barrier angles that already minimize the hth torque harmonic.

For the combination in pairs with (6.14), there are two options:

1. Find flux-barrier angles that minimize the hth torque harmonic, and find the minimum of the kth torque harmonic using optimal combination of the angles previously found.
2. Find flux-barrier angles that minimize the kth torque harmonic, and find the minimum of the hth torque harmonic using optimal combination of the angles previously found.

Since most of the ripple is due to the first four slot harmonics of the electric loading, which cause the first two slot harmonics of the torque, it is advisable to take k as the first torque slot harmonic, and $h = 2k$ as the second one, or vice versa. Note, however, that the two criteria differ greatly. In fact, the number of $2k$-combinations is higher than the k-ones due to the higher number of minima found minimizing $2k$th torque harmonics. So the first criterion leads to a higher probability of finding proper combinations, and it is the one which is here adopted.

6.5.3 Combinations for one flux-barrier rotor and validation

The combinations in pairs of the angles in Table 6.4 could be ranked based on the predicted 12th torque harmonic by (6.14). The result of such procedure is reported in the first column of Table 6.5. Since there were 66 combinations, only the first 10 are reported.

Many of these combinations were tested by FEA to see whether the ripple is lower than in the respective symmetric rotors and to check if the ranking is effectively sorted.

The FEA results are reported in Table 6.5. It can be noticed that the ranking is not perfectly sorted. This is due to the effect of local saturation that virtually changes the position of the flux-barrier end; hence it changes the flux-barrier angle. However, some results are excellent because the ripple is less than half the original one (the improvement can be seen from the example of Figure 6.14(a) and from [55]).

Figure 6.14 shows the torque waveforms and spectra of one suitable combination of the flux-barrier angles found earlier. The chosen combination is the fourth of Table 6.5. labeled as "10 and 8," whose pair of flux-barrier angles is $\vartheta_{\rm b}^{\rm e} = 78.7°$ and $\vartheta_{\rm b}^{\rm e} = 63.7°$ (see Table 6.4). The expected average torque is about 1.56 Nm while the

Table 6.5 FEA results of average torque and ripple achieved adopting the predicted combinations of flux-barrier angles (first column)

i* and *j	**Torque [Nm]**	**Ripple [%]**	**Twelfth harm. [Nm]**	**Twenty-fourth harm. [Nm]**
11 and 9	1.54	26.1	0.095	0.140
11 and 10	1.50	41.8	0.289	0.007
9 and 7	1.56	36.9	0.234	0.068
10 and 8	1.56	21.7	0.035	0.124
8 and 6	1.52	47.5	0.281	0.087
7 and 5	1.44	28.4	0.112	0.097
6 and 4	1.31	21.3	0.024	0.107
5 and 3	1.11	29.1	0.084	0.089
10 and 7	1.54	25.3	0.048	0.132
9 and 8	1.55	38.4	0.258	0.036

Rotor with one flux-barrier per pole.

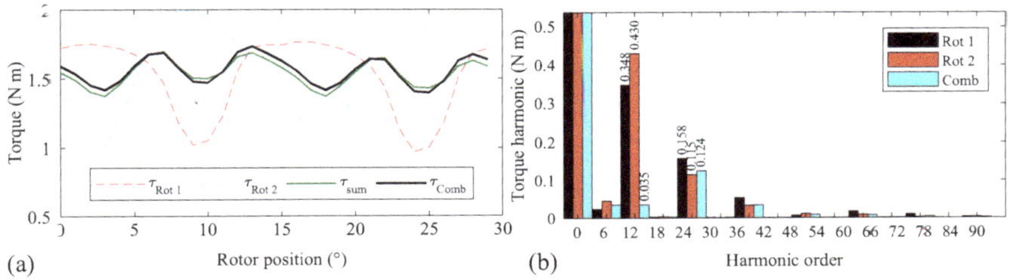

Figure 6.14 FE torque (a) waveforms and (b) spectra of the two single one-barrier rotors and their combination

torque ripple is 21.7%. It can be seen that the two torque waveforms combine together and tend to lower the first torque slot harmonic. In fact, the second harmonic visible in Figure 6.14(a) corresponds to the first torque slot harmonic ($Q/p = 24/2 = 12$) since only a sixth of electrical period is shown. The compensation of this harmonic is evident also from the spectrum in Figure 6.14(b).

6.6 Rotor with two flux-barriers per pole

The torque harmonic equation (6.2) is slightly modified. In particular, there is an additional term that depends on the second flux-barrier angle. Hence the torque contribution of the electric loading harmonic ν results in

$$T_\nu = k_T^* \Delta U_r \frac{\hat{K}_\nu}{\nu} \sin(\nu \tfrac{\pi}{2})[\sin \nu\vartheta_{b1}^e + \sin \nu\vartheta_{b2}^e] \tag{6.15}$$

The amplitude of the torque harmonic of order h is still obtained through (6.2) or (6.13), taking into account the corresponding electric loading harmonics.

Figure 6.15 reports the torque behaviors due to the first four slot harmonics versus the two electrical barrier angles, $(\vartheta_{b1}^e, \vartheta_{b2}^e)$. The 2D maps are again obtained through a complete model of the two flux-barriers SyR machine. On the other hand, the black dots and the red crosses identify the barrier angles that correspond to minima and maxima, respectively, found by the simple model. This model is well suited to predict these points with great accuracy.

In the following, the minima found through (6.13) are used to combine the rotor poles. Similarly to the single flux-barrier rotor, the minima obtained through (6.13) are the same of (6.2) for large angles, while they could be different for small angles. These angles are reported in Table 6.6. The combination procedure for a two flux-barriers rotor is exactly the same as the one previously developed, so the problem to solve is again expressed by (6.14). Since now there are 35 solutions of minima, about 600 asymmetric combinations are available to be tested. They were sorted based on the predicted 12th-order harmonic and then the first solutions were further analyzed.

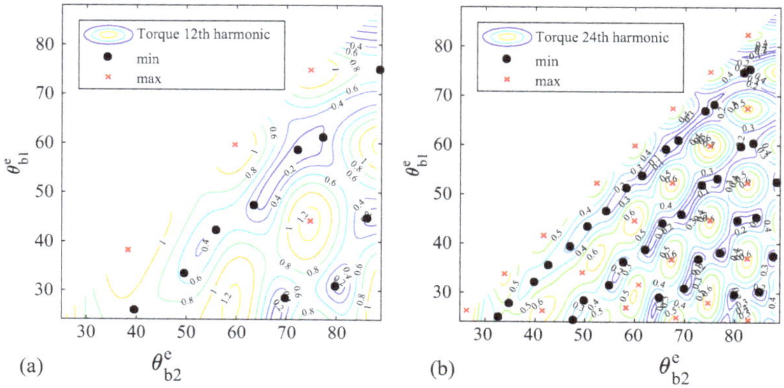

Figure 6.15 Analytical torque maps of the first slot harmonics for a motor with $Q = 24$, $2p = 4$ in a two-barriers machine: (a) $\nu_1 = 11$, $\nu_2 = 13$; (b) $\nu_1 = 23$, $\nu_2 = 25$. Black dots show the minima found through the proposed model

Table 6.6 Set $S_h = \{(\vartheta^e_{b1}, \vartheta^e_{b2})_i\}_{i=1}^n$ of flux-barrier angles corresponding to minima of the 24th torque harmonic (Figure 6.13(b)) for a rotor with two flux-barriers per pole

#	1	2	3	4	5	6	7	8
ϑ^e_{b1} [°]	24.9	24.3	27.7	28.3	29.0	30.2	30.8	29.6
ϑ^e_{b2} [°]	32.7	47.7	34.9	49.9	65.0	85.0	70.0	80.0
#	9	10	11	12	13	14	15	16
ϑ^e_{b1} [°]	32.1	31.5	36.8	35.5	36.2	38.0	37.4	39.3
ϑ^e_{b2} [°]	39.9	54.9	72.8	42.7	57.8	77.1	87.8	47.1
#	17	18	19	20	21	22	23	24
ϑ^e_{b1} [°]	38.7	43.4	44.0	45.2	46.5	44.6	45.9	51.2
ϑ^e_{b2} [°]	62.1	50.6	65.6	84.4	54.3	80.6	69.3	58.4
#	25	26	27	28	29	30	31	32
ϑ^e_{b1} [°]	51.8	53.7	53.1	60.3	59.0	59.6	60.9	66.8
ϑ^e_{b2} [°]	73.4	61.5	76.5	83.7	66.2	81.2	68.7	74.0
#	33	34	35					
ϑ^e_{b1} [°]	68.1	74.7	75.3					
ϑ^e_{b2} [°]	75.9	81.9	83.1					

Table 6.7 reports the first ten combinations. It can be observed again that the ranking is not perfectly sorted when they are analyzed by means of FEA. In addition, the achievable torque ripple results in almost always lower than that of one flux-barrier rotors.

Table 6.7 FEA result of the predicted combinations of two flux-barriers rotors

i and *j*	Torque [Nm]	Ripple [%]	Twelfth harm. [Nm]	Twenty-fourth harm. [Nm]
33 and 26	1.61	12.1	0.031	0.031
18 and 5	1.52	23.2	0.092	0.096
34 and 29	1.56	20.1	0.074	0.081
28 and 23	1.61	12.0	0.059	0.022
16 and 10	1.43	24.7	0.118	0.014
29 and 22	1.63	20.5	0.152	0.045
27 and 15	1.63	22.4	0.144	0.050
22 and 6	1.62	23.0	0.174	0.029
28 and 22	1.62	19.0	0.134	0.023
26 and 23	1.59	21.9	0.122	0.022
22 and 10	1.53	13.5	0.089	0.021

6.6.1 Validations

As an example, two combinations minimizing (6.14) are studied. They are the "33 and 26" and the "28 and 23" combinations. The first one corresponds to the two pairs of barrier angles $\vartheta_{b1}^{e} = 68.1°, \vartheta_{b2}^{e} = 75.9°$ and $\vartheta_{b1}^{e} = 53.7°, \vartheta_{b2}^{e} = 61.5°$. The geometry is shown in Figure 6.16(a).

The second combination corresponds to the two pairs of barrier angles $\vartheta_{b1}^{e} = 60.3°, \vartheta_{b2}^{e} = 83.7°$ and $\vartheta_{b1}^{e} = 45.9°, \vartheta_{b2}^{e} = 69.3°$. The relative geometry is reported in Figure 6.16(b).

The FEA results are reported in Figures 6.17 and 6.18 for both combinations. In Figures 6.17(a) and 6.18(a), dashed line refers to the torque behavior of a rotor with the first pair of angles, and dash-dotted line refers to the second pair. It is evident that the torque oscillations are out of phase, even though they are not perfectly equal in amplitude. This results from the minimization of (6.14). When the two flux-barrier geometries are combined in the same rotor (Figure 6.16 shows the resulting rotors), the torque behavior is smoother as observed in Figures 6.17(a) and 6.18(a), with the

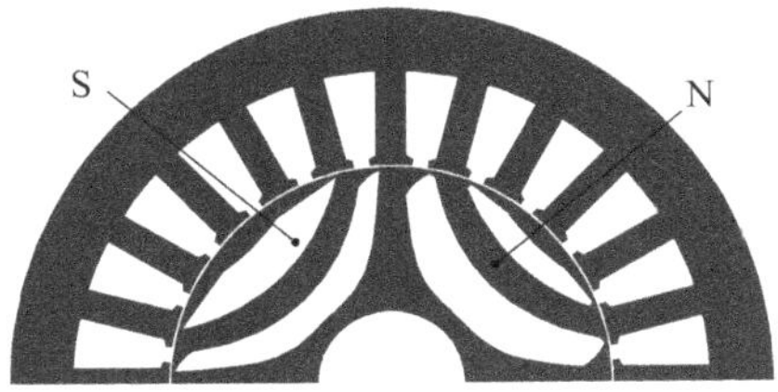

(a) (b)

Figure 6.16 Tested two flux-barriers geometries: (a) first and (b) second

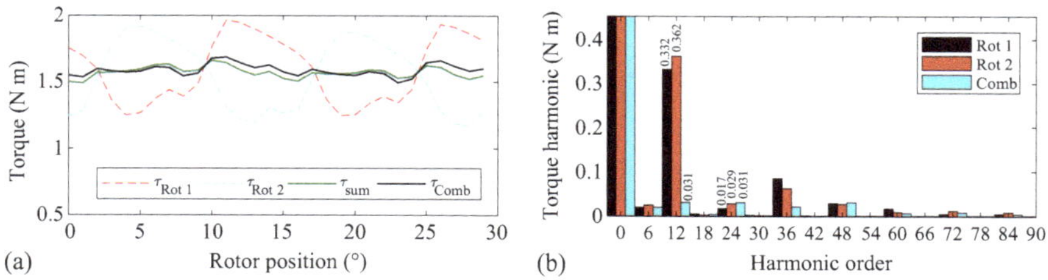

Figure 6.17 FE torque (a) waveforms and (b) spectra of the first combination ("33 and 26") of two single rotors and their combination

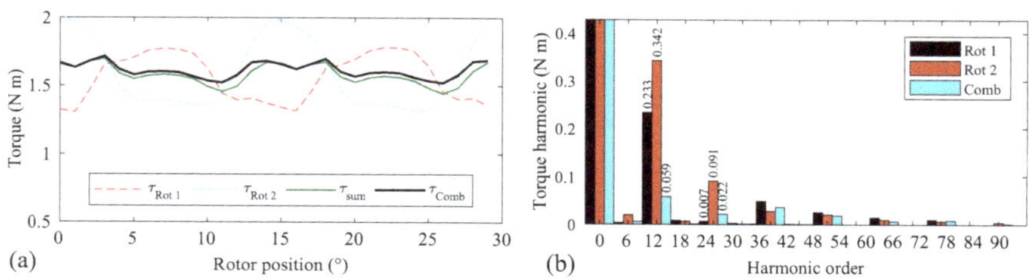

Figure 6.18 FE torque (a) waveforms and (b) spectra of the second combination ("28 and 23") of two single rotors and their combination

solid lines. Furthermore, Figures 6.17(b) and 6.18(b) highlight the reduction of the torque harmonic, especially the 12th that is the highest.

6.7 Experimental measurements

The results of a symmetric and an asymmetric two flux-barriers Interior Permanent Magnet (IPM) prototypes were previously available [34]. The asymmetric geometry was designed with a different technique, searching for flux-barrier angles that minimize and compensate two torque harmonics, as described in [34]. The angles of the flux-barrier ends are $\vartheta^{e}_{b1} = 29.6°$ and $\vartheta^{e}_{b2} = 54.8°$ for the even poles, $\vartheta^{e}_{b1} = 44.4°$ and $\vartheta^{e}_{b2} = 80.4°$ for the odd poles. It should be highlighted that the first two flux-barrier angles are very close to the 10th solution of the set reported in Table 6.6, while the other one is close to the 22nd reported solution. Furthermore, the combinations of such solutions appear in the complete list in position 44, so within the first 10% of all the possible combinations, and it corresponds to the one labeled "22 and 10" in Table 6.7.

The FEA results, shown in Figure 6.19, confirm that the asymmetric rotor (also referred to as "Machaon" in [34]) is able to generate a smoother torque waveform. It is also evident from Figure 6.19(b) that the harmonic content of the Machaon rotor is very low. In particular its average torque results 2.14 Nm, with a ripple of 4.57%,

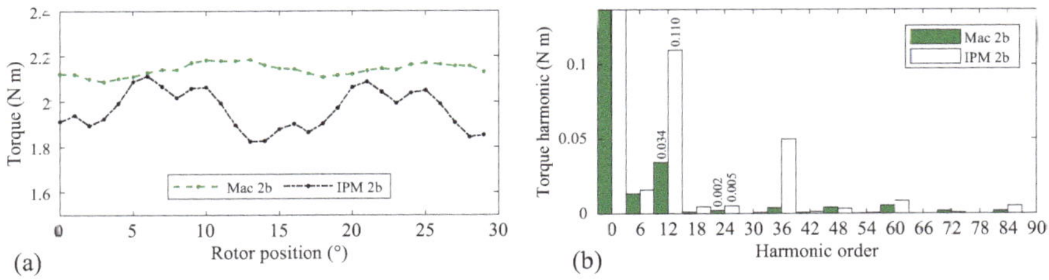

Figure 6.19 Comparison of FE torque (a) waveforms and (b) spectra of the available IPM and Machaon prototypes

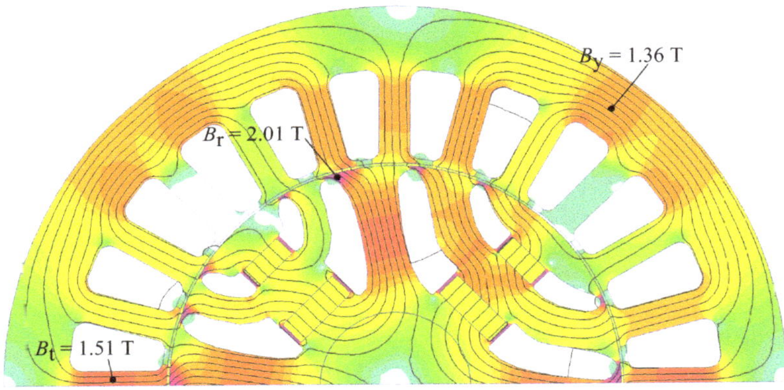

Figure 6.20 Magnetostatic solution of the Machaon two flux-barriers rotor

Table 6.8 Measured torque harmonics in the two prototypes

Harmonic order	Symmetric rotor	Asymmetric rotor
12	5.36	1.61
24	1.56	0.44
Total	13.1	4.76

while the IPM motor has an average torque of 1.97 Nm with a ripple of 14.7%. A magnetostatic FEA solution of the Machaon IPM is also reported in Figure 6.20.

Some measured results are reported in Table 6.8. The torque harmonics are referred to the average torque. From this result, it is evident that the asymmetric rotor allows one to minimize the torque oscillations. In particular the study has been carried out focusing on the higher torque harmonic components (related to the slot harmonics), which are of 12th and 24th order. It is worth noting that both harmonics are strongly reduced. As a consequence, the total torque ripple is reduced to one-third with respect to that of a machine with symmetric rotor.

6.8 Discussion

This chapter dealt with the design procedure for a quick and practical choice of flux-barriers in SyR machines. In particular, a simple but effective model has been derived, useful to select the flux-barrier angles in order to minimize the highest torque harmonics. The adopted assumptions are as follows:

1. Only the first two electric loading slot harmonics are considered because they produce the majority of torque ripple.
2. The rotor magnetic scalar potentials are considered constant.

According to that, the model reduces to the numerical solution of a single equation (or system of equations). This fact ensures a great convenience of the presented model, so the designer can opt for the minimization of any torque harmonic of interest.

The results have been compared with the full torque harmonic functions, derived by the complete analytical model. A perfect match between the minima has been shown, for many combinations of number of flux-barriers and slots per pole.

Furthermore, the combination of proper flux-barrier angles into an asymmetric rotor structure has been presented. This strategy is based on the following two-step design procedure:

1. At first, a set of flux-barrier angles is identified so as to minimize a torque harmonic of given order.
2. Then, pairs of geometries belonging to this set are combined together so that a specific torque harmonic is compensated.

After that, many combinations of these minima have been tested to try compensating the first, and highest, slot harmonics. One of these combinations corresponds to an available prototype previously optimized by a series of FEA simulations, which effectively compensated the first slot harmonic. In conclusion, this model becomes a practical tool for the design of SyR motors.

Chapter 7
Structural analysis with GetDP

7.1 Brief presentation of GetDP

GetDP is a free open-source scientific software for the numerical solution of integrodifferential equations [60,61]. Its name is the acronym of *General environment for the treatment of Discrete Problems*. The main features of GetDP are as follows:

- It deals with one-dimensional models, two-dimensional models (2D), both planar and axisymmetric, and three-dimensional models.
- It deals with different time domains, such as static, transient, harmonic and eigenvalue problems.
- It allows us to select several numerical methods for the problem solutions (e.g. Finite Element (FE), boundary element and volume integral methods).
- It allows one to solve coupled physical models involving electromagnetic phenomena (e.g. magnetostatics, magnetodynamics, wave propagation, lumped electric circuits), acoustic phenomena, thermal phenomena and mechanical phenomena (elasticity, rigid body movement).
- It can be controlled by command prompt and run dedicated script written in ASCII data file.
- The problem definition syntax is very close to the symbolic mathematical expressions.

Figure 7.1 shows some examples of the problems that GetDP can handle.

7.2 Mathematical formulation for structural analysis

The structural analysis of a body is based on the linear elastic theory. So, it describes the deformation and inner stress that occurs in a solid body when a determined load condition is applied. The adjective *linear* specifies that the deformation (or strain) are sufficiently small so that there is a linear relationship between the stress components and strain. A further assumption adopted in the following is that the body is *isotropic,* which means that its constitutive laws are independent from the reference frame used to indicate the stress components. Such analysis is selected since it suits with the characteristic of a Synchronous Reluctance (SyR) lamination.

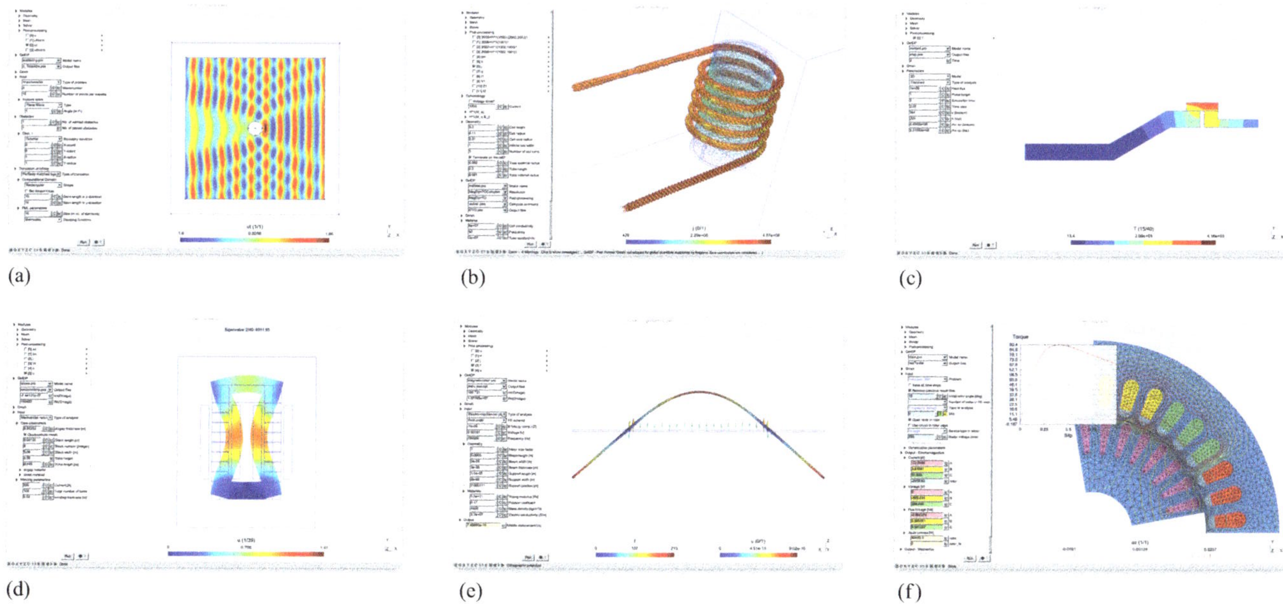

Figure 7.1 Examples of geometries and problems that can be analyzed with GetDP: (a) acoustic scattering, (b) magneto-thermal model, (c) thermal conduction, (d) magnetostriction, (e) structural analysis, (f) electric machines

7.2.1 Basic equations

The basic equations that describe a linear elasticity model are the equation of motion, the strain displacement equations and the constitutive equations [62,63].

7.2.1.1 Equation of motion

The equation of motion represents the dynamic equilibrium of the body and is expressed as follows:

$$\nabla \cdot \overleftrightarrow{\boldsymbol{\sigma}} + \boldsymbol{F} = \rho \frac{\partial^2 \boldsymbol{u}}{\partial t^2} \tag{7.1}$$

where $\overleftrightarrow{\boldsymbol{\sigma}}$ is the Cauchy stress tensor, $\boldsymbol{F}$ is the body force density, ρ is the mass density, $\boldsymbol{u}$ is the displacement vector and t is the time. The last term is null since the structural problems of interest are time independent. $\overleftrightarrow{\boldsymbol{\sigma}}$ is a tensor that contains the nine components of the state stress classified as *normal*, σ, and *tangential*, τ:

$$\overleftrightarrow{\boldsymbol{\sigma}} = \begin{bmatrix} \sigma_{11} & \sigma_{12} & \sigma_{13} \\ \sigma_{21} & \sigma_{22} & \sigma_{23} \\ \sigma_{31} & \sigma_{32} & \sigma_{33} \end{bmatrix} = \begin{bmatrix} \sigma_x & \tau_{xy} & \tau_{xz} \\ \tau_{yx} & \sigma_y & \tau_{yz} \\ \tau_{zx} & \tau_{zy} & \sigma_z \end{bmatrix} \tag{7.2}$$

thanks to the symmetry property, $\boldsymbol{\sigma} = \boldsymbol{\sigma}^{\mathrm{T}}$, the number of distinct stress components is six. The SI unit is $\mathrm{N/m^2}$; however, the MPa is usually adopted in engineering.

7.2.1.2 Strain–displacement equation

As the name suggests, this equation relates the strain tensor, $\overleftrightarrow{\boldsymbol{\varepsilon}}$, to the displacement vector, $\boldsymbol{u}$. The expression of $\overleftrightarrow{\boldsymbol{\varepsilon}}$ is

$$\overleftrightarrow{\boldsymbol{\varepsilon}} = \begin{bmatrix} \varepsilon_{11} & \varepsilon_{12} & \varepsilon_{13} \\ \varepsilon_{21} & \varepsilon_{22} & \varepsilon_{23} \\ \varepsilon_{31} & \varepsilon_{32} & \varepsilon_{33} \end{bmatrix} \quad \text{with} \quad \varepsilon_{ij} = \frac{1}{2}\left(\frac{\partial u_i}{\partial x_j} + \frac{\partial u_j}{\partial x_i}\right) \quad i,j = 1,2,3 \tag{7.3}$$

the diagonal contains the *normal* strains while the left elements are the *shear* strains. By definition, $\boldsymbol{\varepsilon}$ is symmetric so that there are only six distinct share components. Equation (7.3) can be written in the following differential form:

$$\overleftrightarrow{\boldsymbol{\varepsilon}} = \frac{1}{2}\left[\nabla \boldsymbol{u} + (\nabla \boldsymbol{u})^{\mathrm{T}}\right] \tag{7.4}$$

7.2.1.3 Constitutive law

The constitutive law that links the stresses and strains of a continuous elastic material is Hooke's law expressed as follows:

$$\begin{bmatrix} \sigma_{11} \\ \sigma_{22} \\ \sigma_{33} \\ \sigma_{23} \\ \sigma_{13} \\ \sigma_{12} \end{bmatrix} = \frac{E}{(1+\nu)(1-2\nu)} \begin{bmatrix} 1-\nu & \nu & \nu & 0 & 0 & 0 \\ \nu & 1-\nu & \nu & 0 & 0 & 0 \\ \nu & \nu & 1-\nu & 0 & 0 & 0 \\ 0 & 0 & 0 & \frac{1-2\nu}{2} & 0 & 0 \\ 0 & 0 & 0 & 0 & \frac{1-2\nu}{2} & 0 \\ 0 & 0 & 0 & 0 & 0 & \frac{1-2\nu}{2} \end{bmatrix} \begin{bmatrix} \varepsilon_{11} \\ \varepsilon_{22} \\ \varepsilon_{33} \\ 2\varepsilon_{23} \\ 2\varepsilon_{13} \\ 2\varepsilon_{12} \end{bmatrix}$$

where E is Young's modulus and ν is Poisson's ratio. In the case of plane stress conditions, it is verified that

$$\sigma_{31} = \sigma_{13} = \sigma_{32} = \sigma_{23} = \sigma_{33} = 0 \tag{7.5}$$

so that, it results in

$$\begin{bmatrix} \sigma_{11} \\ \sigma_{22} \\ \sigma_{12} \end{bmatrix} = \frac{E}{1-\nu^2} \begin{bmatrix} 1 & \nu & 0 \\ \nu & 1 & 0 \\ 0 & 0 & \frac{1-\nu}{2} \end{bmatrix} \begin{bmatrix} \varepsilon_{11} \\ \varepsilon_{22} \\ 2\varepsilon_{12} \end{bmatrix} \tag{7.6}$$

Some basic applications of the linear elasticity model are solved with GetDP in the following. The results reliability is confirmed by the analytical solutions. Finally, a complete example of the procedure adopted to analyze an SyR rotor is presented.

7.3 Cantilever beam with single load at the end

The cantilever beam represents one of the simplest body to test the linear elasticity equations. In this case, it is clamped at one end while a force, $F = 100$ N, is applied to the other extreme, as shown in Figure 7.2. The beam dimensions and the material properties are reported in Table 7.1.

7.3.1 Analytical solution

The beam moment of inertia is [64]

$$J = \frac{1}{12} w\, d = \frac{10^6}{12} \text{ mm}^4 \tag{7.7}$$

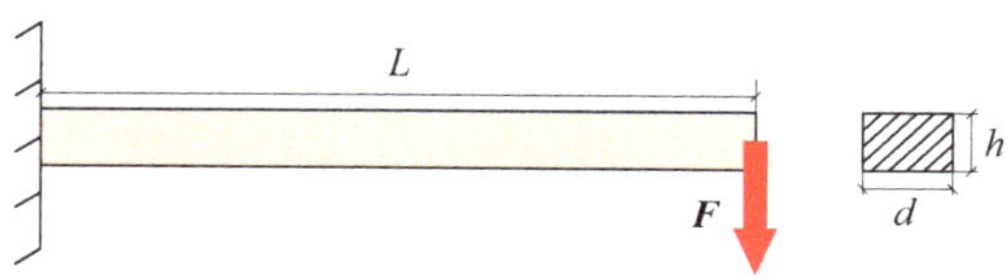

Figure 7.2 Cantilever beam with single load at the end

Table 7.1 Beam dimension and material properties

Name	Symbol	Value
Beam length	L	100 mm
Beam height	h	10 mm
Beam deep	d	1 000 mm
Young's modulus	E	210 MPa
Poisson's ratio	ν	0.3

the maximum deflection at the loaded extreme is

$$u_{\max} = \frac{1}{3}\frac{F\,L^3}{E\,J} = 1.9\ \mu\text{m} \tag{7.8}$$

the maximum bending stress is given by

$$\sigma_x = \frac{6\,F\,L}{d\,h} = 6 \cdot 10^5\ \text{Pa} \tag{7.9}$$

while the maximum shear stress is

$$\tau_{xy} = \frac{3}{2}\frac{F}{d\,h} = 1.5 \cdot 10^4\ \text{Pa} \tag{7.10}$$

finally, the von Mises stress is

$$\sigma_{\text{vm}} = \sqrt{\sigma_x^2 + 3\tau_{xy}^2} \approx 6 \cdot 10^5\ \text{Pa} \tag{7.11}$$

7.3.2 GetDP solution

7.3.2.1 Problem data

The file `data.pro` contains the model dimensions, the physical group labels, the material properties and the force amplitude. It is called both in pre- and post-processing by Gmsh and GetDP, respectively. Its content is reported in Listing 7.1.

Listing 7.1 File `data.pro`

```
// beam dimensions
unit = 1e-3;
L = 100*unit; // length (m)
h = 10*unit; // height (m)
lc = 1*unit; // characteristic length (m)
// physical labels
beam = 100;
left = 10;
bottom = 20;
right = 30;
top = 40;
// material properties
E = 210e9; // Young's modulus (Pa)
v = 0.3; // Poisson's ratio (-)
// force amplitude
Load = 100/h; // (N)
```

7.3.2.2 Model creation

The geometry creation and the assignment of the physical groups (necessary for GetDP) have been done with Gmsh [65]. Then physical group names are reported in Figure 7.3. The left segment, labeled as "`left`," is used to assign the boundary conditions $u_x = 0$ and $u_y = 0$, the right segment, labeled as "`right`," is used

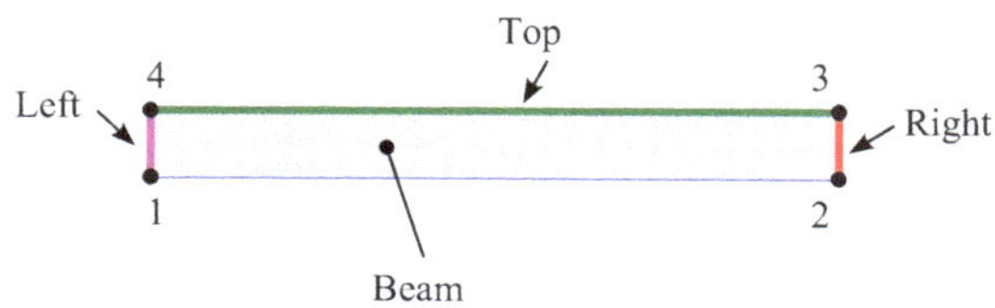

Figure 7.3 Beam geometry and physical group names

to assign the concentrated force F while the beam area is used to set the material properties E and ν and to define the region in which apply (7.6). The Gmsh code (saved in a `.geo` text file) to draw the beam geometry is reported in Listing 7.2.

Listing 7.2 Gmsh code (`.geo`) to draw the beam geometry and assign the physical groups

```
Include "data.pro";

// add points
Point(1) = {0,0,0,lc};
Point(2) = {L,0,0,lc};
Point(3) = {L,h,0,lc};
Point(4) = {0,h,0,lc};
Line(1) = {1, 2};
Line(2) = {2, 3};
Line(3) = {3, 4};
Line(4) = {4, 1};

// create beam surface
Line Loop(5) = {3, 4, 1, 2};
Plane Surface(6) = {5};

// assign physical groups
Physical Surface(beam) = {6};
Physical Line("left", left) = {4};
Physical Line("bottom", bottom) = {1};
Physical Line("right", right) = {2};
Physical Line("top", top) = {3};
```

7.3.2.3 Groups, materials and boundary condition properties

The GetDP code (saved in a `.pro` text file) to define the group, the material properties and the boundary conditions is reported in Listing 7.3.

After the creation of the geometry it is necessary to define the material properties, the boundary conditions and assign them to the proper physical group. First, the group relatives to the beam area, the clamped segment and the load segment are declared. It is worth noticing that the force is distributed along the right segment. For this reason, it is defined as F/h in the file `data.pro`.

Then, the beam material is described by (7.6) and it is assigned to the group `beam` while the force is defined with only the y component and it is applied to the group `load`.

Listing 7.3 GetDP code (`.pro`) to set the model groups, the material properties and the boundary conditions to the cantilever beam model

```
Include "data.pro"; // load physical group labels

// define the model groups
Group{
  beam = Region[{beam}]; // beam region
  fixed = Region[{left}]; // clamped side
  load = Region[{right}]; // loaded side

  Domain_Force_Sur = Region[ { } ];
  Domain_Force_Lin = Region[ {load} ];
  Domain_Force     = Region[ {Domain_Force_Sur ,
    Domain_Force_Lin}];
  Domain_Disp      = Region[ {beam} ];
}

Function{
  // set the beam material function property
  C_m[ beam ] = E/(1-v^2)*TensorSym[ 1,   v,    0,
  1,    0,
  (1-v)/2 ];

  // apply Force on group 'load'
  F[load] = Load*Vector[0,-1,0];
}

// Define boundary conditions
Constraint{
  // d ux = 0
  { Name DeplacementX ; Type Assign ;
    Case {
      { Region fixed; Value 0 ; }
    }
  }
  // d uy = 0
  { Name DeplacementY ; Type Assign ;
    Case {
      { Region fixed; Value 0 ; }
    }
  }
```

```
}

// load the formulation
Include "Elasticity_2D.pro"

// post processing operations
PostOperation Map_u UsingPost Mec2D_u {
  Print [ u,  OnElementsOf Domain_Disp, File "u.pos"] ; //
    displacement
  Print [ um, OnElementsOf Domain_Disp,  File "um.pos"] ; //
    displacement
  amplitude
  Print [ F,  OnElementsOf Domain_Force,  File "F.pos"] ; //
    load
  Print [ Vm , OnElementsOf Domain_Disp, File "Vm.pos"] ; //
    von Mises stress
}
```

After that, the boundary conditions `DeplacementX` and `DeplacementY` are declared within the `Constraint` field. They are assigned to the groups `fixed` with the value `0`.

Finally, the formulation file `Elasticity_2D.pro`, reported in Appendix D, is invoked and the problem is solved. Some post-processing operations on the resulting ***u*** field are performed to get the Force field and the von Mises stress.

7.3.2.4 Remark on the GetDP formulation

The GetDP code relative to the elasticity model formulation is reported in Listing 7.4. Without describing in detail the GetDP syntax, it is possible to see that (7.1) is applied to the group `Domain_disp` (i.e. the beam area) without the term ***F***. The latter is imposed to the group `Domain_force` that, in this example, is the left side of the beam.

Listing 7.4 GetDP formulation of the 2D elasticity problem

```
Formulation{
  { Name Mec2D_u ; Type FemEquation ;
    Quantity {
      { Name u ; Type Local ; NameOfSpace H_u_Mec2D ; }
    }
    Equation {
      Galerkin { [ C_m[] * Dof{D1 u}, {D1 u} ] ;
        In Domain_Disp ; Jacobian Vol ; Integration GradGrad ;
     }
      Galerkin { [ -F[] , {u} ] ;
        In Domain_Force; Jacobian SurLinVol; Integration
   GradGrad; }
    }
  }
}
```

7.3.3 Results

The displacement plot and of the beam is shown in Figure 7.4(a). It is possible to see that the maximum deflection occurs on the right side and its amplitude is equal to the analytical one computed in (7.8). The von Mises stress plot in Figure 7.4(b) shows that the maximum value is in good agreement with the analytical one and is near the clamped side of the beam. In fact, in that region the bending stress is higher and reach the maximum at the clamped side.

7.4 Cantilever beam with distributed load

In this case, the beam is clamped at one end while a uniform distributed load, $q = 1$ N/mm, is applied in vertical direction on the top side as illustrated in Figure 7.5. The beam dimensions and the material properties are reported in Table 7.1.

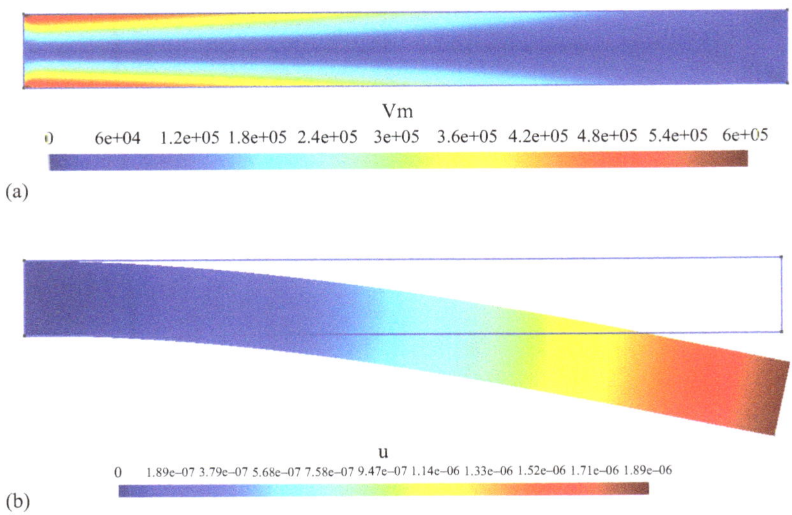

Figure 7.4 Resulting plots of the beam with single load at the end (the graphical displacement is amplified): (a) displacement plot and (b) von Mises plot

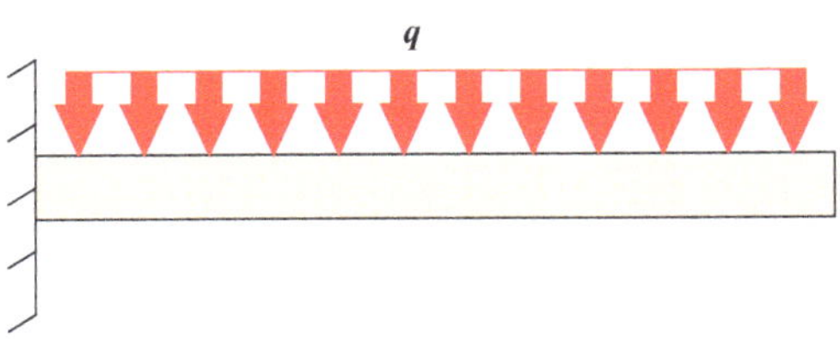

Figure 7.5 Cantilever beam with distributed load

7.4.1 Analytical solution

The moment of inertia has been computed in (7.7). The maximum deflection is

$$u_{\max} = \frac{1}{8}\frac{q\,L^4}{E\,J} = 0.71\,\mu\text{m} \tag{7.12}$$

The maximum bending moment is

$$M_{\max} = \frac{q\,L^2}{2} = 5\ \text{N mm} \tag{7.13}$$

hence, the relative stress is

$$\sigma_x = \frac{6M_{\max}}{d\,h^2} = 3\cdot 10^5\ \text{Pa} \tag{7.14}$$

the maximum shear stress is

$$T_{\max} = q\,L = 100\ \text{N} \tag{7.15}$$

so, the relative stress has been already computed in (7.10). Finally, with (7.11), the von Mises stress is $\sigma_{\text{vm}} \approx \sigma_x = 3\cdot 10^5$ Pa.

7.4.2 GetDP solution

7.4.2.1 Code changes

Only few changes on the GetDP code of the previous example allows one to solve the current one. The Gmsh code does not require any modification since the geometry is the same.

First, the `Load` variable in the file `data.pro` has to be set to 1. Then, the region assigned to the group `load` (Listing 7.5) has to be set to `top` to indicate that the load is applied on the top side (Figure 7.3) of the beam.

Listing 7.5 Setting of the group `load`

```
// continue ...
load = Region[{top}]; // loaded side
// continue ...
```

7.4.2.2 Results

The displacement plot of the beam with uniform distributed load is shown in Figure 7.6(a). It can be see that the maximum deflection is 0.71 μm as predicted analytically by (7.12). Also the von Mises plot in Figure 7.6(b) confirms the analytical results achieving a maximum stress value of $3\cdot 10^5$ Pa.

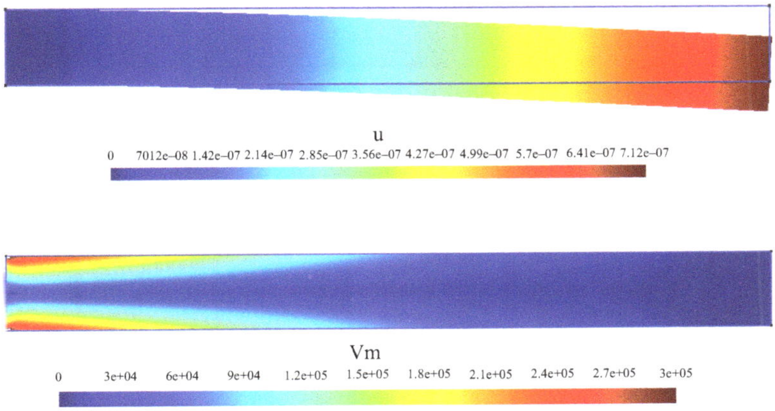

Figure 7.6 Resulting plots of the beam with distributed load on top (the graphical displacement is amplified): (a) displacement plot and (b) von Mises plot

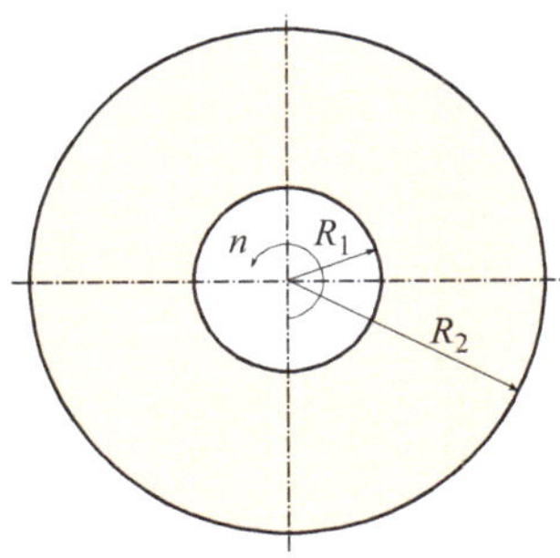

Figure 7.7 Hollow disk geometry

7.5 Rotating hollow disk

The rotating hollow disk is a very simple structural problem whose behavior is similar to the more complex SyR rotor geometry. Furthermore, the displacement and stress can be computed analytically and so it is possible to verify GetDP results reliably.

The hollow disk, shown in Figure 7.7, rotates at speed n and presents a mass density ρ. The inner and the outer diameter are free to move. The unique load is the centrifugal force due to the rotating disk mass. The main problem data are reported in Table 7.2.

Table 7.2 Hollow disk problem data

Name	Symbol	Value
Beam length	L	100 mm
Outer radius	R_1	250 mm
Inner radius	R_2	100 mm
2 Angular speed	n	3 000 rpm
Mass density	ρ	7 800 kg/m^3
Young's modulus	E	210 MPa
Poisson's ratio	ν	0.3

7.5.1 Analytical solution

Since the geometry is axisymmetric, it is convenient to use the polar reference frame. In this way, the stress components are *radial*, σ_r and *tangential*, σ_t.

The analytical formulation of the stress components and displacement are as follows [66,67]:

$$\sigma_r(r) = \frac{3+\nu}{8}\rho\omega^2\left[R_1^2 + R_2^2 - r^2 - \left(\frac{R_1R_2}{r}\right)^2\right]$$

$$\sigma_t(r) = \frac{3+\nu}{8}\rho\omega^2\left[R_1^2 + R_2^2 - \frac{1+3\nu}{3+\nu}r^2 + \left(\frac{R_1R_2}{r}\right)^2\right]$$

$$u(r) = \frac{3+\nu}{8}\rho\omega^2\frac{1-\nu}{E}r\left[R_1^2 + R_2^2 - \frac{1+\nu}{3+\nu}r^2 + \frac{1+\nu}{1-\nu}\left(\frac{R_1R_2}{r}\right)^2\right]$$

The radial stress is zero for $r = R_1$ and $r = R_2$ and is maximum for $r^* = \sqrt{R_1R_2} =$ 158.1 mm, which yields

$$\sigma_{r,\max} = \sigma_r(r^*) = 7.15 \text{ MPa} \tag{7.16}$$

the maximum tangential stress occurs in the inner radius $r = R_1$:

$$\sigma_{t,\max} = \sigma_t(R_1) = 41.05 \text{ MPa} \tag{7.17}$$

while is minimum for $r = R_2$:

$$\sigma_{t,\min} = \sigma_t(R_2) = 14.77 \text{ MPa} \tag{7.18}$$

The maximum displacement occurs for $r = R_1$ and is

$$u_{\max} = u(R_1) = 19.54 \text{ mm} \tag{7.19}$$

while is minimum for $r = R_2$:

$$u_{\min} = u(R_2) = 17.6 \text{ mm} \tag{7.20}$$

7.5.2 GetDP solution

7.5.2.1 Problem data

The main data of the problem are declared in the text file `data.pro` reported in Listing 7.6. In particular, it contains the models dimensions, the physical groups number, the material properties and the input angular speed.

Listing 7.6 File `data.pro` of the hollow disk problem

```
// beam dimensions
unit = 1e-3;
R = 250*unit; // outer radius (m)
r = 100*unit; // inner radius (m)
cl = 2.5*unit; // characteristic length (m)
// physical labels
body = 100;
ux0 = 1001;
uy0 = 1002;
// material properties
MassDensity = 7800; // mass density (kg/m3)
E = 210e9; // Young's modulus (Pa)
v = 0.3; // Poisson's ratio (-)
// input
MechSpeedRPM = 3000; // angular speed (rpm)
```

7.5.2.2 Model creation

The Gmsh code to draw the model in Figure 7.8 is reported in Listing 7.7. The physical lines, labeled as `ux0` and `uy0`, are necessary to set the constraints $u_x = 0$ and $u_y = 0$, respectively. The physical surface of the model is labeled as `body`.

7.5.2.3 Groups, functions and constraints

In the first part of the GetDP code in Listing 7.8, the model groups, functions and constraints are defined. In particular, the disk region is assigned to the group `body` while the physical lines `ux0` and `uy0` are assigned to the homonym groups.

After that, the material behavior, described by (7.6), and the centrifugal force density are defined in the function model field. The centrifugal density force of the jth mesh element is computed as follows:

$$f = \rho\omega^2 r_j \tag{7.21}$$

where r_j is the radius of the jth mesh element center of gravity, automatically computed by GetDP with the function `XYZ[]`.

A zero horizontal displacement constraint is then assigned to the group `ux0`. Similarly, the zero vertical displacement is set to the group `uy0`.

Then, the GetDP formulation is invoked and the problem is solved. It is worth noticing that the formulation code is the same one used for the cantilever beam problem.

Listing 7.7 Gmsh code (.geo) to draw the hollow disk and assign the physical groups

```
Include "data.pro";

Point(1) = {0, 0, 0, cl}; // disk center

// add outer circle points
Point(2) = {R, 0, 0, cl};
Point(3) = {0, R, 0, cl};
Point(4) = {-R, 0, 0, cl};
Point(5) = {0, -R, 0, cl};
// add outer circle arcs
Circle(1) = {2, 1, 3};
Circle(2) = {3, 1, 4};
Circle(3) = {4, 1, 5};
Circle(4) = {5, 1, 2};
// add inner circle points
Point(6) = {r, 0, 0, cl};
Point(7) = {0, r, 0, cl};
Point(8) = {-r, 0, 0, cl};
Point(9) = {0, -r, 0, cl};
// add inner circle arcs
Circle(5) = {6, 1, 7};
Circle(6) = {7, 1, 8};
Circle(7) = {8, 1, 9};
Circle(8) = {9, 1, 6};
// add lines for boundary conditions
Line(9) = {6, 2};
Line(10) = {7, 3};
Line(11) = {8, 4};
Line(12) = {9, 5};

// add disk surfaces
Line Loop(1) = {9, 1, -10, -5};
Plane Surface(1) = {1};
Line Loop(2) = {2, -11, -6, 10};
Plane Surface(2) = {2};
Line Loop(3) = {3, -12, -7, 11};
Plane Surface(3) = {3};
Line Loop(4) = {4, -9, -8, 12};
Plane Surface(4) = {4};

// assign physical groups
Physical Surface("body", body) = {1, 2, 3, 4};
Physical Line("uy=0", uy0) = {9, 11};
Physical Line("ux=0", ux0) = {12, 10};
```

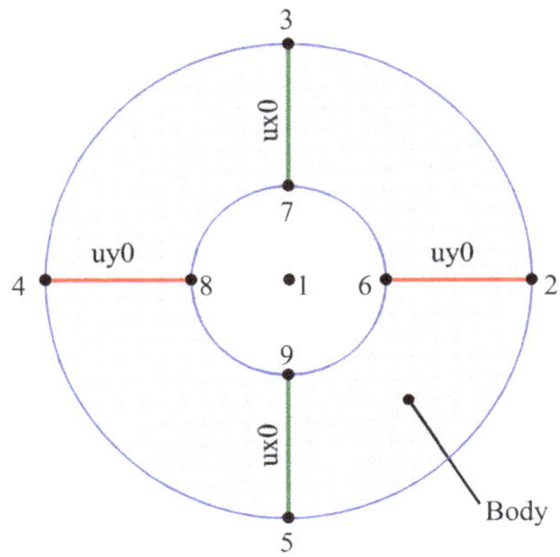

Figure 7.8 Hollow disk geometry and physical group names

At the end, some post-processing computations on the displacement field are performed to get some additional outputs.

7.5.2.4 Results

The displacement plot of the problem is reported in Figure 7.9(a). The displacement distribution and amplitude confirm the analytical results. In fact, at the inner and the outer radius, the displacement is 19.5 mm and 17.6 mm, respectively.

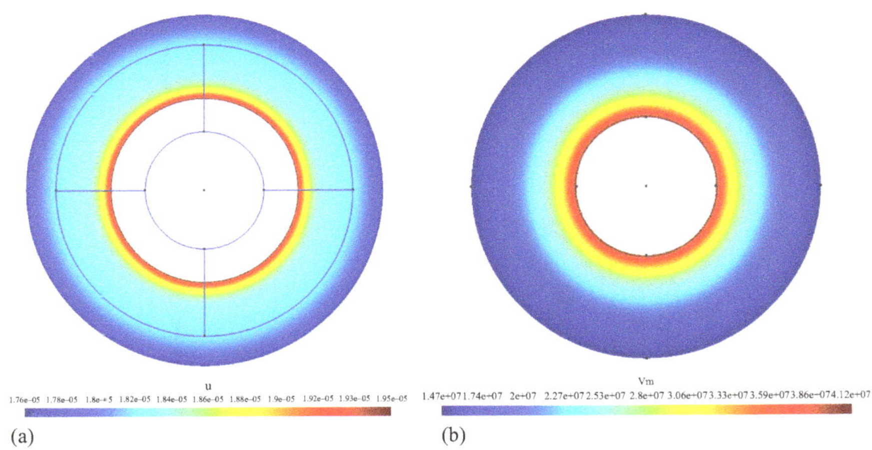

Figure 7.9 Resulting plots of the rotating hollow disk (the graphical displacement is amplified): (a) displacement plot and (b) von Mises plot

Listing 7.8 GetDP code (`.pro`) to set the model groups, the material properties and the boundary conditions to the hollow disk model

```
Include "data.pro";

// define the model groups
Group {
  body = Region[body];
  ux0 = Region[ux0];
  uy0 = Region[uy0];

  Domain_Force_Sur = Region[ {body } ];
  Domain_Force_Lin = Region[ {} ];
  Domain_Force     = Region[ {Domain_Force_Sur ,
    Domain_Force_Lin}];
  Domain_Disp      = Region[ {body} ] ;
}

Function {
  // set the material behavior
  C_m[ disk ] = E/(1-v^2)*TensorSym[ 1,   v,    0,
  1,    0,
  (1-v)/2 ];
  // compute the speed
  omega  = 2*Pi*MechSpeedRPM/60; // angular speed (rad/s)
  rho[]  = MassDensity; // mass density (kg/m^3)
  // assign a radial centrifugal force for each element
  // its amplitude depends on the speed and element mass and
    distance from
  the center
  F[body] = omega^2 * XYZ[] * rho[] ; // centrifugal density
    force (N/m^3)
}

// set the constraints
Constraint{
  { Name DeplacementX ; Type Assign ;
    Case {
      { Region ux0; Value 0 ; }
    }
  }
  { Name DeplacementY ; Type Assign ;
    Case {
      { Region uy0; Value 0 ; }
    }
  }
}
```

```
// load the formulation
Include "Elasticity_2D.pro"

// post processing operations
PostOperation Map_u UsingPost Mec2D_u {
  Print [ u,  OnElementsOf Domain_Disp, File "u.pos"] ; //
    displacement
  Print [ um, OnElementsOf Domain_Disp,  File "um.pos"] ; //
    displacement
  amplitude
  Print [ F,  OnElementsOf Domain_Force,  File "F.pos"] ; //
    load
  Print [ Vm , OnElementsOf Domain_Disp, File "Vm.pos"] ; //
    von Mises stress
}
```

7.6 Rotating SyR rotor geometry

The FE structural analysis of an SyR rotor is similar to the hollow disk problem, where the presence of the flux-barriers represents the only change. In this case, an exact analytical solution does not exist but, as described in Chapter 14, only an estimation of the average stress in the radial rib regions is possible.

For these reason, this section reports only few steps of the problem definition while the results validation and discussion are carried out in Chapter 14.

The complete four poles SyR rotor geometry is reported in Figure 7.10(a). For simplicity, only the reduced one pole model is analyzed in the following. This reduces the model complexity and does not require any change on the GetDP code in Listing 7.8. The main problem data are reported in Section 14.5.

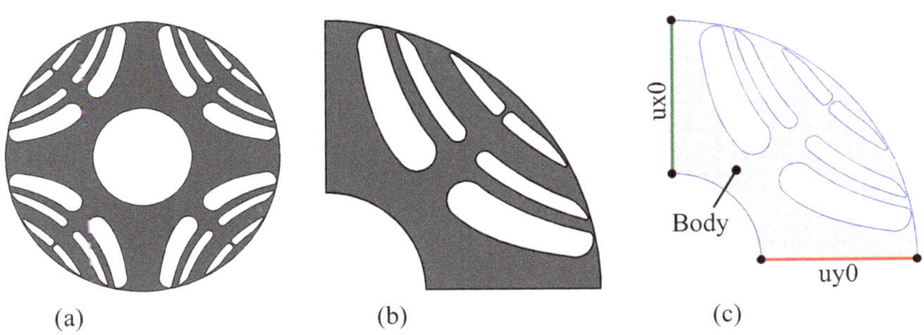

Figure 7.10 SyR rotor geometry sketches and physical group names: (a) complete model, (b) one-pole model, (c) Gmsh model

7.6.1 Model geometry

The rotor geometry is not drawn directly with the Gmsh code because of its complexity but is imported from a `dxf` file. The Gmsh model and the physical group names are reported in Figure 7.10(c).

7.6.2 Results

The displacement and von Mises plots of the SyR rotor structural analysis are shown in Figure 7.11(a) and (b), respectively. The maximum von Mises stress within the radial ribs are in good agreement with the analytical estimation.

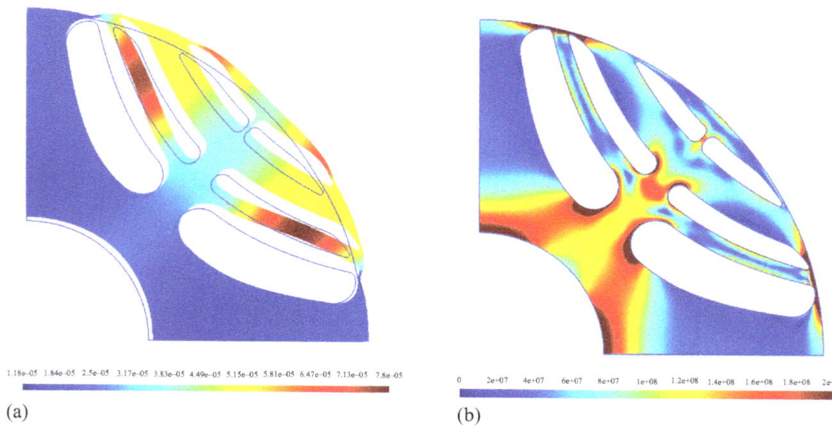

Figure 7.11 Resulting plots of the rotating SyR rotor (the graphical displacement is amplified): (a) displacement plot and (b) von Mises plot

Chapter 8
Efficiency map computation

8.1 Introduction

One of the most effective graphical ways to describe the performance of an electric machine is the efficiency map. The efficiency map is a constant-efficiency contour plot in the torque, or power, versus speed plane [68–74]. Each point of such a map corresponds to a specific current supply condition that exhibits the maximum efficiency for a given torque and speed loadpoint. An example is reported in Figure 8.1.

Numerically speaking, it is a 2D matrix with R rows and C columns; i and j are row and column indices, respectively. The number of rows and columns depends on the size of the torque and speed arrays selected to build the evenly spaced grid. The maximum torque depends on the operating condition and current limit. In general, it can be set equal to the rated or overload torque. The maximum speed depends on the mechanical constraint of the machine. The voltage is automatically taken into account in the efficiency map computation.

Referring to Figure 8.1, the two arrays are defined as follows:

$$T_{\mathrm{m}} = \begin{pmatrix} T_{\mathrm{m}}^{(1)} \\ T_{\mathrm{m}}^{(2)} \\ \vdots \\ T_{\mathrm{m}}^{(R)} \end{pmatrix} \qquad \omega_{\mathrm{m}} = \left(\omega_{\mathrm{m}}^{(1)}, \omega_{\mathrm{m}}^{(2)}, \ldots, \omega_{\mathrm{m}}^{(C)}\right) \tag{8.1}$$

which means that the (i,j) element of the efficiency map, $\boldsymbol{\eta}$, contains the maximum efficiency achievable with the torque, $T_{\mathrm{m}}^{(i)}$, at the mechanical speed, $\omega_{\mathrm{m}}^{(j)}$.

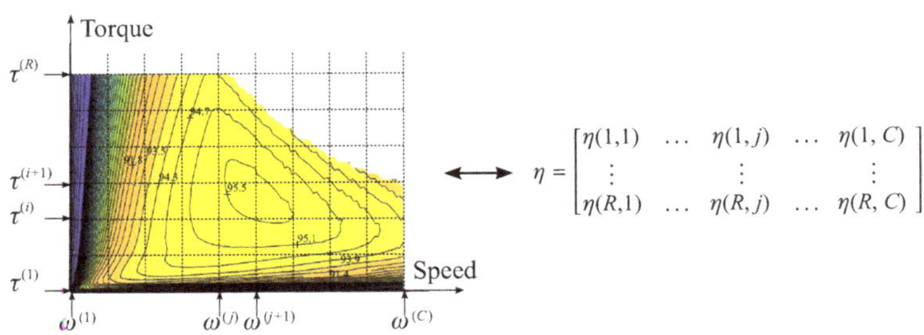

Figure 8.1 Efficiency map in the torque versus speed plane example

In general, for synchronous machine drives, there are infinite combinations of I_d and I_q currents that produce the torque $T_m^{(i)}$ at the speed $\omega_m^{(j)}$ within the current and voltage limits. The hard point of the efficiency map computation is finding the best I_d and I_q combination that yields the maximum efficiency. This chapter reports a detailed description of the algorithm used to compute the efficiency map.

In the following, unless otherwise stated, all the operations performed with the arrays or matrices are *element-wise* operations. For example, let us define matrices $\boldsymbol{A}$ and $\boldsymbol{B}$ as follows:

$$\boldsymbol{A} = \begin{pmatrix} A_{11} & A_{12} \\ A_{21} & A_{22} \end{pmatrix} \qquad \boldsymbol{B} = \begin{pmatrix} B_{11} & B_{12} \\ B_{21} & B_{22} \end{pmatrix} \tag{8.2}$$

the element-wise product is

$$\boldsymbol{C} = \boldsymbol{A} \odot \boldsymbol{B} = \begin{pmatrix} A_{11} \cdot B_{11} & A_{11} \cdot B_{12} \\ A_{21} \cdot B_{21} & A_{22} \cdot B_{22} \\ & , \end{pmatrix} \tag{8.3}$$

while the element-wise reciprocal of $\boldsymbol{A}$ is

$$\frac{1}{\boldsymbol{A}} = \begin{pmatrix} 1/A_{11} & 1/A_{12} \\ 1/A_{21} & 1/A_{22} \end{pmatrix} \tag{8.4}$$

$$\eta = \begin{bmatrix} \eta(1,1) & \dots & \eta(1,j) & \dots & \eta(1,C) \\ \vdots & & \vdots & & \vdots \\ \eta(R,1) & \dots & \eta(R,j) & \dots & \eta(R,C) \end{bmatrix} \tag{8.5}$$

8.2 Input data

The main input data are the magnetizing curves and iron losses matrices obtained for an evenly spaced grid of I_q and I_q currents, as reported in Figure 8.2.

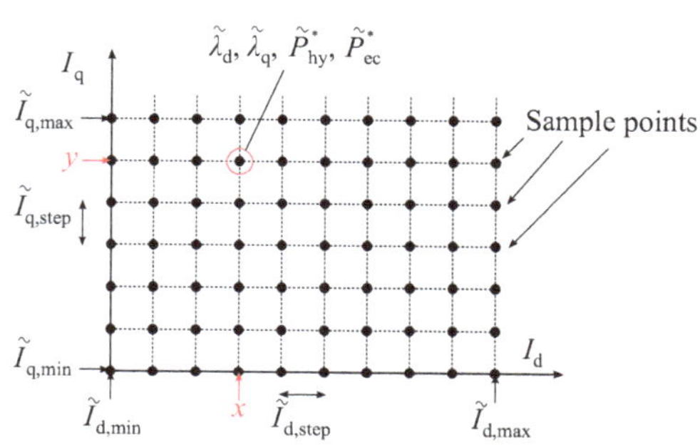

Figure 8.2 I_d–I_q evenly spaced grid. For each point of the grid, the dq flux linkages and the hysteresis and eddy current iron losses have to be computed. These matrices allow to describe entirely the machine behavior with a proper data processing

The iron losses can be computed in different ways. In the case of Finite Element (FE) results, it is convenient to compute the hysteresis and eddy current losses for every grid point at a given reference speed, ω_{m}^{*}, yielding the matrices P_{hy}^{*} and P_{ec}^{*}.

8.3 Constant torque loci current points

The next step is to find the I_{d} and I_{q} current combinations that describe the constant torque levels of the array T_{m}.

In the proposed method, the MATLAB function `contourc` is adopted. For a given constant torque level $T_{\mathrm{m}}^{(i)}$ (ith element of $\boldsymbol{T}_{\mathrm{m}}$), the function returns a certain number of points. Figure 8.3(a) shows qualitatively the constant torque curves relative to $T_{\mathrm{m}}^{(i)}$ and $T_{\mathrm{m}}^{(i+1)}$. In particular, the first and second loci are described by k (i.e. 4) and h (i.e. 10) points, respectively.

8.4 Constant torque number of points homogenization

For a reason that will be clarified later on, it is necessary to describe all the constant torque curves with the same number of points, N. To do that, a 1D interpolation has to be performed.

First, a new set of N evenly spaced I_{d} currents is defined setting the minimum and maximum values of the initial data set as extreme values. From the new set of d-axis currents, I_{d}, the respective q-axis currents, I_{q}, are then interpolated. This step is described qualitatively in Figure 8.3(b). The filled circles on the I_{d}-axis represent the I_{d} values, while the star markers are the interpolated I_{q} currents. The empty circles on the plot represent the initial data set obtained in Section 8.3. This operation has to

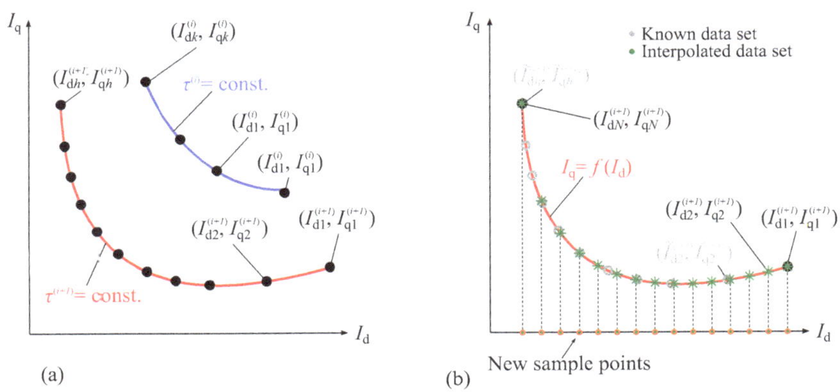

Figure 8.3 Computation and interpolation of the constant torque loci in the (I_d, I_q) plane: (a) constant torque locus points and (b) interpolation of the $T_m^{(i)}$ constant torque locus points

be performed for each element of the array $\boldsymbol{T}_{\text{m}}$ in (8.1). At the end of the procedure, R different I_{d} and I_{q} arrays, of size $N \times 1$, will be created, as reported in (8.6):

$$I_{\text{d}}{}^{(i)} = \begin{pmatrix} I_{\text{d},1}^{(i)} \\ \vdots \\ I_{\text{d},N}^{(i)} \end{pmatrix} \qquad I_{\text{q}}{}^{(i)} = \begin{pmatrix} I_{\text{q},1}^{(i)} \\ \vdots \\ I_{\text{q},N}^{(i)} \end{pmatrix} \qquad i = 1, 2, \ldots R \tag{8.6}$$

8.5 Creation of the 3D matrices

For more clarity, let us first consider the data relative to the torque $T_{\text{m}}^{(i)}$. The column arrays $I_{\text{d}}{}^{(i)}$ and $I_{\text{q}}{}^{(i)}$ are replicated C times along the column direction yielding the following matrices:

$$I_{\text{d}}^{(i)} = \begin{bmatrix} \begin{pmatrix} I_{\text{d},1}^{(i)} \\ \vdots \\ I_{\text{d},N}^{(i)} \end{pmatrix}_1 \cdots \begin{pmatrix} I_{\text{d},1}^{(i)} \\ \vdots \\ I_{\text{d},N}^{(i)} \end{pmatrix}_j \cdots \begin{pmatrix} I_{\text{d},1}^{(i)} \\ \vdots \\ I_{\text{d},N}^{(i)} \end{pmatrix}_C \end{bmatrix}_i$$

$$I_{\text{q}}^{(i)} = \begin{bmatrix} \begin{pmatrix} I_{\text{q},1}^{(i)} \\ \vdots \\ I_{\text{q},N}^{(i)} \end{pmatrix}_1 \cdots \begin{pmatrix} I_{\text{q},1}^{(i)} \\ \vdots \\ I_{\text{q},N}^{(i)} \end{pmatrix}_j \cdots \begin{pmatrix} I_{\text{q},1}^{(i)} \\ \vdots \\ I_{\text{q},N}^{(i)} \end{pmatrix}_C \end{bmatrix}_i$$

After that, the speed matrices have to be created. The construction is quite simple and consists of replicating N times the row array ω_{m} along the row dimension. The result is the matrix $\boldsymbol{\omega}_{\text{m}}^{(i)}$:

$$\boldsymbol{\omega}_{\text{m}}^{(i)} = \begin{bmatrix} \left(\omega_{\text{m}}^{(1)} \quad \cdots \quad \omega_{\text{m}}^{(C)}\right)_1 \\ \vdots \\ \left(\omega_{\text{m}}^{(1)} \quad \cdots \quad \omega_{\text{m}}^{(C)}\right)_n \\ \vdots \\ \left(\omega_{\text{m}}^{(1)} \quad \cdots \quad \omega_{\text{m}}^{(C)}\right)_N \end{bmatrix}_i \tag{8.7}$$

Then, matrices (8.5), (8.6) and (8.7) have to be replicated R times along the third dimension (i.e. *layer* in MATLAB syntax) as shown schematically in Figure 8.4. The current amplitude 3D matrix is

$$[\boldsymbol{I}] = \sqrt{[\boldsymbol{I}_{\text{d}}]^2 + [\boldsymbol{I}_{\text{q}}]^2}, \tag{8.8}$$

while the hysteresis and eddy current losses, relative to $[\boldsymbol{I}_{\text{d}}]$ and $[\boldsymbol{I}_{\text{q}}]$, can be computed interpolating the initial data set obtaining the matrices $[\boldsymbol{p}_{\text{hy}}^{*}]$ and $[\boldsymbol{p}_{\text{ec}}^{*}]$.

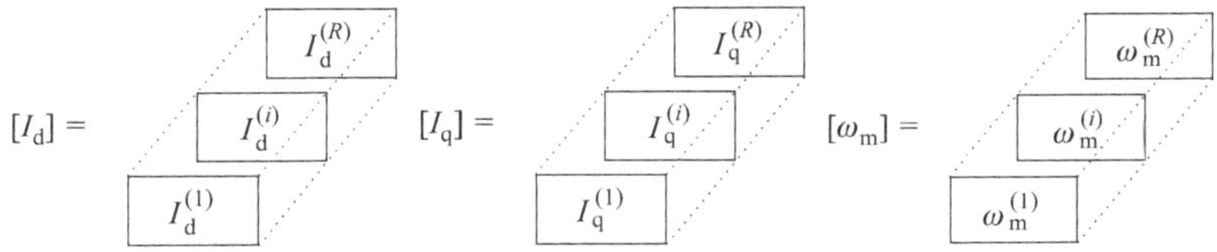

Figure 8.4 3D current and speed matrices. Their dimension is $N \times C \times R$ where N is the number of points for each constant torque curve, while C and R are the length of ω_m and T_m, respectively

8.6 Computation of the derived quantities

Once the current and speed 3D arrays are built, it is possible to compute all the derived quantities. The first step is computing the dq flux linkages for each current combination in $[I_d]$ and $[I_q]$ interpolating the initial data set discussed in Section 8.2. The results are the 3D arrays $[\mathbf{\Lambda}_d]$ and $[\mathbf{\Lambda}_q]$ with the same size and indices meaning of $[I_d]$.

The losses values, $[P_{hy}]$ and $[P_{ec}]$, relative to the currents $[I_d]$ and $[I_q]$ are computed interpolating the initial data set. Then, they are rescaled according to the actual speed $[\boldsymbol{\omega}_m]$. Such an operation depends on the iron losses equation used for the computation of $p_{hy}^{\sim *}$ and $p_{ec}^{\sim *}$. Using the standard Steinmetz equation, they can be computed as follows:

$$[\boldsymbol{p}_{hy}] = [\boldsymbol{p}_{hy}^*] \odot \left(\frac{[\boldsymbol{\omega}_m]}{\omega_m^*}\right)^{\alpha} \quad \text{and} \quad [\boldsymbol{p}_{ec}] = [\boldsymbol{p}_{ec}^*] \odot \left(\frac{[\boldsymbol{\omega}_m]}{\omega_m^*}\right)^{2} \tag{8.9}$$

where α is the hysteresis iron losses coefficient, usually set to 1.

The 3D array of the Joule losses is simply computed as follows:

$$[\boldsymbol{P}_{Js}] = \frac{3}{2} R_w \sqrt{[I_d]^2 + [I_q]^2} \tag{8.10}$$

where R_w is the phase winding resistance, assumed to be constant for all the working points neglecting the temperature effect on its value. A rule of thumb is to compute the resistance at the maximum temperature allowed by the winding insulation class.

The 3D array of the torque is computed as follows:

$$[\boldsymbol{T}_m] = \frac{3}{2} p \left([\mathbf{\Lambda}_d] \odot [\boldsymbol{I}_q] - [\mathbf{\Lambda}_q] \odot [\boldsymbol{I}_d]\right) \tag{8.11}$$

It is worth noting that, according to the adopted indices meaning, the ith layer of $[\boldsymbol{T}_m]$ presents all the elements equal to $T_m^{(i)}$.

The mechanical power is given by

$$[\boldsymbol{P}_m] = [\boldsymbol{T}_m] \odot [\boldsymbol{\omega}_m], \tag{8.12}$$

while the efficiency is

$$[\eta] = \frac{[\boldsymbol{P}_{\rm m}]}{[\boldsymbol{P}_{\rm m}] + [\boldsymbol{P}_{\rm Js}] + [\boldsymbol{P}_{\rm Fe}]}. \tag{8.13}$$

The terminal voltage components and the total current have to be computed. Their computation depends on how the iron losses equivalent resistance is connected to the standard *dq* electric model of the machine. In general, the iron losses resistance can be connected in parallel with the machine terminals or in series with the machine impedance as reported in Figure 8.5(a) and (b), respectively.
For parallel and series connection, the equations are

$$\text{parallel:} \begin{cases} U_{\rm d} = U'_{\rm d} \\ U_{\rm q} = U'_{\rm q} \\ I_{\rm Fe} = \frac{p_{\rm Fe}}{\sqrt{U_{\rm d}^2 + U_{\rm q}^2}} \\ I_{\rm d} = I'_{\rm d} + I_{\rm Fe} \\ I_{\rm q} = I'_{\rm q} + I_{\rm Fe} \end{cases} \qquad \text{series:} \begin{cases} I_{\rm d} = I'_{\rm d} \\ I_{\rm q} = I'_{\rm q} \\ U_{\rm Fe} = \frac{p_{\rm Fe}}{\sqrt{I_{\rm d}^2 + I_{\rm q}^2}} \\ U_{\rm d} = U'_{\rm d} + U_{\rm Fe} \\ U_{\rm q} = U'_{\rm q} + U_{\rm Fe} \end{cases} \tag{8.14}$$

where $U'_{\rm d}$ and $U'_{\rm q}$ are the inner voltage *dq* components given by

$$\begin{cases} U_{\rm d} = R_{\rm w} I_{\rm d} - \omega \Lambda_{\rm q} \\ U_{\rm q} = R_{\rm w} I_{\rm q} + \omega \Lambda_{\rm d} \end{cases} \tag{8.15}$$

Setting $[I_{\rm d}]$, $[I_{\rm q}]$, $[P_{\rm hy}]$ and $[P_{\rm ec}]$ in (8.14), we obtain the 3D arrays $[\boldsymbol{U}_{\rm d}]$ and $[\boldsymbol{U}_{\rm q}]$, while the voltage amplitude is given by

$$[\boldsymbol{U}] = \sqrt{[\boldsymbol{U}_{\rm d}]^2 + [\boldsymbol{U}_{\rm q}]^2} \tag{8.16}$$

At this point, the element indices of $[\boldsymbol{U}]$ and $[\boldsymbol{I}]$ that are not compatible with the voltage and current limits are found. Then, the corresponding elements in all the

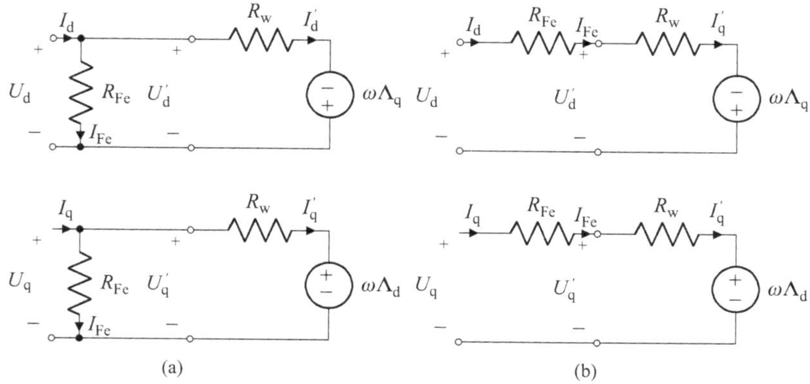

Figure 8.5 dq Electric model of the machine with different connection of the iron losses resistance: (a) parallel and (b) series

previously computed matrices are replaced by NaN in order to exclude these elements from the maximum efficiency computation.

8.7 Maximum efficiency computation

Thanks to the matrix indexing order adopted, it is possible to observe that the jth column of the ith layer of $[\eta]$ contains the efficiencies computed for all the current combinations relative to $T_{\mathrm{m}}^{(i)}$ at the speed $\omega_{\mathrm{m}}^{(j)}$. This means that the maximum efficiency of the jth column and ith layer represents a point of the efficiency map, in the torque versus speed plane, coordinates of which are $\omega_{\mathrm{m}}^{(j)}$ and $T_{\mathrm{m}}^{(i)}$. Furthermore, the set of column maximum efficiencies of the ith layer is a row of the efficiency map corresponding to the torque $T_{\mathrm{m}}^{(i)}$. So, applying the function **maxcol** for each layer of $[\eta]$, we obtain the complete efficiency map, $\hat{\eta}$, as shown in Figure 8.6.

All the other quantities, indicated with the hat symbol, can be obtained selecting the elements which corresponds to the maximum efficiency position. For example,

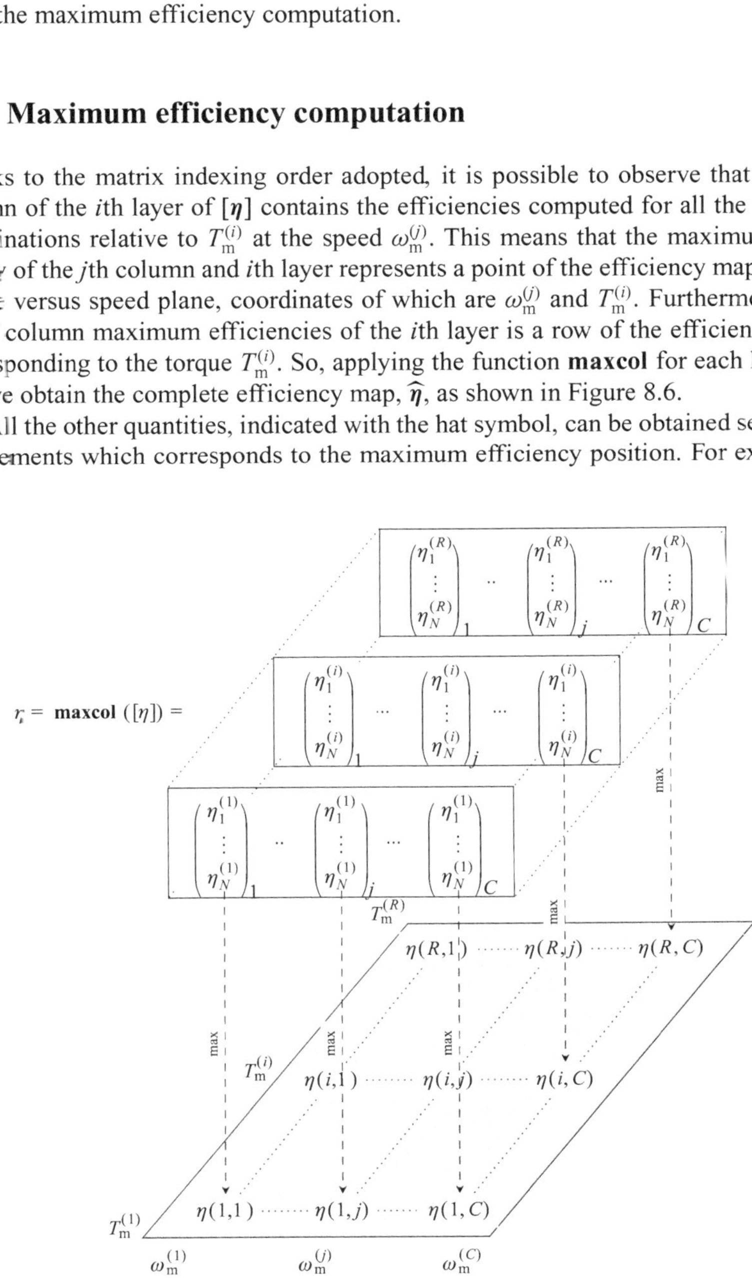

Figure 8.6 Scheme for the computation of the maximum efficiency from the 3D array $[\eta]$. *The function* **maxcol** *returns the maximum of each column*

the ith row and jth column of the matrix $\widehat{\boldsymbol{P}_{Js}}$ will contain the Joule losses computed at the speed $\omega_{\mathrm{m}}^{(j)}$ and torque $T_{\mathrm{m}}^{(i)}$ obtained with the current–voltage combination which yields the maximum efficiency $\widehat{\eta}(i,j)$.

The same approach can be carried out for the induction motor with the addition of the rotor slip and Joule losses. Their computation is outside the scope of the present work, but it is described in detail in [75,76].

8.8 Efficiency map, including thermal limit

The proposed efficiency map algorithm can be further improved taking into account the thermal limit and resistance variation with the temperature. As usual, the performances are computed at steady-state condition.

It is worth noticing that the resistance dependence on the temperature affects the Joule losses computation as well as the terminal voltage computation by means of (8.10) and (8.15), respectively. The influence of the core temperature on the eddy current losses is not considered. Also the temperature effect on the PMs is neglected.

8.8.1 Efficiency map at different temperatures

On the basis of the iron and Joule losses maps computed in (8.9) and (8.10), it is possible to estimate the actual winding temperature by means of a thermal analysis. Such analysis can be carried out by means of a lumped parameters network [77] or FE analysis. In any cases, the actual winding temperature, T_{w}, is described by the following relationship:

$$T_{\mathrm{w}} = T_{\mathrm{w}}\left(P_{\mathrm{Js}}(\widetilde{T_{\mathrm{w}}}), p_{\mathrm{hy}}(\omega_{\mathrm{m}}), p_{\mathrm{ec}}(\omega_{\mathrm{m}}^{2})\right) \tag{8.17}$$

where $\widetilde{T}$ is the assumed winding temperature used for the resistance computation.

Let us first define the assumed winding temperature vector:

$$\widetilde{T_{\mathrm{w}}} = \left(\widetilde{T_{\mathrm{w}}}^{(1)}, \widetilde{T_{\mathrm{w}}}^{(2)}, \ldots, \widetilde{T_{\mathrm{w}}}^{(M)}\right) \tag{8.18}$$

where $\widetilde{T_{\mathrm{w}}}^{(1)}$ and $\widetilde{T_{\mathrm{w}}}^{(M)}$ can be set equal to the environment temperature (i.e. 20°C) and the maximum insulation class temperature, respectively.

For each element of $\widetilde{T_{\mathrm{w}}}$, it is possible to compute the Joule losses and efficiency maps by means of (8.10) and (8.13) obtaining the 3D arrays reported in Figure 8.7.

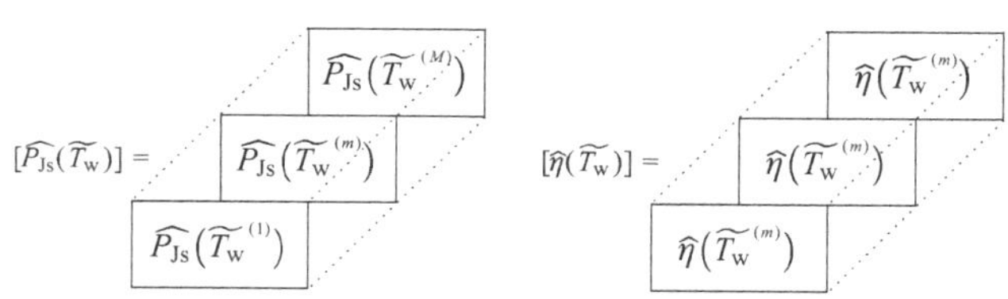

Figure 8.7 Efficiency and Joule losses 3D maps computed for different winding temperature

Since the iron losses do not depend on the temperature, the matrices $\widehat{\boldsymbol{p}_{\mathrm{hy}}}$ and $\widehat{\boldsymbol{p}_{\mathrm{ec}}}$ can be replicated M times along the third dimension obtaining $[\widehat{\boldsymbol{p}_{\mathrm{hy}}}]$ and $[\widehat{\boldsymbol{p}_{\mathrm{ec}}}]$.

8.8.2 Thermal analysis

Once the losses maps are know, it is possible to perform a thermal analysis for each of their element. In particular, the Joule and iron losses are set as heat source in the thermal model. This computation provides M different winding temperature matrices as shown in Figure 8.8. In particular, the ith row, jth column and mth layer of the matrix $[\widehat{\boldsymbol{T}_{\mathrm{w}}}]$ contain the winding temperature obtained at the speed $\omega_{\mathrm{m}}^{(j)}$ and torque $T_{\mathrm{m}}^{(i)}$ with an assumed winding temperature of $\widetilde{T_{\mathrm{w}}}^{(m)}$. All the elements that present a value higher than the maximum allowed temperature are replaced by a `NaN` value, since they do no represent a valid steady-state solution.

It is worth noticing that after a preliminary estimation of the maximum iron and Joule losses, it is possible to carry out the thermal analyses separately from the electromagnetic ones so that their computational time is not altered. Furthermore, for a given electromagnetic map, it is possible to couple different thermal analyses with different material properties (e.g. material conductivity), boundary conditions (e.g. heat exchange), computational accuracy (2D or 3D FE analysis or CFD analysis) and other settings.

The mth layer of the $[\widehat{\boldsymbol{T}_{\mathrm{w}}}(\widetilde{T_{\mathrm{w}}})]$ is shown qualitatively in the bottom right $(\omega_{\mathrm{m}}, T_{\mathrm{m}})$ plane in Figure 8.9. Among the map elements, only the ones that belong to the constant winding temperature locus $\widehat{T_{\mathrm{w}}}^{(m)}$ are computed correctly since the actual winding temperature is equal to the assumed one $\widetilde{T_{\mathrm{w}}}^{(m)}$. The efficiency of each of this nodes is then computed interpolating the relative efficiency map layer $\widehat{\boldsymbol{\eta}}(\widetilde{T_{\mathrm{w}}}^{(m)})$ and finally, only these nodes are stored for the final efficiency map assembling. The Mth layer of the matrix $[\widehat{\boldsymbol{T}_{\mathrm{w}}}(\widetilde{T_{\mathrm{w}}})]$ refers to the maximum allowed temperature. The nodes that belong to the constant winding temperature locus $\widehat{T_{\mathrm{w}}}^{(M)}$ describe the maximum thermal limit; hence, it delimits the actual $(\omega_{\mathrm{m}}, T_{\mathrm{m}})$ plane that can effectively be used at steady-state condition.

Repeating such a procedure for each layer of $[\widehat{\boldsymbol{T}_{\mathrm{w}}}(\widetilde{T_{\mathrm{w}}})]$, the correct efficiency map $\widehat{\boldsymbol{\eta}}^{*}$ shown qualitatively on the right of Figure 8.9 is obtained. It is possible to see that the efficiency map is different from the estimated one and the actual maximum torque is lower than the one provided by the voltage and current limits.

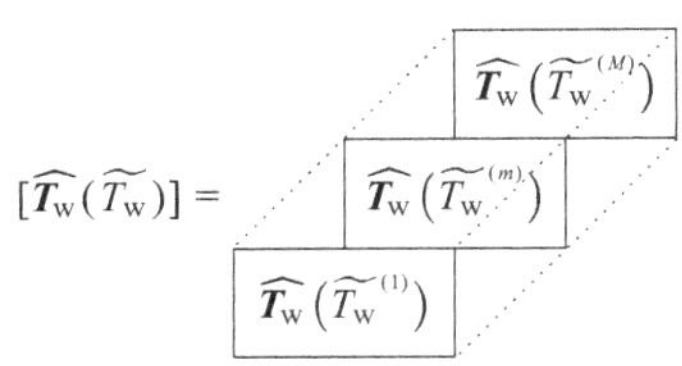

Figure 8.8 Winding temperature 3D matrix obtained from the thermal analysis

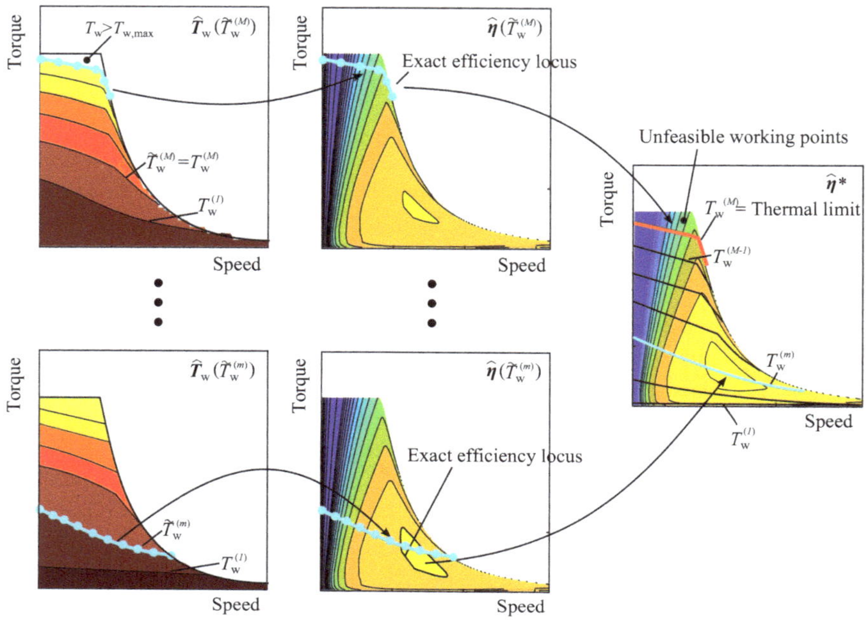

Figure 8.9 Proposed procedure for the identification of the thermal limits

Chapter 9
Multi-objective optimization

This chapter presents a very brief introduction to multi-objective optimization algorithms, in particular, focusing on the Differential Evolution (DE) algorithm.

9.1 Problem statement

The design of electric machines is an inverse problem, it is a synthesis. These are among the most difficult problems to solve, because there is no direct way to go through it. But all these kinds of problems can be formulated in terms of minimizing an objective function:

$$\min_{x \in X} f(x)$$

Actually, we search for the value of x, within a certain interval X, which gives the global minimum of the function f. So what we are really looking for is solving

$$\arg\min_{x \in X} f(x) \tag{9.1}$$

Even though the problem is rather simple to state, there is actually no meaning in minimum of a vector function, that is, when we have more than one objective. And the variable x is often a vector variable, so a set of parameters (which may be diameters, lengths, number of poles, etc.). Then, the subspace (subdomain) X can be defined not only by simple bounds but also by some functional relationships among some quantities. This is not such an uncommon thing, because often there are some feasibility checks on dependent variables, such as the slot opening width, or the yoke height, and so on. For all these reasons, the most simple way to solve this problem is to employ an automatic algorithm, namely, an optimizer. In fact, global optimization has become widespread in solving problems in many fields of science, engineering, economics and more. Many practical problems have objective functions that are non-differentiable, noncontinuous, nonlinear, noisy, with many local minima, etc. While these kinds of problems are very difficult to solve analytically, if not impossible, they can be quite easily implemented into an optimization routine.

A very successful type of optimization algorithms is the Evolutionary Algorithm (EA) class. This class tries to find the minimum of a function borrowing ideas from natural evolution (reproduction, mutation, recombination and selection). EAs belong to a broader class of optimizers, called stochastic, where there is a substantial amount of randomness in the search process. This family is set against the deterministic algorithms. Among the stochastic optimization algorithms, we may have the following [78]:

- Simulated annealing: It aims at emulating the annealing heat treatment process of steel.
- Particle swarm optimization: It aims at emulating the social behavior of birds flocking.
- Game-theory-based optimization: It aims at emulating the evolution of a game in which different players try to fulfill their own objectives. They are based on the game theory from Nash.
- Evolutionary Algorithms (EAs): These aim at emulating the evolution of species by natural selection according to Darwin's theory. It is the most important category of stochastic optimization together with Genetic Algorithms (GAs).
- Genetic Algorithms (GAs): Alike EAs, these aim at emulating the evolution of species. For this reason at times they are considered a subcategory of EAs. However, EAs and GAs, in practice, have different approaches to the emulation of evolution and can also be considered two different categories of stochastic optimization algorithms.

The source of inspiration of many randomized search methods comes from the observation of nature. Concepts from biology, physics, geology or some other fields of investigation are borrowed and implemented in a simplified model of some natural phenomena. Most of these methods are population-based algorithms, in which a set of initial samples evolves (or moves) up to convergence. Stochastic optimization methods are the most innovative and advanced approaches to optimization. Compared to deterministic optimization methods, they have both advantages and drawbacks [78]:

- simpler mathematical background;
- randomness included in the algorithm;
- slower convergence toward the optimum;
- wider search in the design space, so less chance to get stuck at a local minimum;
- adaptable to any kind of function.

One of the most effective optimization algorithms for engineering problems is the DE. The flowchart of the optimization scheme is reported in Figure 9.1. Its algorithm works through three basic principles:

1. mutation: expands the search space
2. recombination: reuses previously successful individuals
3. selection: mimics survival of the fittest

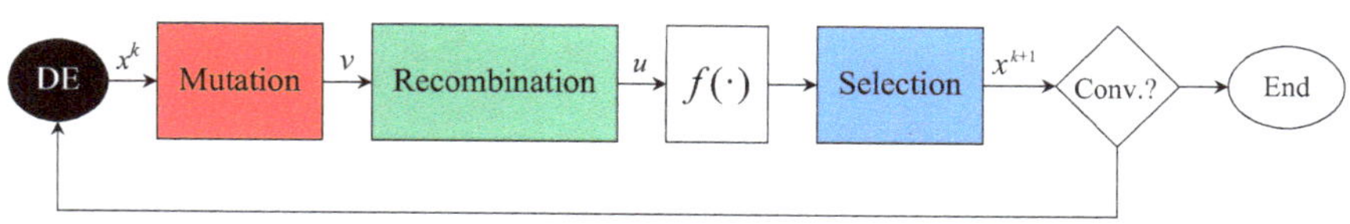

Figure 9.1 Flowchart of the optimization scheme employed by the DE

9.1.1 Mutation

There may be many ways to perform mutation on an individual, and here are some examples:

- DE/rand/1 $v_i = x_{r_1} + F(x_{r_2} - x_{r_3})$
- DE/best/1 $v_i = x_{\text{best}} + F(x_{r_2} - x_{r_3})$
- DE/rand2best/1 $v_i = x_{r_1} + F_1(x_{r_2} - x_{r_3}) + F_2(x_{\text{best}} - x_{r_1})$
- DE/curr2best/1 $v_i = x_i + F_1(x_{r_2} - x_{r_3}) + F_2(x_{\text{best}} - x_i)$
- DE/rand/2 $v_i = x_{r_1} + F(x_{r_2} - x_{r_3} + x_{r_4} - x_{r_5})$
- DE/best/2 $v_i = x_{\text{best}} + F(x_{r_2} - x_{r_3} + x_{r_4} - x_{r_5})$
- ...

What is common among these techniques is that we add one (or more) difference vector(s) to a base individual in order to move the individual and explore more the design space. x_i is the ith *individual* of the current population. After a mutation happens, we have a new individual referred to as v, called *donor*. The factors F_x are called mutation factors, and they can be constants chosen at the beginning of the optimization, or they can change their value during the optimization run, either deterministically (according to a certain law) or stochastically (included among the optimization parameters).

9.1.2 Recombination

The recombination step mixes successful individuals from the previous generation with current donors, to try to keep the best features:

$$u_{i,j} = \begin{cases} v_{i,j} & \text{if } \text{rand}_{i,j} \leq \text{CR} \ \vee \ j = i_{\text{rand}} \\ x_{i,j} & \text{otherwise} \end{cases} \tag{9.2}$$

The individual u_i is called *trial*, CR is the crossover ratio, and with j we are referring to the jth parameter of the considered individual.

9.1.3 Selection

The selection block is what makes the DE really effective, since it mimics a "greedy" natural selection where only the fittest can "survive." In fact, the objective function is evaluated on the trials and, if the objective value results better than the previous generation individual, the trial survives and becomes the new-generation individual. Mathematically, it translated to

$$x_i^{k+1} = \begin{cases} u_i^k & \text{if } f(u_i^k) < f(x_i^k) \\ x_i^k & \text{otherwise} \end{cases} \tag{9.3}$$

9.2 Missing pieces

There are a lot of other missing details that have not been described in this chapter and are essential to perform a proper optimization, though. Nonetheless, they are out of scope of this work, so in the following there will be only a brief reference to these aspects.

- How to choose the population size?
- How to choose the total number of generation?
- How to choose the mutation factor and the crossover ratio?
- Which is the best mutation strategy?
- How to deal with stagnation?
- How to handle constraints?
- How to deal with multi-objective optimization?

Some possible enhancements:

- Asynchronous strategy: use good solutions immediately without waiting for the next generation to be created.
- Elitism: compare trial not only with target but also with the worst individual, therefore keeping the most promising.
- Coevolution: use several populations, each working on a different subset of the design space.

Some of these items are explained and developed in [78–80], where a more thorough bibliographic section is also present.

Some applications and recent advancements about optimization in electric machine can be found in [81,82].

Chapter 10
Design and optimization of a PMaSyR motor for low-voltage E-scooter applications

Nowadays, the increasingly evident effects of the transportation pollution on the climate change and human health [83–85] are inducing many countries to employ policies to encourage the diffusion of electric vehicles [86]. All these aspects and the technological advances of the automotive industry in the recent years stimulated the market of electric vehicles with promising results in China, United States and Europe [87,88].

Among the automotive products, the electric motorcycle is one of the most effective solutions for the mobility, and its market is anticipated to grow in the next years. The main benefits are the high efficiency, low weight, compactness, ease of use and relative low cost with respect to an electric car.

The brushless DC and Interior Permanent Magnet (IPM) motors have been proposed in the past with different geometry layouts (i.e. outer or inner rotor) and transmissions (i.e. belt or direct) [89–93]. The main advantages of these machines are the high-torque density and efficiency due to use of rare-earth Permanent Magnets (PMs) with high remanence flux. However, the sharply fluctuations of the rare-earth PMs price in the last years and their non-negligible environmental impact encouraged the research activities to find alternative solutions with cheaper PMs (e.g. ferrite) or, more drastically, without any PM. The switched reluctance motor has been proposed for its design simplicity, low-cost and fault-tolerant reliability but, to date, is still no attractive for the automotive application due to its high noise and vibrations [94].

In the last years, the Synchronous Reluctance (SYR) motor has been proposed in several applications for its good torque density, low cost, robustness and overload capability [44,95]. The main drawbacks of such a machine are the torque ripple, the low-power factor and the limited Constant Power Speed Range (CPSR). The first point can be reduced effectively with a proper design of the rotor geometry [27,96] while the other two items are intrinsic properties of the machine and are hard to improve with the motor geometry design only. Regarding the last two aspects, the PM-assisted SYR (Permanent Magnet assisted Synchronous Reluctance (PMASYR)) machine represents a valid solution [97–99].

In such a machine, PMs with poor magnetic properties (e.g. ferrite) or a small amount of rare-earth PMs are inset in the rotor flux-barriers to saturate the iron ribs, improve the power factor and extend the CPSR. The rotor PMs addition represents an additional cost of the motor but allows one to limit the volt–ampere ratings of the inverter yielding a lower overall cost of the electric drive with respect to the SYR one. All these features make the PMASYR machine an optimal solution for the automotive application in which high-torque density and a wide CPSR are required [100].

The multi-objective design of such a machine has been deeply investigated in literature [101,102]. In these works, the average torque, the torque ripple and the CPSR are set as objective functions and only the transversal motor geometry is optimized. Furthermore, the number of winding turns and the motor stack length are properly adapted after the optimization to meet the rated current and voltage required by the application.

However, this is possible if the number of turns is high enough that a relatively small modification of the number of turns or stack length does not involve significant changes on the motor performance. Conversely, if the voltage limit and the maximum target speed yield a very low number of turns (i.e. less than 3) any modification of the number of turns or the stack length may involve completely different machine inputs or, more drastically, compromise the feasibility of the drive system.

This chapter proposes a multi-objective PMASYR motor design procedure that (a) evaluates the feasibility of the drive system in overload and flux-weakening condition and (b) computes the optimization cost functions in the same operating condition to allow easy and direct considerations of the Pareto fronts, without any further computations and checks after the optimization [103].

The proposed procedure is used to design the motor for an actual E-scooter application, shown in Figure 10.1. In the present case study, the selected cost functions are (a) motor cost, (b) torque ripple and (c) losses per outer surface in a given steady-state flux-weakening condition. Finally, a prototype of the most promising design solution is manufactured and tested to prove the effectiveness of the proposed design procedure.

Figure 10.1 Render of the E-scooter prototype

10.1 Design requirements and data

The main design requirements and constraints are reported in Table 10.1. The current and voltage limits are imposed by the inverter and battery cells provided for the E-scooter.

The motor has to produce the overload torque $T_{m,OL}$ (taken as reference) in the speed range between 0 rpm (standstill condition) and n_B (base condition). The upper limit chosen for the base speed is 2 000 rpm but can be omitted if all the other requirements are met. The HS torque $T_{m,HS}$, equal to one-third of the overload one, has to be guaranteed up to the maximum speed, n_{HS}, of 9 000 rpm. The maximum axial and radial dimensions are imposed by the available space on board. The maximum motor stack length is L_{stk} 150 mm while the maximum diameter is fixed and equal to D_e 150 mm. The slot number is $Q = 36$ while the pole number is $2p = 6$. The PMs have to be rectangular in shape and their number per pole does not have to exceed four. These constraints are imposed to ease the PMs insertion in the rotor flux-barriers and to reduce their total cost. Sintered ferrite PMs are adopted, their working temperature

Table 10.1 Design requirements and constraints

Definition	Symbol	Value	Unit
Current limit (RMS)	I_{inv}	300	A
Voltage DC bus	V_{DC}	54	V
Torque (overload)	$T_{m,OL}$	1	p.u.
Torque (HS)	$T_{m,HS}$	0.33	p.u.
Base speed	n_B	$< 2\,000$	rpm
Maximum speed	n_{HS}	9 000	rpm
External diameter	D_e	150	mm
Stack length	L_{stk}	¡150	mm
PM pieces per pole*	–	≤ 4	–
Air-gap length	g	0.4	mm
Slot number	Q	36	–
Pole number	$2p$	6	–
No parallel connections			
	Objectives		
Maximum torque ripple	$r_{\%}$	¡5	%
Maximum total cost	C_{tot}	1	p.u.
Losses/$(\pi D_e L_{stk})$	k_{loss}	< 20	kW/m^2
	Materials		
Element		**Name**	**Temperature**
Lamination		M330-35A	–
Winding		Copper	145°C
Permanent magnet		Sintered ferrite	120°C

(*) Rectangular shape is required.

is assumed to be 120°C. The motor air-gap length is set to $g = 0.4$ mm according to the manufacturing limitations. Copper is selected for the winding and the assumed working temperature, for the Joule losses computation, is 145°C according to the selected insulation class. No winding parallel connections are allowed to simplify the coil terminals welding. A standard M330-35A lamination is used for the stator and rotor core. The torque ripple percentage, computed as the relative standard deviation percentage of the torque waveform, has to be lower than 5%. The total materials cost of the motor has to be lower than a reference cost C_{tot}. The coefficient k_{loss} is the ratio between the admitted total losses and the stator stack external surface (i.e. $\pi D_e L_{stk}$). It strongly affects the design of the machine cooling. Typical values are 3 kW/m^2 for natural cooled system, 10 kW/m^2 for forced ventilation and up to 40 kW/m^2 for liquid cooling and potted end windings. A maximum limit of 20 kW/m^2 has been set preliminary.

10.2 Stator and rotor geometries

This section reports the stator and rotor geometries adopted for the design of the PMASYR motor. The parameters and relative variation ranges set for the optimization are also presented.

10.2.1 Stator

The stator slot geometry and its main parameters are reported in Figure 10.2(a). The outer diameter, D_e, is fixed while the ratio between the inner and outer diameter has been set between 0.5 and 0.7. The tooth width, w_t, and slot height, h_s, reference

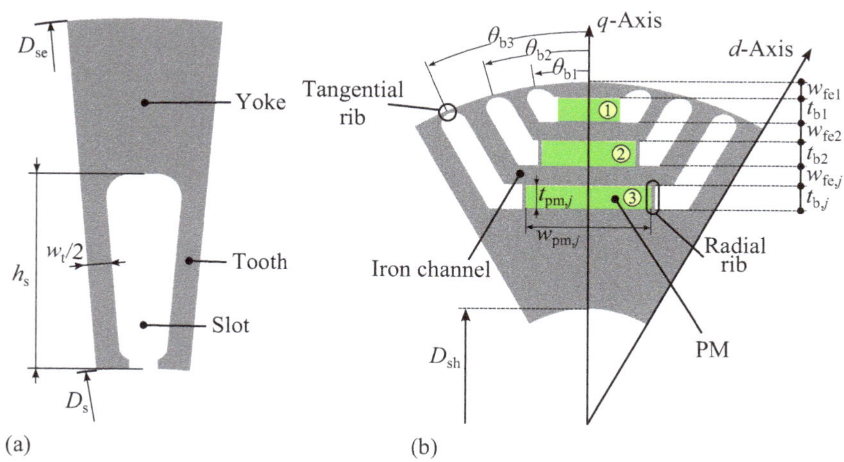

Figure 10.2 Main geometrical parameters of the PMASYR motor: (a) stator and (b) rotor

values have been selected considering a standard stator lamination with the same outer diameter. The variation range of these two parameters has been set between 0.8 and 1.2.

10.2.2 Rotor

In the present case study, the segmented flux-barrier shape, shown in Figure 10.2(b), has been selected for the simple flux-barriers geometry and the possibility of achieving a high filling with standard rectangular shaped PMs. The number of three flux-barriers per pole has been selected according to the results provided in the literature [44,95]. The geometrical parameters used for the optimization are the flux-barrier thicknesses $t_{b1}, \ldots t_{b3}$ and four iron channel widths $w_{c1}, \ldots, w_{c4}$. Each flux-barrier angle, ϑ_{bj} (with $j = 1, 2, 3$), is free to move between 0.05 and 0.95 of the angle defined by the jth outer and inner flux-barrier edges. All the three flux-barriers are filled by a magnet in the middle. The thickness of the jth magnet, t_{pmj}, is equal to the corresponding flux-barrier, t_{bj}, mainly to limit the total number of free parameters and second to use the same matrix for the magnet production. In the optimization, the width of each magnet is free to move between 50% and 100% of the available flux-barrier space. The tangential iron ribs thicknesses are set equal to the lamination width, as usually required. The radial iron ribs have been used for all the flux-barriers and their thicknesses have been set equal to the tangential rib one.

10.3 Design aspects regarding the number of turns

The given current, I_{inv}, and voltage, $V_N = V_{DC}/\sqrt{3}$, limits yield the following constraints:

$$\begin{cases} I_{slot,OL} \le n_{cs} I_{inv} \\ V_N \ge p\omega_{FW}\, k_w \dfrac{n_{cs} Q}{6}\, \phi_{FW} \end{cases} \tag{10.1}$$

where n_{cs} is the number of series conductor per slot, $I_{slot,OL}$ is the overload current in a slot, k_w is the winding factor and ϕ_{FW} is the magnetic flux in the target HS working point (HSWP). From (10.1), the upper and the lower bound of n_{cs} can be found:

$$\begin{cases} n_{cs} \ge \dfrac{I_{slot,OL}}{I_{inv}} \\ n_{cs} \le 6\, \dfrac{V_N}{p\omega_{FW} k_w Q \phi_{FW}} \end{cases} \tag{10.2}$$

The value of $I_{slot,OL}$ is unknown a priori but can be estimated as follows:

$$I_{slot,OL} = \frac{\hat{K}_s \pi D_{s,min}}{k_w Q} \tag{10.3}$$

where $\hat{K}_s$ is the electric loading, typically set between 40 kA/m and 60 kA/m, while $D_{s,min}$ is the minimum feasible stator diameter. Setting $\hat{K}_s = 60$ kA/m and $D_{s,min} =$ 85 mm, from the first of (10.2), it results in

$$I_{slot,OL} = 463\text{A} \qquad n_{cs} \geq 2 \tag{10.4}$$

The upper n_{cs} limit is computed assuming the geometric configurations with the minimum stator diameter and stack length so that the lowest value of magnetic flux can be estimated. In overload condition, the air-gap flux density, B_g, can be set between 1 T and 1.2 T. Hence, the air-gap magnetic flux is given by

$$\phi_{OL} = B_g \frac{D_{s,min} L_{stk,min}}{p} \approx 2.4 \text{ mWb} \tag{10.5}$$

where $L_{stk,min}$ is the minimum stack length equal to 85 mm. The High-Speed (HS) flux is then computed with the following relationship:

$$\phi_{FW} = \frac{n_{OL}}{n_{FW}} \phi_{OL} = 0.53 \text{ mWb} \tag{10.6}$$

where n_{OL} is the overload speed set equal to the minimum n_B. Finally, according to the HSWP speed, the second of (10.2) gives $n_{cs} \leq 3$. Since no parallel connections are allowed, the limit is rounded to the lower even number so that results in $n_{cs} \leq 2$. The results suggest that the number of turns has to be $n_{cs} = 2$.

10.4 Individual evaluation procedure

The flowchart in Figure 10.3 summarizes the main steps performed for each individual evaluation during the optimization.

10.4.1 Geometry feasibility check

At first, a set of parameters is generated by the optimizer algorithm and the geometric feasibility of the motor is evaluated. If no problems occur, the procedure goes on the next step; otherwise, the individual is rejected.

10.4.2 Overload capability

In this step the overload capability, in terms of average torque and PMs demagnetization, is investigated. The current angle, α_i^e, is fixed to 55° while the current amplitude is estimated in an iterative way by means of Algorithm 1. To simply the computation, the PM torque's contribution is neglected so that the torque is assumed to be a quadratic function of the current only. The motor torque is evaluated with a single-step static Finite Element (FE) analysis. To reduce the estimation error due to the torque ripple, the torque is computed as follows:

$$T_{dq} = \frac{3}{2} p (\lambda_d I_q - \lambda_q I_d) \tag{10.7}$$

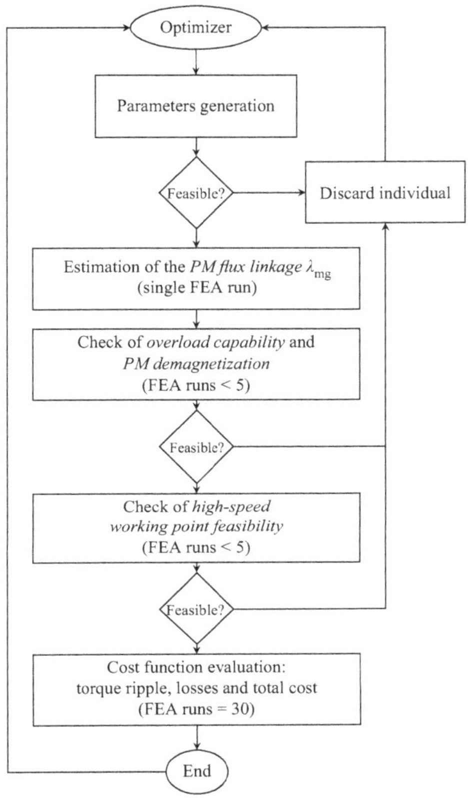

Figure 10.3 Flowchart of the evaluation procedure performed for each individual during the optimization

---

Algorithm 1: Evaluation of the overload capability and PMs demagnetization

---

$k = 0$
$k_{\max} = 10$
$I_1 = I_{\text{inv}}$
$\alpha_i^{\text{e}} = 55$
Run single step static FEA
Compute T_{FEA} by means of (10.7)
Evaluate PMs demagnetization
while $|T_{\text{FEA}} - T_{\text{m,OL}}| > \varepsilon_{\text{T}}$ **and** $k \leq k_{\max}$ **do**
    $k = k + 1$
    $I_k = I_{(k-1)} \cdot \sqrt{\frac{T_{\text{m,OL}}}{T_{\text{FEA}}}}$
    Run single step static FEA
    Compute T_{FEA} by means of (10.7)
    Evaluate PMs demagnetization
end while

In the same simulations, the minimum flux density within the PMs is checked to evaluate the demagnetization risk. If after some iterations the target torque, $T_{\mathrm{m,OL}}$, is not reached, because of current limitation or PMs demagnetization, the considered individual is rejected; otherwise, the evaluation of the HSWP feasibility is performed.

10.4.3 High-speed working point feasibility check

In this step, the algorithm determines the feasibility of the HSWP estimating, with an iterative procedure, the current vector that yields the required torque, $T_{\mathrm{m,HS}}$, at the given mechanical speed n_{HS} within the current and voltage limits.

If the voltage margin is very low, as in this work, it is essential to take into account the voltage drop due to resistive and end-winding inductance effects. The computation of the winding resistance at the maximum admissible working temperature allows one to have, in the worst case, an estimation of the maximum voltage drop. It is given by

$$R_{\mathrm{s}} = \rho_{145^\circ\mathrm{C}} \frac{(L_{\mathrm{stk}} + \ell_{\mathrm{ew}})\, n_{\mathrm{cs}}^2 Q}{3 S_{\mathrm{slot}} k_{\mathrm{fill}}} \tag{10.8}$$

where $\rho_{145^\circ\mathrm{C}}$ is the winding material resistivity at 145°C, ℓ_{ew} is the end-winding length, S_{slot} is the slot cross-section and k_{fill} is the slot fill factor. The end-winding length can be estimated analytically as follows:

$$\ell_{\mathrm{ew}} = 2.5 \frac{D_{\mathrm{s}}}{p} \tag{10.9}$$

while the relative inductance is

$$L_{\mathrm{ew}} = 2pq^2 \mu_0\, n_{\mathrm{cs}}^2 \lambda_{\mathrm{ew}} \ell_{\mathrm{ew}} \tag{10.10}$$

where q is the number of slots per pole per phase, μ_0 is the vacuum magnetic permeability and λ_{ew} is a coefficient that depends on the winding disposition (i.e. $\lambda_{\mathrm{ew}} = 0.35$) [104,105].

The d- and q-axis components of the current vector can be found solving the following system of nonlinear equations:

$$\begin{cases} I_d = \dfrac{2}{3} \dfrac{T_{\mathrm{m,HS}}}{p\left[\lambda_{\mathrm{mg}} + (L_d - L_q) I_q\right]} \\[2ex] V_{\mathrm{N}}^2 = R_{\mathrm{s}}^2 (I_d + I_q)^2 + \omega_{\mathrm{HS}}^2 \left[(L_q I_q - \lambda_{\mathrm{mg}})^2 + (L_d I_d)^2\right] \end{cases} \tag{10.11}$$

where L_d and L_q are the apparent inductances of the d- and q-axis computed as follows:

$$L_d = \frac{\lambda_d}{I_d} + L_{\mathrm{ew}} \qquad L_q = \frac{\lambda_q + \lambda_{\mathrm{mg}}}{I_q} + L_{\mathrm{ew}} \tag{10.12}$$

Equation (10.11) represents the intersection between the constant torque locus, $T_{\mathrm{m,HS}} = T(I_d, I_q)$, and the constant voltage locus, $V_{\mathrm{N}} = V(I_d, I_q)$. The working point is not feasible if (10.11) does not have any solution that means that the constant torque curve does not intersect the constant voltage locus as reported in Figure 10.4(a).

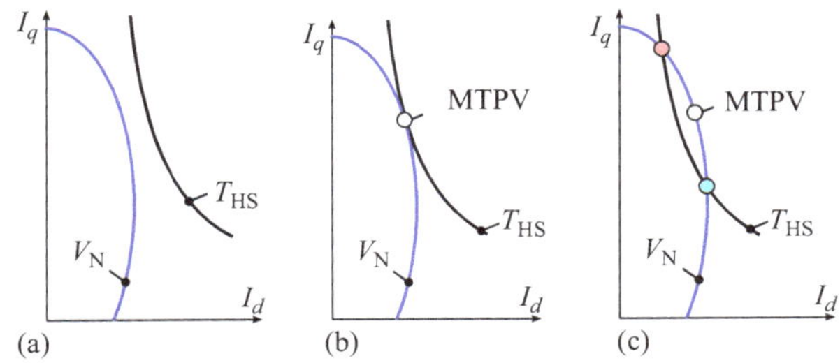

Figure 10.4 Computation of the working point feasibility for different solutions: (a) unfeasible, (b) one solution, (c) two solutions

Otherwise, one or two solutions may exist. In particular, the first case, shown in Figure 10.4(b), corresponds to the Maximum Torque Per Voltage (MTPV) condition in which the constant torque locus is tangent to the constant voltage one. In the case of two solutions, shown in Figure 10.4(c), only the one with the lowest current amplitude has to be selected since it yields the same flux linkage (i.e. same iron losses) with lower Joule losses. Since the inductances strongly depend on the motor saturation, the solution of (10.11) provides accurate results only locally. For this reason, some iterations have to be carried out to reach proper accuracy.

Figure 10.5 shows qualitatively the procedure used to estimate the feasibility of the HSWP in the $I_d - I_q$ plane. Figure 10.5(a) shows the starting situation of the algorithm in which only the current vector, $\boldsymbol{I}_{\text{OL}}$, and the relative inductances are known from the overload capability test carried out in the previous step. The dash dotted and dashed lines represent the constant torque and constant voltage loci in (10.11), respectively. The red and blue lines are the actual constant torque and constant voltage loci (unknown a priori) obtained sampling the $I_d - I_q$ plane with several FE analyses. The difference between the actual and the estimated curves is due to aforementioned saturation effect. The current vector $\boldsymbol{I}_{\text{HS}}$ is the result of (10.11) and represents the first estimation of the HSWP. The vector $\varepsilon_{\mathbf{I}}$ is the difference between the initial and the estimated current vectors and is used to stop the algorithm when its value is lower than a given tolerance $\varepsilon_{\mathbf{I},\text{min}}$. In the current example, shown in Figure 10.5(a), the amplitude of ε_I is higher than $\varepsilon_{\mathbf{I},\text{min}}$ (shaded circle near the estimated working point) so that a further iteration has to be performed.

In the next iteration, Figure 10.5(b), a single FE analysis is carried out with the estimated current vector $\boldsymbol{I}_{\text{HS}}$ and the inductances are updated. Then, a new current vector is estimated by means of (10.11) and the relative error is computed. This procedure is repeated until $\varepsilon_{\mathbf{I},\text{min}} \leq \varepsilon_{\mathbf{I}}$. If no convergence is achieved after a certain number of iterations, the individual is rejected. Otherwise, the procedure continues to the final step. In Figure 10.5(b), the previous and the estimated working points are

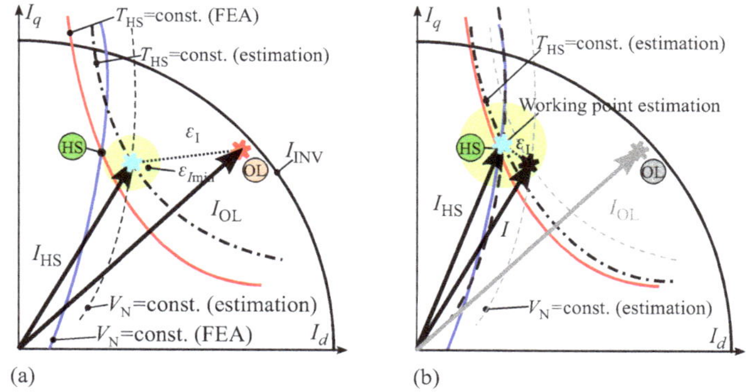

Figure 10.5 Estimation of the HSWP. The estimated *curves are computed with (10.11) while the Finite Element Analysis (FEA) ones are obtained sampling the $I_d - I_q$ plane with several FE analyses: (a) first iteration and (b) final iteration and convergence*

within the shaded circle, which means that the convergence is reached. Algorithm 2 reports the main steps needed to implement such a procedure.

Algorithm 2: Estimation of the HS working point

Estimation of R_s and L_{ew}
$j = 0, j_{\max} = 10$
$\mathbf{I} = \boldsymbol{I}_{OL}$
while $\varepsilon_I > \varepsilon_{I,\min} >$ **and** $j < j_{\max}$ **do**
  if $j > 0$ **then**
    Run single step static FEA
    Inductances computation by means of (10.12)
  end if
  Estimation of $\mathbf{I}_{HS}$ by means of (10.11)
  $\varepsilon_I = |\mathbf{I} - \mathbf{I}_{HS}|$
  $\mathbf{I} = \mathbf{I}_{HS}$
  $j = j + 1$
end while

10.4.4 Evaluation of the optimization objectives

In this last step, the optimization cost functions are evaluated. The total motor cost is computed as the summation of the PMs, copper and iron costs determined by their weight and the relative specific costs.

Since k_{loss} depends on the sum of winding and core losses, it is essential to have an accurate estimation of their individual values to allow proper design solutions. In fact, a wrong estimation of their value could yield solutions with unbalanced electric and magnetic loading [97].

Between the copper and the core losses, the second one are surely more affected by uncertainty. In general, the Steinmetz equation is used for the estimation of the stator teeth and back-iron losses as follows:

$$p_{Fet} = \left(c_{hy} f B_t^2 + c_{ec} f^2 B_t^2\right) G_t \tag{10.13}$$

$$p_{Fey} = \left(c_{hy} f B_y^2 + c_{ec} f^2 B_y^2\right) G_{bi} \tag{10.14}$$

where B_t (B_y) is the maximum flux density in the teeth (back-iron), f is the electric frequency, G_t (G_{bi}) is the teeth (back-iron) weight; c_{hy} and c_{ec} are the hysteresis and eddy-current losses coefficients, respectively. Usually, their values can be computed with a curve fitting of the lamination specific losses reported in the data sheets. The actual flux density behavior in the motor core contains many harmonics due to several effects (e.g. saturation, slotting, PWM supply) and their contribution is not taken into account in (10.14). An improved method for computing the iron losses in FE analysis is to compute the fast Fourier transform of the flux density behavior in all the mesh elements and to apply (10.14) for each harmonic. Recently, such a method has been implemented also in FEMM [106]. Table 10.2 reports the iron losses computed by means of the simplified Steinmetz Method (SM), the fast Fourier method (FFTM) and a commercial software (CSW). For the comparison, the optimal individual, selected in Section 10.5, has been analyzed. The coefficients c_{hy} and c_{ec} and the BH curve have been set equal in all the approaches. Several operating points at different electric frequencies have been evaluated. The first column reports the per unit torque while the second column reports the per unit speed. The relative reference values are the overload torque, $T_{m,OL}$, and the maximum mechanical speed, n_{HS}. For the losses comparison, the FFTM losses in the HSWP are taken as reference value. No additional losses coefficients have been considered in the comparison.

There is a good agreement between the FFTM and the CSW results while SM underestimates the losses in all the conditions. Furthermore, the losses computed with the SM do not vary significantly with the operating point unlike the other two methods in which a losses increase of 47% is observed from the lowest to the highest speed. The good agreement between the FFTM and the CSW results induced the authors to adopt the FFTM for the iron losses computation. However, such a choice involved a longer computational time since, in general, at least one electric period has to be simulated to get a reliable harmonic analysis of the flux-density harmonics. Thanks to the symmetry of the machine, only an half of the electric period has been simulated.

Table 10.2 Iron losses computation: comparison between different computational approaches: Steinmetz method (SM), fast Fourier transform method (FFTM) and commercial software (CSW)

Torque (p.u.)	**Speed (p.u.)**	**Iron losses (p.u.)**		
		SM	**FFTM**	**CSW**
0.33	1	0.20	1	1.03
0.48	0.79	0.20	0.87	0.93
0.62	0.67	0.21	0.86	0.87
0.77	0.59	0.21	0.77	0.77
0.92	0.52	0.22	0.63	0.65
1	0.49	0.22	0.57	0.59

The highest FFTM iron losses, n_{HS} and $T_{m,OL}$ are selected as reference values.

10.5 Optimization results

This section reports the results of the optimization performed with the parameters and the evaluation procedures discussed earlier. A multi-objective differential evolution algorithm has been used for the optimization. The total number of parameters is 11, as seen in Section 10.2, while the number of objectives is three. The maximum number of generations was set to 300 with 55 individuals each. However, the optimization has been interrupted after 266 generations since no improvements were achieved after the 245th generation.

The objective planes of the optimization are shown in Figure 10.6. Figure 10.6(a) reports the torque ripple versus k_{loss}. The gray markers indicate the total individuals of the optimization while the other markers represent the individuals that yield the cost functions within the required ranges (see Table 10.1). The red circle markers indicate the Pareto front of the considered plane (PF1) while the blue circles represent the Pareto front of the cost versus k_{loss} plane (PF2). In the first plot, the dimension of the markers depends on the cost so that a bigger marker means a higher cost. In the second plot the size is related to the torque ripple analogously. Figure 10.6(b) reports the total cost versus k_{loss} plane. Similarly to the previous plot, the blue markers are the PF2 individuals while the red ones refer to the PF1 front.

It can be noted that the number of the PF2 individuals (25) is higher than PF1 (9). This means that many combinations of motor geometry (i.e. motor cost) are possible but a very limited part of them satisfy the conditions to achieve a low-torque ripple. The decrease of the total cost for a higher value of k_{loss} is due to the reduction of the machine volume, the stack length, in particular.

Being the cooling system unknown at the time of the writing, the authors preferred to select the individual that exhibits a good compromise in terms of low k_{loss} and torque ripple, leaving to the total cost a minor relevance for the realization of a first prototype.

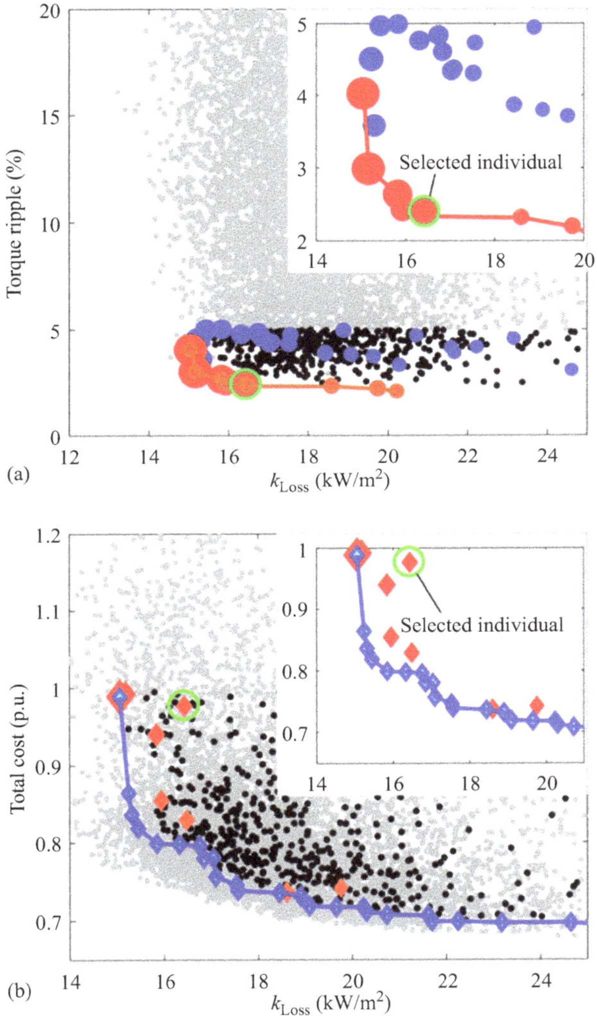

Figure 10.6 *Objectives planes of the optimization: (a) and (b). The gray markers indicates all the population, the black markers are the individuals whose cost functions are within the design requirement limits, the red markers are the non-dominated individuals in the torque ripple versus k_{loss} plane and the blue markers are the non-dominated individuals in the total cost versus k_{loss} plane.*

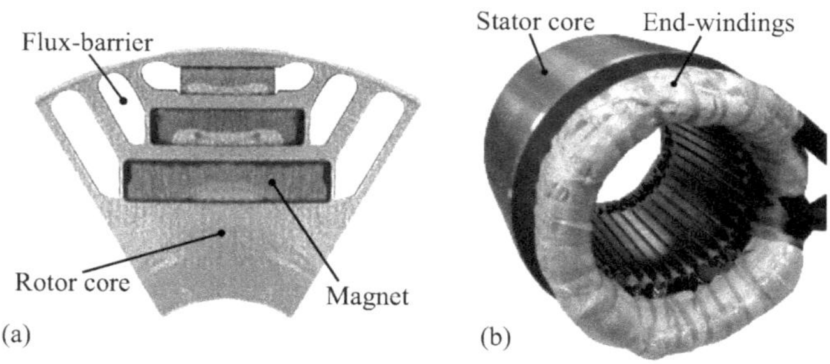

Figure 10.7 Pictures of the motor prototype: (a) rotor pole and (b) stator

The selected individual is highlighted in the plots by green circle. Obviously, many other choices can be done on the basis of the two Pareto fronts.

10.6 Experimental results

This section reports the preliminary experimental results carried out on the motor prototype whose rotor and stator are shown in Figure 10.7(a) and (b), respectively.

Only low-speed tests are carried out for the lack of a test bench with a suitable power. This means that only the torque versus rotor position behaviors and the magnetization characteristics (i.e. $\lambda_d(I_d, I_q)$ and $\lambda_q(I_d, I_q)$) were actually tested. The power versus speed curves are obtained processing the available data with the same algorithm used for the FE results.

Figure 10.8(a) reports the FE and experimental torque versus rotor position behaviors, obtained with different current amplitudes in Maximum Torque Per Ampere (MTPA) condition, the reference torque value is $T_{\mathrm{m,OL}}$. It can be noted that there is a good agreement between the FE and experimental curves both in terms of average torque and torque ripple. In particular, the discrepancy between FE and experimental average torque is always lower than 4%. The power versus speed curves, for different current amplitudes, are reported in Figure 10.8(b). It can be noted that the HSWP, highlighted in the same figure, is always below the maximum power profile, which suggests that the motor drive is capable of delivering the rated power at the given maximum speed.

10.7 Discussion

This chapter dealt with the design of a PMASYR motor for an E-scooter application by means of FE-based multi-objective optimization procedure. In particular, such a procedure is able to find solutions that meet determined overload and flux-weakening capability with a given inverter volt–ampere rating. According to the optimization

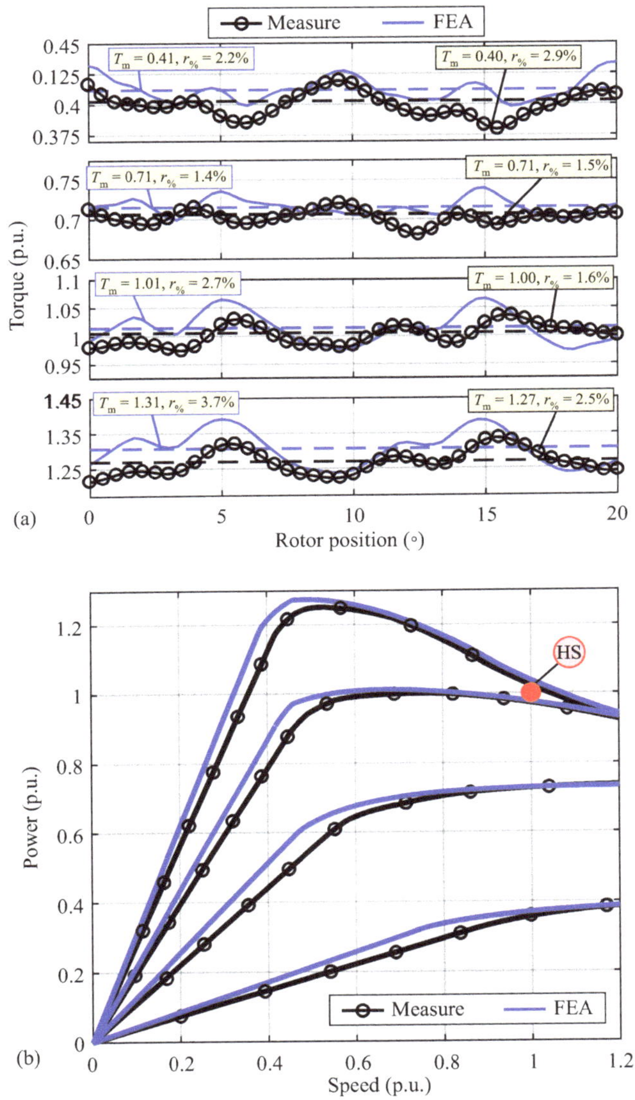

Figure 10.8 Comparison between FEA and experimental results for different current amplitudes: (a) torque versus rotor positions in MTPA condition and (b) power versus speed (indirect results)

results, a prototype has been manufactured and tested to investigate the reliability of the proposed design procedure. The preliminary experimental results have shown a good agreement with the FE predictions in terms of average torque and torque ripple. Promising results are also obtained in terms of power capability.

Chapter 11
Synchronous reluctance motor optimization for pumping application

A multi-objective optimization of a Synchronous Reluctance (SyR) motor is carried out in this chapter. The aim is to substitute some permanent magnet servomotors with SyR motors for variable-speed pumping applications. The requirements are listed in Table 11.1(a). In particular, a torque of at least 90 N m was required at 2 000 rpm. The overload torque is double the rated one, which represents quite a demanding requirement for this kind of machine. Lastly, a FW speed range of at least two to one was required.

11.1 Rotor optimization

The optimization has been performed using the provided stator. The most common way to optimize the SyR machine is to maximize the average torque and minimize the torque ripple. Since the stator is fixed, the efficiency is directly proportional to the average torque, so it is not necessary to include it among the optimization objectives.

The optimization parameters were the three flux-barrier angles, and the magnetic insulation ratio k_{air}.

11.1.1 Optimization results

The objectives plane is reported in Figure 11.1(a). This is the plane made by the first two objectives (average torque and torque ripple), while the dots dimension is proportional to the efficiency.

Obviously, being the stator fixed (constant Joule and stator iron losses), the higher the average torque, the higher the efficiency. It can be seen that all the Pareto individuals are able to reach and exceed the required torque, but a clear conflicting relationship between the average torque and the torque ripple can be observed. The lowest achievable ripple is about 8%, and the individual with the lowest ripple on the Pareto front is selected for further investigations.

11.2 Chosen motor for prototype

As written earlier, the selected individual is the one showing the lowest ripple belonging to the Pareto front. The parameters and the expected results of the selected

Table 11.1 Specifications, requirements and first results of the motor under investigation

(a) Specifications and requirements of the motor

Quantity	Symbol	Value	u.m.
Rated torque	T_B	90	N m
Overload torque	T_{max}	180	N m
Rated speed	n_B	2 000	rpm
Maximum speed	n_{max}	4 000	rpm
Supply voltage	U_n	380	V
Low torque ripple			
Stator outer diameter	D_e	180	mm
Stator inner diameter	D_s	110	mm
Rotor shaft diameter	D_{sh}	40	mm
Stack length	L_{stk}	288	mm
Air-gap thickness	g	0.4	mm
Number of slots	Q	36	
Number of poles	$2p$	4	
Iron ribs thickness	$t_{rib,}$	0.8	mm
Forced air cooling			

(b) Parameters and expected results of the selected individual in the first optimization

Par.	Value	u.m.
$\vartheta_b[1]$	18.813	°
$\vartheta_b[2]$	27.953	°
$\vartheta_b[3]$	35.568	°
k_{air}	0.516	

Obj.	Value	u.m.
$\langle T_m \rangle$	92.305	N m
ΔT_m	8.07	%
η	93.12	%

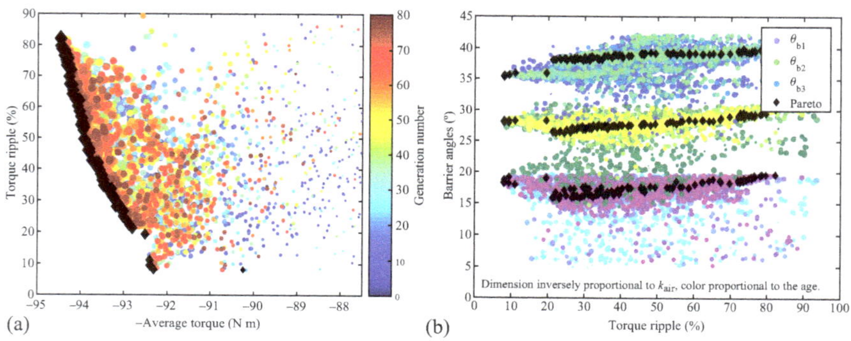

Figure 11.1 Optimization result: (a) objectives plane and (b) flux-barrier angles versus torque ripple. The colors show the evolution of the individuals, which try to reach the Pareto front, represented with black diamonds. In the first plot the size of the points is directly proportional to the efficiency, while in the second one it is inversely proportional to the magnetic insulation ratio.

individual are reported in Table 11.1(b). The results are referred to as expected because during the optimization run the electromagnetic analysis was made quick with a coarser mesh and fewer rotor rotation.

A more detailed electromagnetic finite-element simulation has then been performed, and the final results are reported in Table 11.2. Comparing it with Table 11.1,

Table 11.2 Results of the detailed simulation on the selected individual

Peak slot current	$\hat{I}_{slot}$	575	A
Current angle	α_i^e	68	°
Conductor current density	J	7.025	A/mm^2
Flux link./conductor	Λ/n_{cs}	0.0745	Vs
Average torque	$\langle T_m \rangle$	92.195	N m
Torque ripple	ΔT_m	10.06	%
Joule losses	p_{Cu}	1 093.6	W
Iron losses	p_{Fe}	198.19	W
Efficiency	η	93.14	%

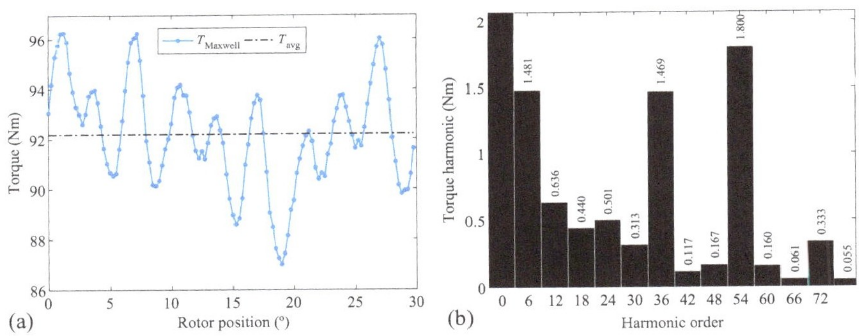

Figure 11.2 (a) Torque waveform versus rotor position and (b) torque harmonic spectrum of the selected candidate

it can be seen that the average torque prediction is quite close to the final result, while the torque ripple is slightly higher than expected (due to the aforementioned simplifications). The efficiency is just a little bit higher than 93%.

The detailed electromagnetic torque behavior for a sixth of electrical period is shown in Figure 11.2(a), while the spectrum in Figure 11.2(b). Since the motor has 36 slots and 2 pole pairs, the highest torque harmonic is expected to be the first slot harmonic, which is the 18th in this case. It can be seen from Figure 11.2(b) that this torque harmonic has been properly minimized during the optimization. In addition, the 36th torque slot harmonic is lower than the 54th, which is unusual, meaning that this torque harmonic has also been minimized by the automatic algorithm.

Figure 11.3 reports the final geometry. The only notable things are that the first flux-barrier angle is quite large, and then the angular step between one flux-barrier and the other is quite small. The second thing is that there is more air than iron along the rotor q-axis, even though it is still larger than the stator back-iron. So a higher iron saturation is expected to be found in the stator rather than in the rotor.

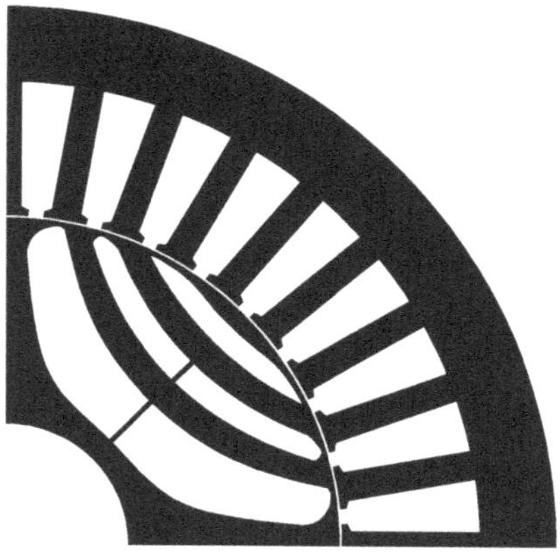

Figure 11.3 Geometry of the optimized selected individual

Table 11.3 Winding quantities for the selected motor after the winding design

Flux link./conductor	$\hat{\Lambda}/n_{cs}$	0.0745	V s
N. series cond./slot	n_{cs}	9	1
N. cond./phase	N_c	108	1
N. parallel paths	n_{pp}	To be decided	1
Flux linkage	$\hat{\Lambda}$	0.6705	V s
Peak BEMF	$\hat{E}$	281	V
RMS BEMF	E	199	V
Winding current	I_w	45.2	A

11.2.1 Winding design

Starting from the steady-state on-load simulation, it is possible to design the winding (Table 11.3). The winding is star-connected; therefore the RMS available phase voltage results in

$$U_w = 220\,\text{V}$$

Assuming a voltage drop on the winding resistance and end-winding reactance of about 10%, the BEMF should result about

$$E_w = 200\,\text{V}$$

Knowing that

$$\hat{E}_w = \omega\hat{\Lambda} = 2\pi f\,\hat{\Lambda}$$

and using the computed value of λ/n_{cs} it results in the fact that n_{cs} should be 9.

Now it is also possible to compute the winding resistance.

$$L_{\mathrm{stk}} = 288\,\mathrm{mm}$$
$$L_{\mathrm{ew}} = 138\,\mathrm{mm}$$
$$L_{\mathrm{c}} = 426\,\mathrm{mm}$$
$$S_{\mathrm{slot}} = 145\,\mathrm{mm}^2$$

The number of parallel paths is chosen equal to $n_{\mathrm{pp}} = 2$, so that

$$n_{\mathrm{c}} = 18$$
$$S_{\mathrm{c}} = 3.2\,\mathrm{mm}^2$$

The closest commercial conductor size is the one with diameter equal to 2 mm. Therefore, the conductor cross-section area should be recomputed accordingly, and so the slot fill factor:

$$d_{\mathrm{c}} = 2\,\mathrm{mm}$$
$$S_{\mathrm{c}}^{\star} = 3.14\,\mathrm{mm}^2$$
$$S_{\mathrm{c,eq}} = 6.28\,\mathrm{mm}^2$$
$$R = \frac{\rho_{\mathrm{Cu}}^{@T} N_{\mathrm{c}} L_{\mathrm{c}}}{S_{\mathrm{c,eq}}} = 138\,\mathrm{m}\Omega$$
$$k_{\mathrm{fill}}^{\star} = 0.39$$

Once the winding resistance is known, the vector diagram of the motor can be drawn (Figure 11.4). Thus, it is possible to derive the value of the external power factor, equal to

$$\cos\varphi = 0.732$$

11.2.2 Motor mapping

The selected motor has been further simulated mapping the $(I_{\mathrm{d}}, I_{\mathrm{q}})$-plane. The map is reported in Figure 11.5(a). The black circle represents the rated torque limit. The highlighted green torque contour is the rated torque (90 N m). The red curve represents the Maximum Torque Per Ampere (MTPA) trajectory, which continues along the green Maximum Torque Per Volt-Ampere (MTPVA) region, and it ends finally with the Maximum Torque Per Voltage (MTPV) trajectory, which is basically a straight line.

The base point is the point lying on the MTPA at the rated current. It can be noted that the corresponding current angle is quite high, meaning that the motor is working under heavy saturation. This is also evident from the shape of the MTPA trajectory: it starts from the origin along a line at 45° and then it curves at low current values and it continues in a straight line.

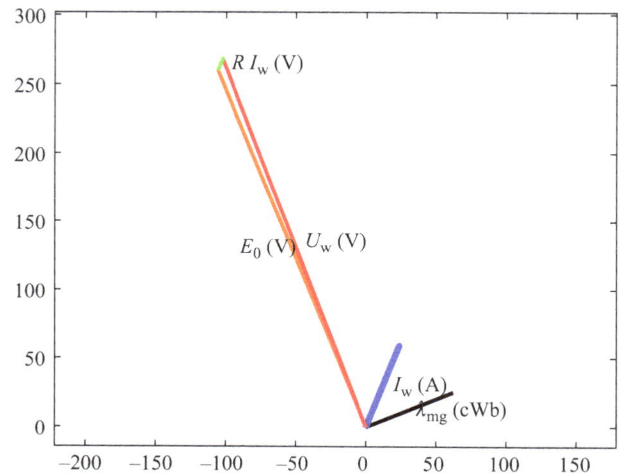

Figure 11.4 Vector diagram in the base point

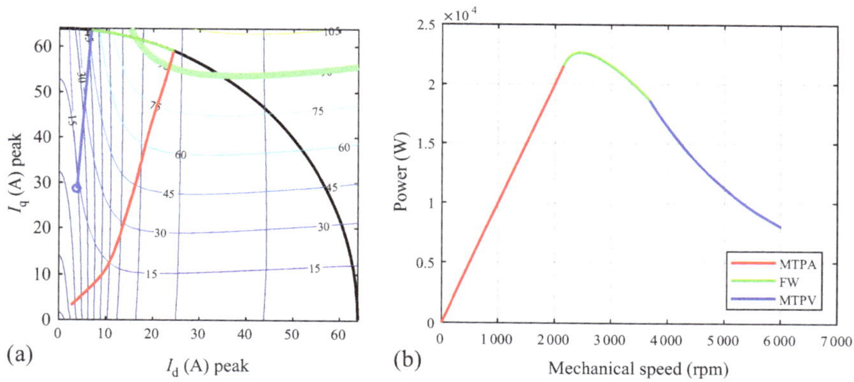

Figure 11.5 (a) Map of the (I_d, I_q)-plane and (b) rated power-versus-speed limit curve for the chosen motor geometry

11.2.3 Structural analysis

The selected geometry has also been analyzed mechanically under centrifugal load. The speed was set to 8 000 rpm, well above the operating speed (Figure 11.6). Despite this fact, the maximum von Mises stress is found in the last radial iron rib, and it results about 85 MPa. This is quite a low stress value for iron alloys, whose ultimate tensile strength usually starts at around 400 MPa. Of course, the corners of the flux-barriers should be properly rounded to avoid stress concentrations. In addition, the maximum

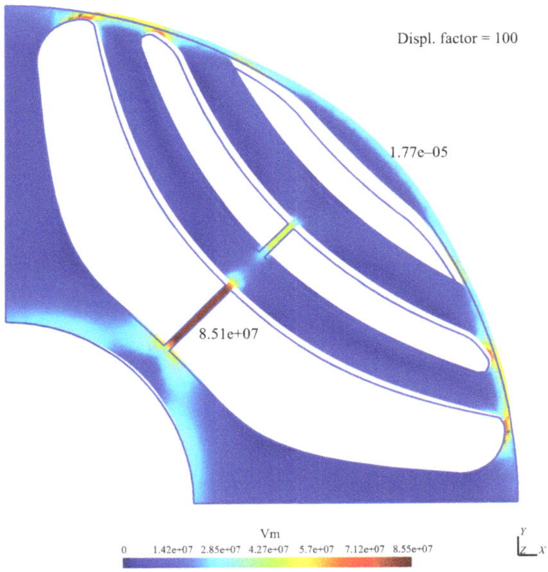

Figure 11.6 *von Mises stress plot at 8 000 rpm of the optimized rotor. The maximum displacement at the air gap is 17.7 μm, while the maximum von Mises stress in the innermost rib is 85.1 MPa*

deformation of the rotor at the air gap results about 18 μm, which is very small compared to the air-gap thickness ($g = 0.4$ mm).

11.3 Experimental measurements

The prototyped motor is a scaled-down version of the actual machine. In particular, the rated torque of the prototype is 20 N m, so 4.5 times lower than the original torque. This results in a 4.5-time lower stack length of the prototype. In addition, to overcome the limitation of the test bench DC bus voltage, current supply and maximum speed, the winding has been rearranged to reach 500 rpm as the rated speed, and 1 000 rpm as the desired FW high-speed operation. Magnetically, if the supplied current is scaled accordingly, nothing changes. A summary of the specifications for the prototype is given in Table 11.4.

Typically, two kinds of test can be conducted on synchronous machines:

- low-speed rotation to acquire the torque waveform and measure the torque map in the (I_d, I_q)-plane;
- high-speed rotation to acquire the flux linkages and measure the respective maps in the (I_d, I_q)-plane.

These kinds of measurements are performed in two different test benches.

Table 11.4 Specifications of the scaled down prototype

Quantity	Symbol	Value	u.m.
Rated torque	T_B	20	N m
Overload torque	T_{max}	40	N m
Rated speed	n_B	500	rpm
Maximum speed	n_{max}	1 000	rpm
Stack length	L_{stk}	64.5	mm

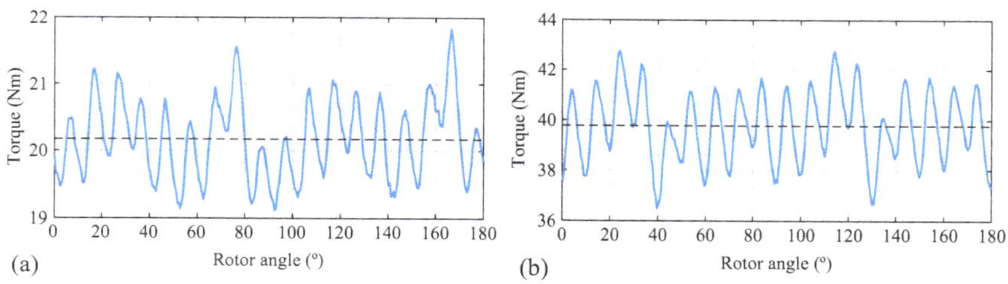

Figure 11.7 Torque waveforms versus the rotor position for two different current levels: (a) rated current and (b) double the rated current

In addition, MTPA and MTPVA operations for a specific current amplitude at the respective speed may be done, to measure external performance indexes as efficiency and power factor.

11.3.1 Torque waveform

The first test is the rotation of the rotor at low speed with the many current amplitudes and the correct MTPA current angle for each amplitude.

Figure 11.7 shows the torque waveforms obtained during this test. It can be noted that the motor is able to reach the target torque, both at rated and at overload condition. However, the torque waveforms show a higher harmonic content than predicted by Finite Element Analysis (FEA). In particular, the torque ripple results about 14%, and the waveform shows a distinctive 18th harmonic, which was not found in FEA. The torque ripple value is consistent for different current values, as it can be verified from the overload current.

11.3.2 Flux maps

The second test, the one done at relatively high speeds, is used to determine the flux linkage maps versus the d- and q-axis currents. Their behavior is reported in Figure 11.8. As expected for this kind of machine, the d-axis flux linkage is always

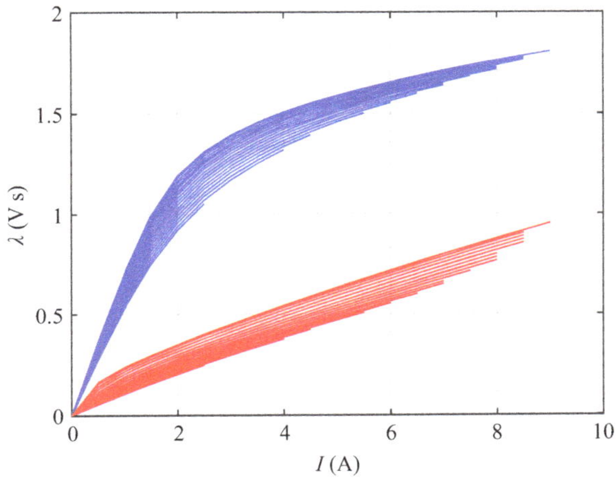

Figure 11.8 Flux linkages behavior as functions of the two axes currents

higher than the q-axis one. There are two families of curves due to the cross-saturation effect: the higher flux linkage is obtained with the respective axis current when the other axis current is kept to zero; as soon as there is also the other axis current, the flux linkage decreases. While a q-axis flux decrease is beneficial, typically the decrease of the d-axis flux linkage is higher than the decrease of the q-axis, leading to worse performance. Another interesting fact that can be appreciated from Figure 11.8 is that the slope of q-axis flux linkage near the origin is quite similar to the q-axis one. This is due to the presence of the iron ribs, which must be saturated for the machine to shown appreciable saliency.

The behavior of the flux linkages can be plotted also in the (I_d, I_q)-plane as contours. They are shown in Figure 11.9. The effect of cross-saturation is evident from the curvature of the isolines.

Even though the torque has been measured during the low-speed test of the motor, it is possible to estimate its value in the (I_d, I_q)-plane using the well-known relationship

$$T_m = \frac{3}{2}p\left(\lambda_d i_q - \lambda_q i_d\right)$$

as done in FEA. The results are shown in Figure 11.10. This map can be directly compared with the one obtained through FEA (considering the scaling factor, of course). As long as the rotational speed is low (therefore, iron losses are negligible), there should be good accuracy between the measurements and finite elements.

11.3.3 Flux-weakening and MTPV

Using the high-speed test bench and imposing the appropriate I_d and I_q, it is possible to reconstruct the whole FW operation, so moving along the MTPVA and MTPV trajectory. The torque and power versus speed characteristics of the prototype at the

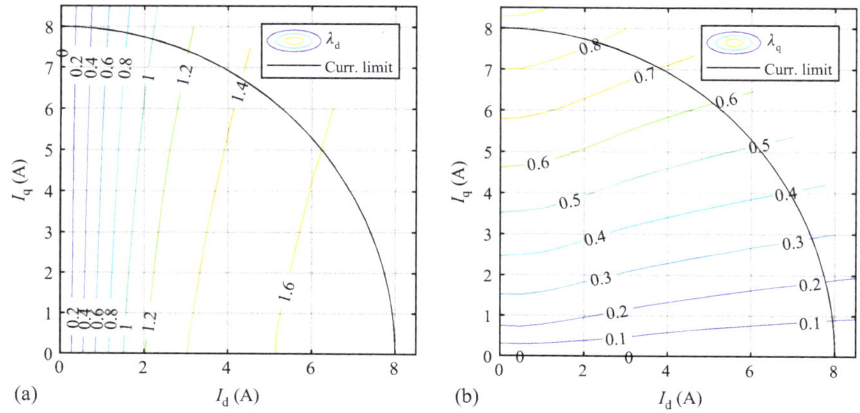

Figure 11.9 Flux linkages maps in the (I_d, I_q)-plane: (a) d-axis flux linkage and (b) q-axis flux linkage

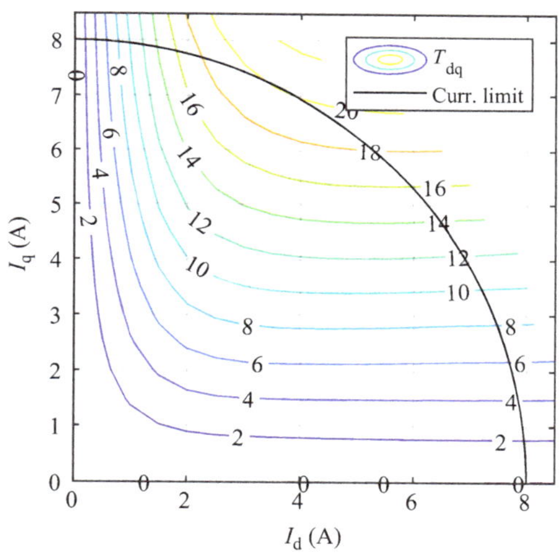

Figure 11.10 Torque map in the (I_d, I_q)-plane

rated current are reported in Figure 11.11. It is noticeable that the torque is decreasing with the speed, despite the fact that the motor is working in MTPA operation. This may be due to the friction and ventilation losses. The corner speed is just above 500 rpm. At this point the machine has to be flux-weakened. The peak power is reached at about 650 rpm, and after that the power rapidly decreases.

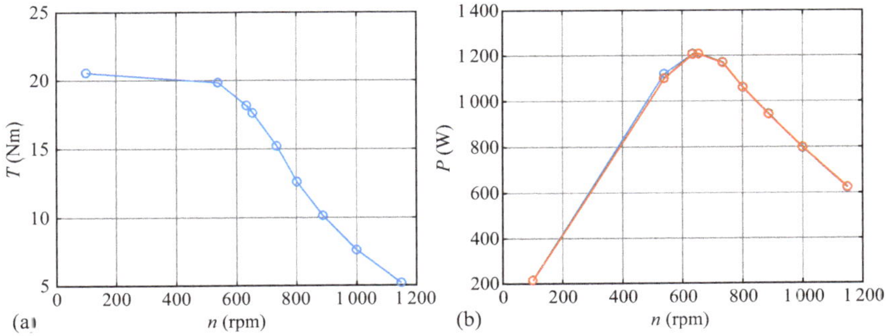

Figure 11.11 (a) Torque and (b) power versus speed characteristics of the prototype at rated current

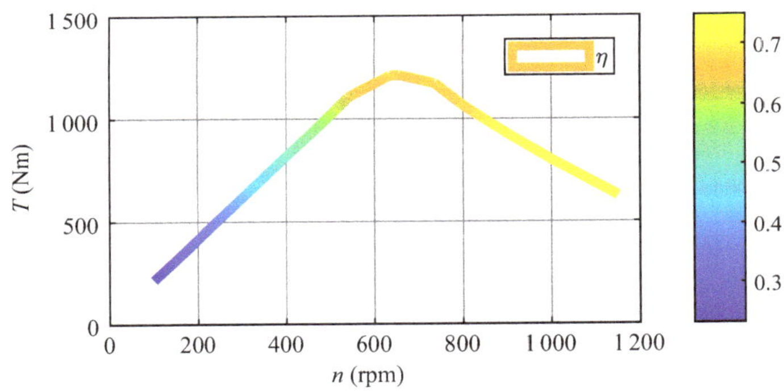

Figure 11.12 Efficiency plot along the power versus speed curve

The rated output power of the prototype is about 1.1 kW, and considering the scaling factor for the stack length (4.5) and the base speed ratio (4), the actual motor power should result in 18 times higher.

Using a wattmeter during this test, it is also possible to compute the efficiency of the motor. The plot is reported as a colored contour in Figure 11.12. Since smaller motors are intrinsically less efficient than large ones, the efficiency values of the prototype are not a good performance index for the actual machine. However, the variation of this quantity in that plane is still a good metric to check the progress of the measurement and a proper motor operation.

11.4 Discussion

In this chapter a multi-objective optimization has been done to design an SyR motor within a defined size. Due to time constraints, the automatic optimization represents the best tool to reach a sound design in the shortest time. The chosen design was able to reach all the requirements, in terms of torque, power and efficiency. The torque ripple was the most critical aspect, though. In fact, the prototyped scaled-down motor displayed higher harmonic content than the expected simulation result.

Anyway, the motor performances in MTPA and FW operations were properly satisfied, so the design was a suitable candidate for the application under investigation. In particular, no new stator lamination was required, minimizing the fixed manufacturing cost.

Chapter 12
High-torque low-speed permanent magnet assisted synchronous reluctance motor design

Direct-drive electric machines represent a valid solution for low-speed applications thanks to the absence of the gear-box. This ensures lower weight, noise, axial length and lower mechanical losses. Furthermore, a higher reliability of the overall system is guaranteed, and this is very important requirement when the machine is installed in harsh environments [107–110]. Such machines find applications in energy production, electrical propulsion, industrial automation, low-head pumps and so on. They typically employs strong Rare-Earth (RE) magnets with high pole numbers. However, in this application ferrite magnets have to be preferred, because of the high cost of RE magnets. This means that the machine should exploit the reluctance torque component as much as possible. Therefore, a Permanent Magnet assisted Synchronous Reluctance (PMASYR) machine is chosen.

The structure of the chapter is as follows: at first the specifications and constraints are listed; after that, some parametric analyses are carried out [27]. Then, multi-objective optimization coupled to finite-element analysis is performed to find the best possible design solutions. Finally, a thorough electromagnetic analysis is performed on one possible design solution.

12.1 Specifications, requirements and hypothesis

The motor outer dimensions are the constraints of the application and they are reported in Table 12.1. In particular, they are 2 500 mm for the outer diameter, and 1 500 mm for the stack length. The power requirement is 1 MW at 22 rpm, which results in a torque of about 434 kN m. To achieve such performance, a machine with a high pole number is required. Furthermore, a modular machine has to be preferred, where all the machine sectors are independent. This limits the choice of the winding arrangement and the number of poles. In fact, either two or three three-phase systems have to be employed for supplying the machine. This means that the number of pole pairs (one sector) should be a multiple of 2, 3 or both.

The slot shape is rectangular to ease the manufacturing of the machine, and the conductors are chosen to be rectangular too. Such conductors do not represent a problem for skin and proximity effects since the supply frequency is extremely low. Iron losses are negligible too, so the iron flux densities could be increased to higher values.

Table 12.1 Motor specifications, constraints and hypothesis

Quantity	Value	u.m.
Power	1	MW
Speed	22	rpm
DC bus voltage	690	V
Air gap	2	mm
Outer diameter	2 500	mm
Stack length	1 500	mm
Torque ripple	1	%
Hypothesis		
Split ratio	0.92	
Slot p. pole p. phase	4	
Power factor	0.875	
Estimated fill factor	0.6	
Packing factor	0.97	
Current density	6	A/mm^2
Air-gap flux density	1.1	T
Yoke flux density	1.6	T
Tooth flux density	1.8	T

To start with an initial design, some hypotheses and choices have been made. They are reported in Table 12.1. In particular, a high split ratio is chosen since the machine has a large external diameter and a high poles number. In order to improve the mmf waveform and reduce torque ripple, the number of slots per pole and per phase is selected to be four.

The choice of the pole number results from a compromise between the minimum slot width (which should not be too small for manufacturing issues), and the iron weight. The reduction of the iron weight obviously implies an overall weight and material cost reductions, as the iron losses do not represent an issue [6]. On the other hand, the slot width is mainly due to the number of slots per pole per phase, which should be relatively high to reduce the torque ripple [34,55,111]. Thus, for the initial design, 36 poles and 4 slots per pole per phase are adopted. Through a preliminary design, it has been found that the efficiency target could be equal to 93%.

12.2 Parametric analyses

At first, some parametric analyses have been carried out in order to better understand the dependency of the performance factors on the design variables (Figure 12.1).

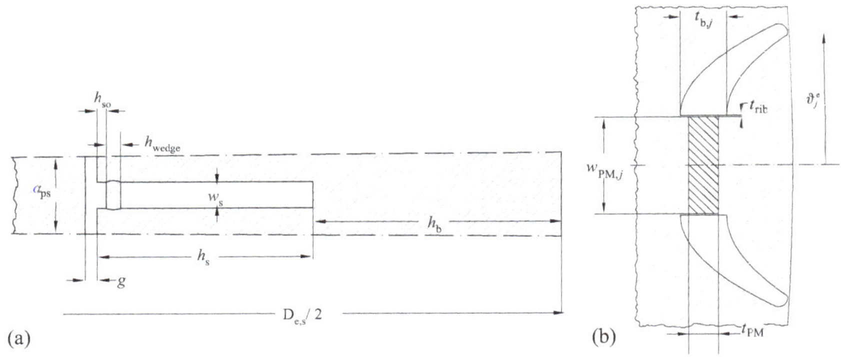

Figure 12.1 Machine geometrical parameters: (a) stator and (b) rotor

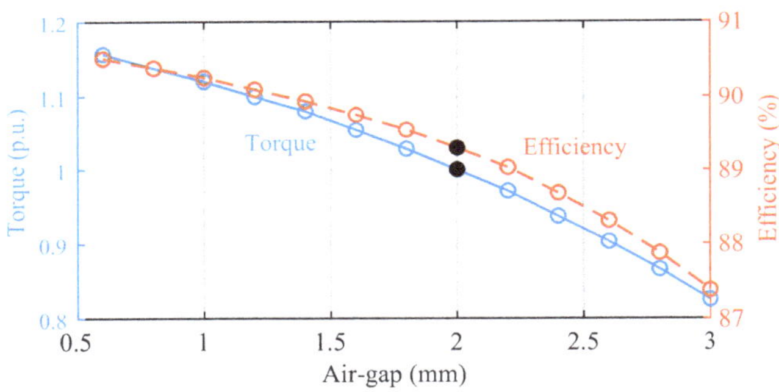

Figure 12.2 Torque and efficiency parametric analysis varying the air-gap thickness

12.2.1 Air-gap thickness

The first one is the variation of the air-gap thickness. Since the torque depends on the air-gap flux density, which is almost inversely proportional to the air-gap thickness with the same Permanent Magnet (PM) and currents, an increase in the air gap reduces the available torque. The results of the parametric analysis are shown in Figure 12.2, while a comparison of the actual flux density maps is reported in Figure 12.3.

As expected, the available torque increases with the decrease of the air gap. This could potentially reduce the current needed to reach the desired torque and, thus, increase the efficiency. However, the minimum air gap has been set to 2 mm according to mechanical constraints, given the large stator diameter. From Figure 12.3, it is evident that a smaller air gap results in a more magnetized machine, with all the other parameters constant.

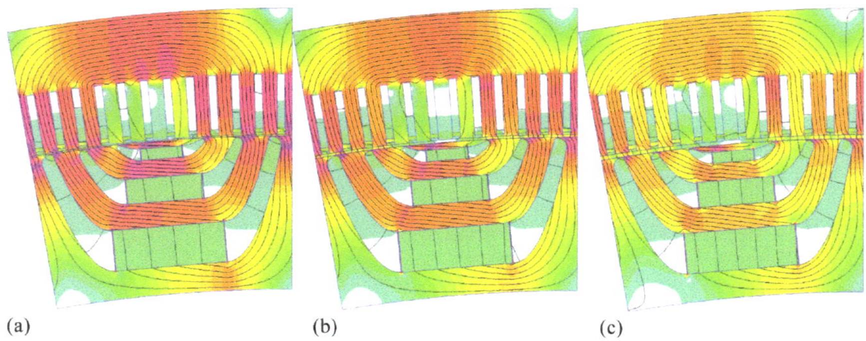

Figure 12.3 Parametric analysis varying air gap: (a) g = 1 mm, (b) g = 2 mm, (c) g = 3 mm

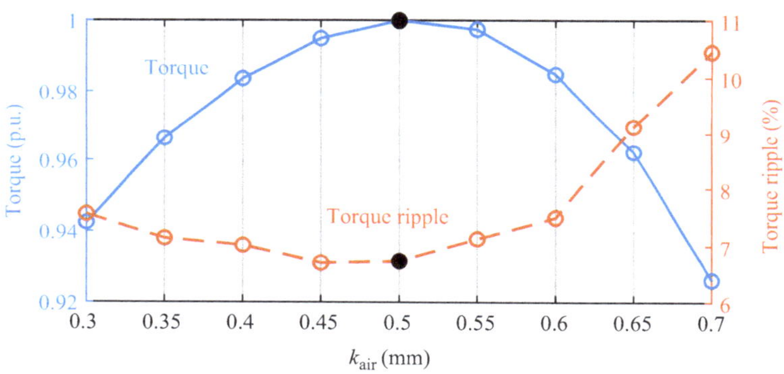

Figure 12.4 Torque and torque ripple parametric analysis varying k_{air}

12.2.2 Magnetic insulation ratio

Another parametric analysis deals with the magnetic insulation ratio, k_{air}. This factor determines how much space is reserved for flux-barriers in the rotor. Generally speaking, this parameter strongly affects the reluctance torque, but less the ripple. These aspects can be observed from Figure 12.4, where the average torque and torque ripple are plotted against this factor. The initial choice of $k_{air} = 0.5$ (so as much air as iron) represents the best compromise for this kind of machine. At this stage of the design, the thickness of the magnets is the same as the one of the corresponding barriers. It can also be noted that the torque ripple remains almost constant up to a value of k_{air} equal to 0.6. After this value, the ripple increases. This is due to the fact that the large flux-barrier thicknesses compromise the actual position of the flux-barrier-end angle, leading to a different air-gap flux density distribution and to a different torque behavior.

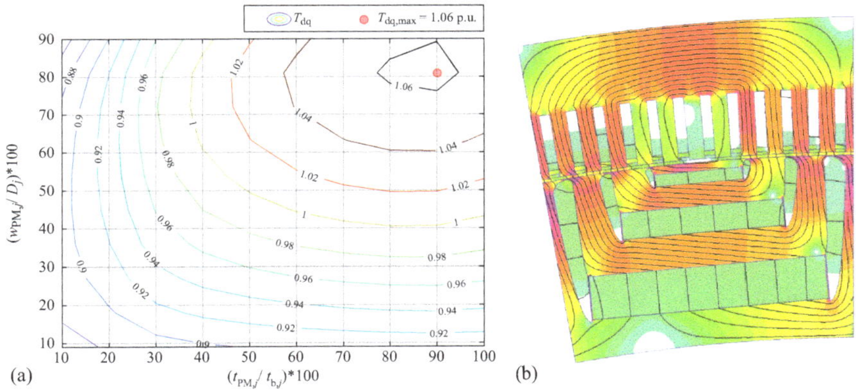

Figure 12.5 Results of the parametric analysis changing the magnet dimensions: (a) torque contours versus magnet thickness and width and (b) magnetostatic solution of the maximum torque geometry after the magnet dimensions parametric analysis

12.2.3 Magnets dimensions

Another parametric analysis that has been carried out deals with the variation of the magnets dimensions. The response variable is the average torque. Since only one rotor position was evaluated, the *dq* torque was used as the best estimate of the average torque.

The results are reported in Figure 12.5(a). The magnet thickness, on the x-axis, is varied between 10% and 100% with respect to the relative flux-barrier thickness, while the magnet width, on the y-axis, between 10% and 90% of the flux-barrier projection on the rotor periphery. It can be seen that there exists a maximum for the torque at a specific combination of magnets thickness and width, and it exhibits a torque about 6% higher than the initial design.

Such a geometry and a magnetostatic simulation is reported in Figure 12.5(b). It can be noted that the magnets are really wide and they cover almost all of the available space, especially comparing it to the initial design of Figure 12.2. Despite the great increase in the magnet size, the torque increase is modest. This is due to the use of ferrite and probably such a solution is not reasonable when the magnet cost is considered.

12.2.4 Number of poles and slots

The most defining parametric analysis is the variation of the number of poles and the number of slots per pole per phase, though. In this case, the geometry and the current change accordingly to the following constraints:

- output torque equal to 440 kN m,
- current density equal to 6 A/mm^2,

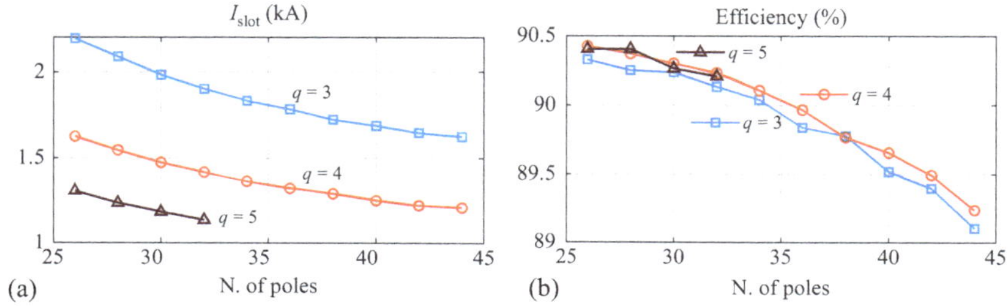

Figure 12.6 Results of the parametric analysis varying the number of pole pairs p and the number of slots per pole per phase q: (a) slot current and (b) efficiency

- flux-barrier electrical angles equal to {28°, 52°, 76°},
- maximum tooth flux density equal to 1.8 T and stator yoke flux density equal to 1.6 T,
- magnetic insulation ratio, k_{air}, equal to 0.5,
- equal magnet volume.

As can be seen from Figure 12.6(a), the required slot current decreases as the number of poles increases, as expected. In fact, the electric loading can be defined as

$$\hat{K}_s = \frac{k_w \hat{I}_{slot}}{p_s} \tag{12.1}$$

where k_w is the winding factor, $\hat{I}_{slot}$ the peak slot current, and p_s the slot pitch. It is evident that as the number of poles increases, the slot pitch decreases and so does the current to keep the same electric loading, thus the same torque. Furthermore, the current required decreases as the number of slots increases. However, the overall efficiency of the motor decreases as the number of poles increases, because of the increase of the copper quantity. Since the current density is kept constant, the overall Joule losses increase. On the other hand, the rotor diameter increases and the iron volume decreases. From these analyses, it is clear that the efficiency requirement cannot be met with such current density. This is also highlighted in Figure 12.6(b), where the maximum achievable efficiency is 90.4% for the machine with the lowest pole pairs number. Therefore, the supply condition of the machine should be modified to better approach the efficiency requirement.

12.3 Optimization

Geometries with 30 poles will be optimized, not to penalize the efficiency too much and to guarantee the feasibility of three three-phase systems (Table 12.2). Only four slots per pole per phase are considered in the following since this combination

Table 12.2 Slots and pole combinations analyzed in the parametric analysis

		2p									
		26	28	30	32	34	36	38	40	42	44
q	3	•	•	•	•	•	•	•	•	•	•
	4	•	•	•	•	•	•	•	•	•	•
	5	•	•	•	•						

Table 12.3 Optimization parameters and ranges

Symb.	Value	u.m.	Symb.	Value	u.m.
h_y	$50 \pm 50\%$	mm	k_b	$0.1 \div 0.7$	1
h_s	$45 \pm 50\%$	mm	w_{pm1}	$45 \pm 25\%$	mm
w_s	$5.7 \pm 50\%$	mm	w_{pm2}	$85 \pm 25\%$	mm
ϑ_{b1}^{e}	$30 \pm 10\%$	°	w_{pm3}	$125 \pm 25\%$	mm
ϑ_{b2}^{e}	$50 \pm 10\%$	°	t_{pm1}	$8 \pm 25\%$	mm
ϑ_{b3}^{e}	$70 \pm 10\%$	°	t_{pm2}	$18 \pm 25\%$	mm
k_{air}	$0.4 \div 0.7$	1	t_{pm3}	$28 \pm 25\%$	mm

achieved the best torque ripple results and it allows a wide enough slot. A multi-objective optimization has been carried out, with 14 parameters and 3 objectives.* The parameters, together with their limits, are reported in Table 12.3. They are the yoke height h_y, the slot height h_s, the slot width w_s, the three flux-barrier angles $\vartheta_{b1}^{e}, \vartheta_{b2}^{e}, \vartheta_{b3}^{e}$, the magnetic insulation ratio k_{air}, the flux-barrier-end shape factor k_b, the three magnets width $w_{pm1}, w_{pm2}, w_{pm3}$, and the three magnets height $t_{pm1}, t_{pm2}, t_{pm3}$.

A total of 70 individuals over 120 generations were evaluated, so a total of 8 470 function evaluations have been performed. Each evaluation lasts for about 150 s, meaning an optimization time of about 6 days with parallel computations. The current was adjusted so as to always reach the target torque.

The three objectives are the active material cost, the torque ripple and the efficiency. One of the possible results plot is shown in Figure 12.7(a). This represents a common and expected result, which highlights the relationship between efficiency and cost. It is evident that newer generations are closer to the Pareto front, shown with black diamonds, so newer generations have better efficiency for the same cost, and it means that the optimizer is actually improving the objective functions. Furthermore, efficiency equal to the target can be reached, but only at higher costs, which means with a heavier machine, generally speaking.

Figure 12.7(b) reports the torque ripple versus the cost. The dimension of the markers is proportional to the efficiency. It can be noted that the torque ripple is

*See Chapter 9.

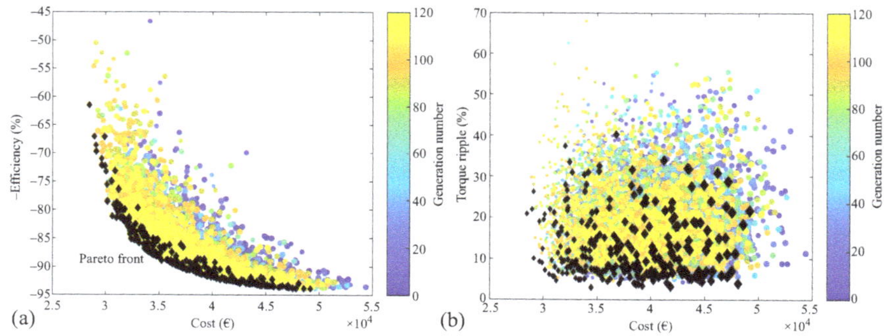

Figure 12.7 Optimization objectives planes: (a) efficiency versus cost plane and (b) torque ripple versus cost plane. Lighter colors represent newer generations, while black diamonds are the points belonging to the Pareto front. The size of the markers is proportional to the objective not shown.

basically independent of the cost of the machine, proving once again that it is mostly due to the rotor geometry. In addition, the achievable torque ripple is lower than 5%, which is a remarkable result for this kind of machine. The target was 1%, but that was very ambitious for a machine relying heavily on the reluctance torque.

Figure 12.8 reports a summary of the optimization results, where each objective is plotted against each parameter. This kind of plot enables the designer to understand the relationships existing between different quantities. The first thing that is evident is that the cost has a linear relationship with the slot height and width (h_s and w_s), due to the amount of copper needed. Also the Pareto points creates a sort of straight line in those planes. As a consequence, as already seen from Figure 12.7(a), the efficiency is proportional to these two parameters, but less than linearly. Furthermore, there is an upper and lower bound for the cost as a function of these two parameters.

Another thing that can be noted is that solutions with small second and third magnets are preferred, thanks to the decrease in cost. In fact, newer generations and Pareto points are close the lower bound of that quantity, meaning that it is convenient to keep these magnets small, both from a performance and from a cost point of view.

At last, low-torque ripple can be achieved only with certain combinations of flux-barrier angles. This can be deduced easily from Figure 12.9(a), where the leftmost points appear only at specific combinations of the three angles (for instance, the one with $\vartheta^e_{b1} \simeq 30°$, $\vartheta^e_{b2} \simeq 56°$, $\vartheta^e_{b3} \simeq 68°$).

One of the optimal solutions belonging to the Pareto front was selected for deeper electromagnetic and thermal investigations. Figure 12.9(b) reports the geometry and the magnetostatic solution of this optimum. It can be noticed that the yoke height is quite large, and the slot is tall and narrow. This would imply a high slot leakage inductance but, thanks to the low frequency, the voltage drop due to it is still quite low. The magnets dimensions were changed as little as possible in order to obtain their widths and heights proportional and so decreasing supply and manufacturing costs.

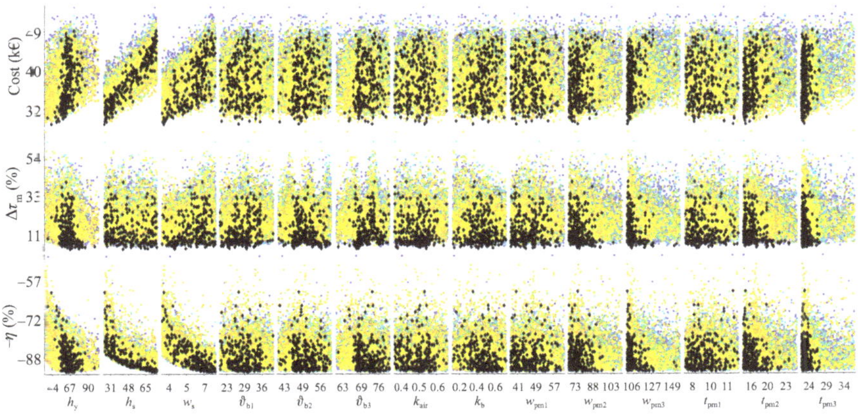

Figure 12.8 Summary of the optimization. Lighter colors represent newer generations, while black diamonds are Pareto points. **Objectives**: cost, torque ripple, efficiency. **Parameters**: yoke height, slot height, slot width, three barrier angles, rotor magnetic insulation ratio, barrier-end shape factor, three magnet widths, three magnet thicknesses.

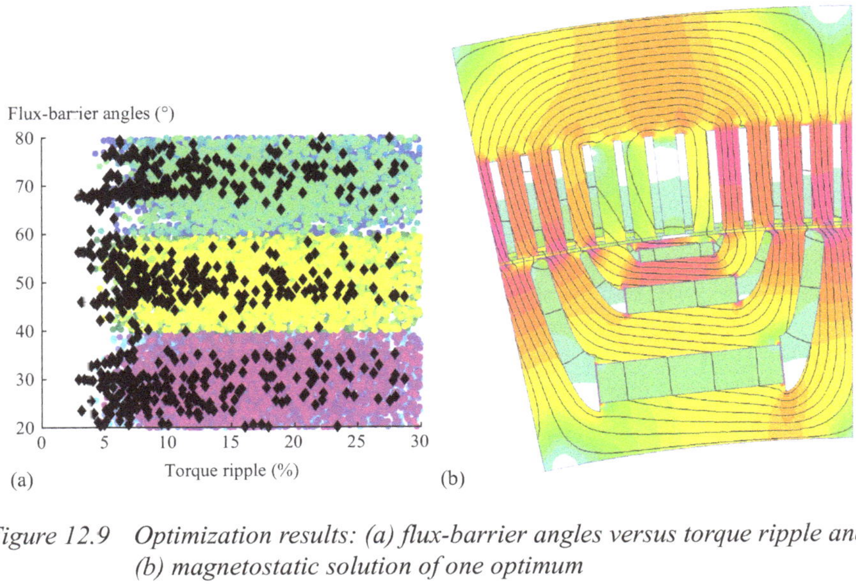

Figure 12.9 Optimization results: (a) flux-barrier angles versus torque ripple and (b) magnetostatic solution of one optimum

The main performances of this machine are reported in Table 12.4. The motor is able to almost reach the efficiency target, reducing the conductor current density, though. It can be noted that iron losses are negligible, thanks to the very low frequency. The power factor is also quite low, which may result in inverter oversizing. Finally,

Table 12.4 Chosen optimal motor performance, weight and cost results

Quantity	**Value**	**u.m.**
Current	2 905	A
Current density	4.4	A/mm^2
Average torque	446.5	kNm
Torque ripple	3.7	%
Joule losses	69.9	kW
Iron losses	4 897	W
Efficiency	92.6	%
Power factor	0.741	1
Material weight	25.5	t
Material cost	45.9	k€
Torque density	17.5	Nm/kg

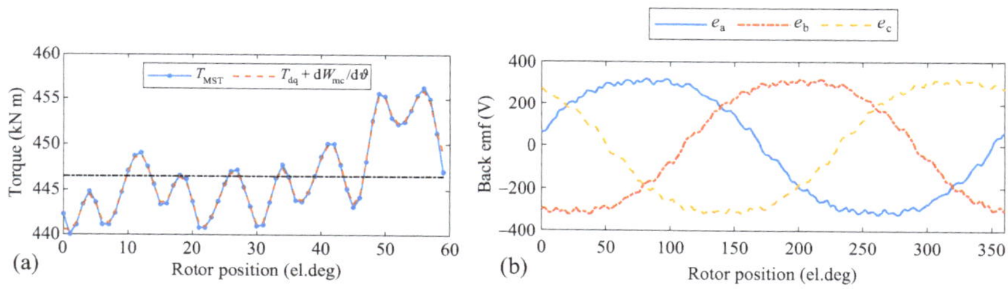

Figure 12.10 Simulation results for the chosen optimum: (a) torque waveform and (b) BEMF waveforms

the torque ripple is very good for this kind of machine. The corresponding torque waveform is reported in Figure 12.10(a). Here, a sixth of an electrical period is shown. It is immediate to notice a first harmonic, which corresponds to the sixth harmonic of torque. After that, the highest visible harmonic is the eighth, which corresponds to the 24th in a full electrical period. This is the second slot harmonic ($2Q/p$), while the first slot harmonic (Q/p) has been minimized during the optimization. Figure 12.10(b) also reports the resulting phase Back Electromotive Force (BEMF). On top of the desired sinusoidal waveform, there are some higher order harmonics. This aspect is quite typical for synchronous reluctance and PMASYR machines.

A thermal analysis has also been carried out to estimate whether the selected cooling system meets the requirement. In particular, the conductor temperature should not exceed 120°C, which is the supposed temperature for the electromagnetic computations. The ambient temperature is 25°C, and the machine is water-cooled on the outer surface. An equivalent air region has been added between the stator iron and the frame to model the nonideal contact between the two parts. The rotor is virtually without losses, so it is not necessary to model it. Figure 12.11 shows the temperature

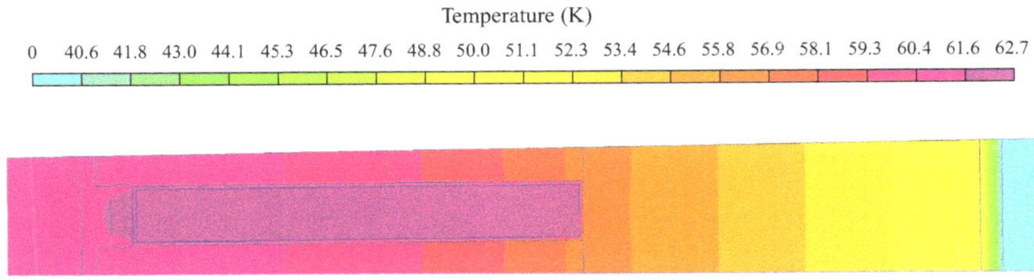

Figure 12.11 Thermal simulation of one stator slot

distribution in one stator slot when the machine is on full load. The conductor relative temperature reaches 62°C, so 87°C absolute, which is far below the supposed one. This means that the Joule losses could be lower, and the efficiency higher. Alternatively, the cooling system could be made smaller. The absolute rotor temperature results about 86°C, which means that a slightly higher PM flux density may be expected.

12.4 Discussion

In this chapter, a comprehensive feasibility study of a PMASYR motor for direct-drive cableway application has been done. The overall performance and limits of this kind of machine have been highlighted. In particular, the motor is able to exhibit good performance, compromising slightly some of the requirements (in particular torque ripple, which was very ambitious). Parametric analysis and multi-objective optimization coupled to finite-element analysis are fundamental tools in the design of these machines, in particular when the design space is wide. In conclusion, the PMASYR machine is suitable to reach the given targets and dimension constraints.

Chapter 13
Bonded magnets in PMaSyR machines

The Permanent Magnet assisted Synchronous Reluctance (PMASYR) machine is well known to mitigate the low-power factor and power capability that characterize the Synchronous Reluctance (SYR) machine by means of a small Permanent Magnet (PM) flux contribute. Depending on the adopted design and economic strategies, such an excitation flux may be produced by a small amount of rare earth PMs or by low magnetic energy ferrites. Both solutions allow us to keep the PMASYR cost very competitive with the classical rare-earth-based permanent magnet synchronous machine; however, in the recent years, the second solution is often preferred to avoid any risk related to the high rare-earth materials cost fluctuation. Thanks to its advantages, the PMASYR machine is adopted in home-appliance, spindle and automotive applications [103,112,113].

The main design keys of the PMASYR machine are the flux-barriers geometry and PM compensation [103,114]. Even though these two aspects are not independent, the first one mainly affects the average torque and torque ripple while the second one affects the high-speed power capability [55,115,116].

As pointed out in Chapter 10, rectangular PMs are usually adopted to simplify as much as possible the rotor geometry and the overall manufacturing cost as well. However, when high PM compensation is required, such a design choice greatly limits the machine performance since, as shown in Figure 13.1(a), only a small part of the effective flux-barriers cross-sectional area can be filled by PMs.

The Bonded Magnet (BMs) represent a valid solution since they can be easily shaped or directly injected into the flux barriers [20–22]. Figure 13.1(b) shows how the BMs can be effectively used for the realization of a PMASYR rotor and highlights the higher amount of PM materials with respect to the classical solution in Figure 13.1(a). The BMs are constituted by a mixture of a magnetic powder and polymeric binder [23]. The magnetic powder may be hard ferrite, NdFeB, SmCo while the binders depend on the typology of the molding process. The normally used polymeric materials are Polyamide (PA) for injection molding and thermoset resins for compression molding [24]. The BMs can be made in very complex shapes using both molding technologies [25]. Finally, the magnetic characteristics of the BMs can be "tuned" through the percentage of the plastic binder. The NdFeB BMs have better magnetic characteristics with respect to ferrites, mainly for the high intrinsic coercivity.

The aims of this chapter are to present some experimental investigation about the impact of different production technologies on the BMs properties and to evaluate their

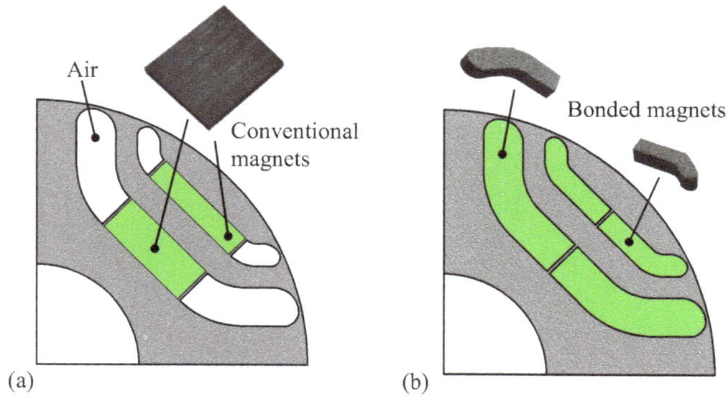

Figure 13.1 PMASYR *motors with different magnet types: (a) sintered magnets bonded magnets and (b) BMs*

Figure 13.2 *Motor prototypes manufactured for the study: one* SYR *machine and two* PMASYR *machines*

suitability for PMASYR machine applications. In the first part a brief description of the motor geometry and its design is presented. Then, the steps for BMs preparation and magnetization are described. In particular, two BM typologies are tested: one prepared with compression molding and the other with injection molding. Next, three motor prototypes, reported in Figure 13.2, are manufactured: one SYR machine and two PMASYR machines with compression and injection molding BMs. These prototypes are tested in laboratory and their performance are finally compared.

13.1 Motor geometry

A standard geometry, whose main data are reported in Table 13.1, has been selected for the stator. The segmented flux-barriers geometry, described in Section 3.2, has been preferred for the rotor because of its simplicity. The number of flux-barriers

Table 13.1 Main data of the SYR *machine*

Number of pole pairs	2	–
Number of slots	36	
Stack length	40	mm
Stator outer diameter	150	mm
Stator inner diameter	96	mm
Rotor outer diameter	95.2	mm
Air gap	0.4	mm
Rotor inner diameter	40	mm
Current density	6	A/mm^2
Flux-barrier angles	{16.05, 33.8}	°
Magnetic insulation ratio	0.46	–

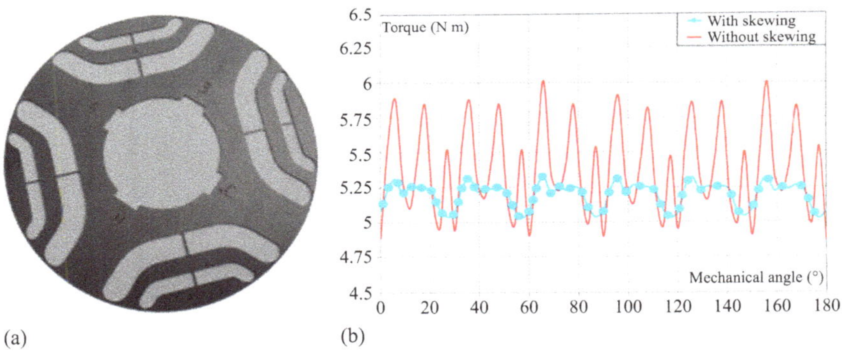

Figure 13.3 (a) SYR *rotor geometry and (b) torque behavior in MTPA condition with and without skewing*

per pole has been set to two to limit the number of BMs to manufacture. It is worth noticing that these design choices are made only to simplify the BMs production in the laboratory. The rotor geometry has been optimized using the procedure described in Chapter 9. The selected rotor parameters were the two flux-barrier angles and the magnetic insulation ratio. The flux-barriers and iron channel dimensions have been computed with the procedure reported in Section 3.3. The optimization objectives were to maximize the average torque and to minimize the torque ripple. The optimization has been performed in 60 generation of 20 individuals each. The optimal rotor parameters are reported in Table 13.1 and a sketch of the rotor geometry is shown in Figure 13.3(a). The Finite Element (FE) torque behavior computed in Maximum Torque Per Ampere (MTPA) condition is reported in Figure 13.3(b), and the motor achieves an average torque and ripple of 5.3 N m and 22%, respectively. A rotor skewing of a slot pitch (i.e. 10°) with eight steps has been selected to reduce the main torque harmonic. With such a solution, the average torque slightly reduces to 5.1 N m while the torque ripple decreases to 5%, as reported in Figure 13.3(b).

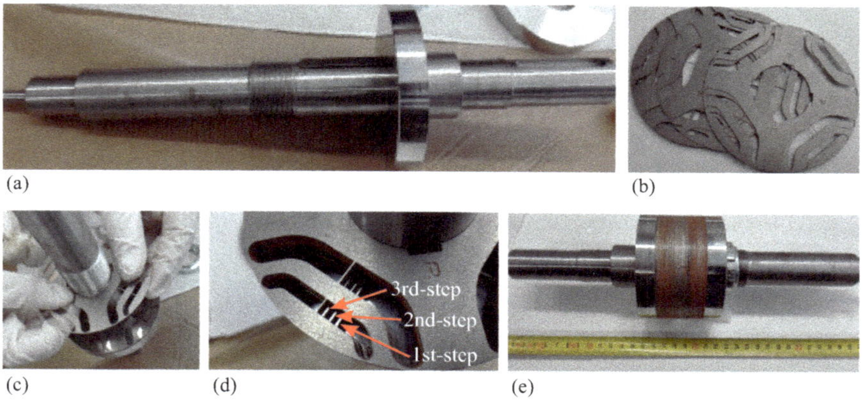

Figure 13.4 SYR *rotor assembling steps. (a) Motor shaft with one end-cap mounted, (b) Rotor lamination sheets, (c) Manual assembling of the lamination sheets, (d) Inside view of the rotor lamination stack with visible step-skewing, and (e) Manual assembling of the lamination sheets*

13.2 Prototype assembling

13.2.1 SyR rotor

The main steps of the SYR rotor assembling are reported in Figure 13.4. The rotor has been assembled in laboratory manually starting from the shaft, Figure 13.4(a), and the single rotor laminations, Figure 13.4(b). The latter have been stacked one by one on the shaft as reported in Figure 13.4(c). The four different rotor keyway slots on the rotor inner diameter have been used as reference for the eight rotor steps. In fact, they have been stacked following the order:

$$A, B, C, D, \bar{D}, \bar{C}, \bar{B}, \bar{A},$$

where the bar indicates the mirrored orientation. Three of these steps are clearly visible in Figure 13.4(d) while the assembled rotor is shown in Figure 13.4(e).

13.2.2 PMASYR *rotor*

The PMASYR assembling is similar to the SYR one, the only difference is the BMs insertion into the rotor flux-barrier slots. After the completion of the first rotor step stack, Figure 13.5(a), the BMs are inserted manually one by one in the flux-barrier slots as reported in Figure 13.5(b). This operation has been carried out for each rotor step. In Figure 13.5(c), it is possible to see the effect of the rotor skewing on the complete rotor.

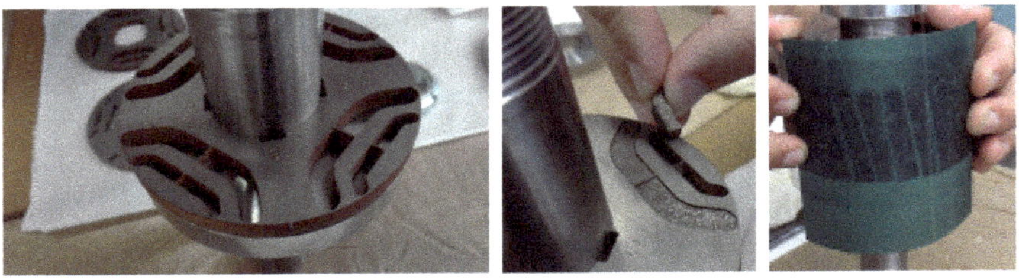

(a) (b) (c)

Figure 13.5 PMASYR *rotor assembling steps. (a) First rotor substack mounted on the rotor shaft, (b) Insertion of the magnets inside the flux-barriers, and (c) Magnet paper used to show the realized step-skewing of the rotor stack*

13.3 Experimental results

The tests carried out on the three prototypes allowed to get their MTPA locus and torque behaviors for different current amplitudes.

13.3.1 Torque versus current angle

The experimental and FE average torque versus current angle curves of the three prototypes are reported in the first column of Figure 13.6. The simulation data are indicated by means of continuous lines while the measurement values are represented by black markers. The MTPA trajectory is reported in red continuous line. In general, the measured average torque values are slightly lower than FE ones. Many aspects can explain such a difference. The most important is of course related to the nonuniform magnetic characteristics in the BM samples. Another aspect is the presence of additional air gaps between the BMs and flux-barrier surfaces determined by the BMs and lamination mechanical tolerances. This is confirmed by the fact that in some cases the BMs insertion was easy and in other cases was very hard. Also the BMs temperature plays an important role on the machine performance. In the FE analyses, it was set to 80°C while there is any indication of the temperature behavior during the tests. Finally, the actual lamination magnetic properties (i.e. *BH* curve) may be not completely coincident to the one used in the FE model. The latter aspect is surely present since a certain error is present in the SYR machine results as well.

With respect to the SYR machine, the average torque increases of 55% and 39% for the PMASYR motor with the compression and injection molding BMs, respectively. This difference is due to the higher remanence of the BMs obtained by compression.

13.3.2 Torque behaviors

The second column of Figure 13.6 reports the FE and experimental torque behaviors of the three motors obtained in MTPA condition with the rated current. From

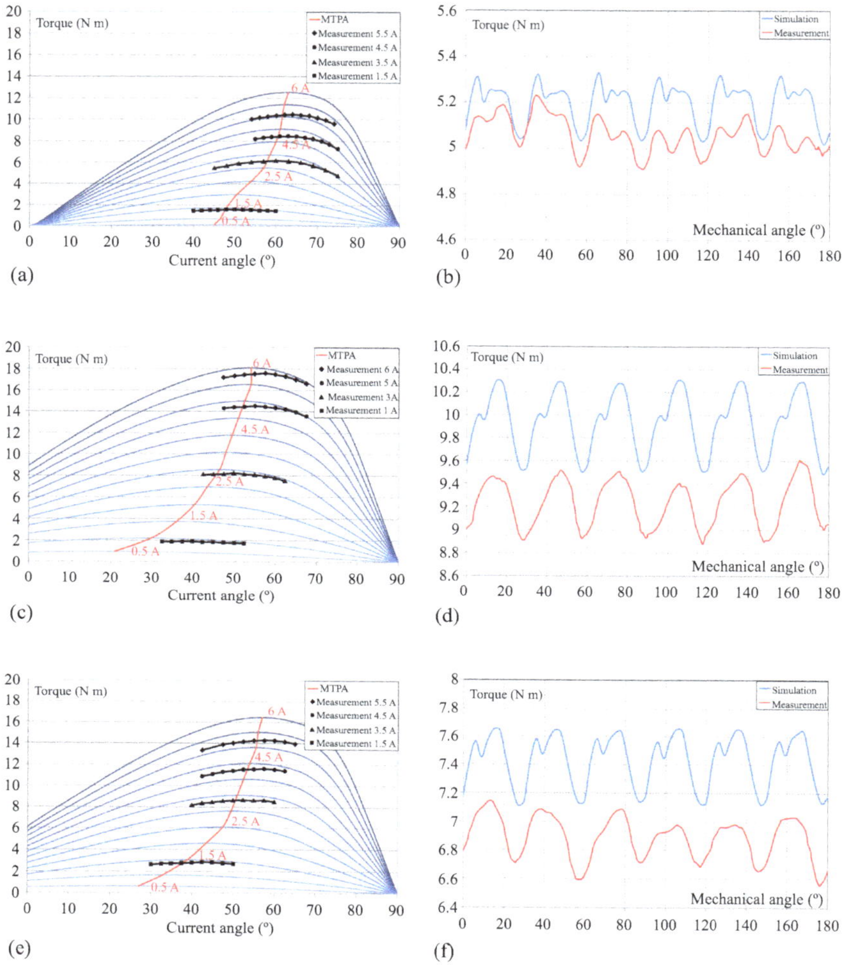

Figure 13.6 Torque versus current angle (first column) and torque versus rotor position curves (second column): comparison between FE and experimental results. The torque behaviors are obtained at the rated current in MTPA *condition: (a)* SYR *machine, (b)* SYR *machine, (c)* PMASYR *machine with compression BM, (d)* PMASYR *machine with compression BM, (e)* PMASYR *machine with injection BM, (f)* PMASYR *machine with injection BM*

Figure 13.6(b), it can be noted that the SYR motor has the best matching between measurement and simulation data. In fact, the average torque error is about 3% while for the ripple it is about 6%. For the PMASYR motor with BMs made through compression molding, Figure 13.6(c), the average torque difference is higher, i.e. 7%,

Table 13.2 Power factor of the prototypes at the rated current (3A) in MTPA *condition*

Motor types	**PF**
SyR motor	0.71
PMASyR motor with PM compression	0.93
PMASyR motor with PM injection	0.87

while the ripple one is 1%. Similar average torque error but higher torque ripple difference, i.e. 9%, has been found for the PMASyR motor with BMs prepared by injection molding, Figure 13.6(e). The use of BMs increases the ripple, respectively, of 145% for the PMASyR motor with compression-molded BMs and of 80% for PMASyR motor with injection-molded BMs.

13.3.3 Power factor

For all prototypes, the power factor has been computed on the basis of the experimental *dq* flux linkages and the winding resistance at the rated speed of 1 000 rpm, and the results are reported in Table 13.2. It can be noted that the highest power factor, 0.93, is achieved by the PMASyR machines equipped with compression-molded BMs. A slightly lower value, i.e. 0.87, is achieved by the other PMASyR machine. Both motors exhibit yield of a quite good power factor improvement with respect to the SyR one (0.71).

13.4 Conclusion

An investigation of the BMs suitability for PMASyR machines has been carried out in this chapter. Two types of BMs based on different production technologies have been proposed: compression and injection. The former allows us to obtain better magnetic properties, while the latter is more suitable for an industrial process, mainly in terms of production rates. A good match between experimental and FE results has been obtained, in particular regarding the torque waveforms. A torque density increment up to 50% has been found for the PMASyR machine prototype with BMs obtained by compression. The same prototype has a power factor of 0.93 at rated condition, much higher than SyR motor (0.71). In general, a noticeable torque ripple increase has been obtained in both PMASyR motors. However, such an increment can be effectively reduced adopting a more complex rotor with almost three-flux barriers per pole and optimizing their parameters accordingly.

Chapter 14
High-speed synchronous reluctance machines

This chapter describes a guideline to properly design Synchronous Reluctance (SyR) machines suitable for High-Speed (HS) applications. The main target is to give some guidelines for an accurate design of the rotor geometry to obtain a robust mechanical structure, a quite high-torque density and a low-torque ripple. A particular care is paid to the rotor rib thicknesses, designed to guarantee the structural integrity of the rotor [42] and to minimize the q-axis magnetic flux. In fact, the thickness of these iron ribs has to be wide enough to mechanically sustain the rotor and it increases as the design speed increases [117]. This is a significant drawback, since a quite large part of the magnetic flux flows through them, reducing the available saliency and, consequently, the machine torque.

14.1 Overview on a high-speed SyR machine

In the last few decades, the interest for HS electrical machines such as vacuum pumps, turbo-compressors and machining spindles arose from the requirement of compact and efficient drive systems [118]. In addition, the lack of gearbox, in the direct-drives systems, involves some important advantages such as high reliability, low cost and weight. The advances in new power electronic components, materials and control methodologies contributed to the HS drive systems success.

During the years, different electrical machine typologies have been proposed for HS applications [119,120]. Despite the universal motor was widely diffused in many civil and industrial applications in the past, the presence of the brushes strongly limited its diffusion in HS applications. The Induction Motor (IM) has been proposed for its intrinsic robustness and mature technology with several design solutions aimed to increase the rotor strength (i.e. solid rotor) and to reduce its inertia (i.e. solid cage, coated conducting sheet) [121–123,123–126] (Figure 14.1).

However, its low-power factor and efficiency limited its suitability for HS applications.

The interest to find a more efficient alternative to IM increased the attention for the Permanent Magnet Synchronous Machine (PMSM) well known for its high torque/power density and efficiency mainly due to the high-energy magnets which, on the other hand, have a strong impact on the machine's total cost.

In general, the magnetic and thermal aspects are taken into account in the PMSM design to avoid any Permanent Magnet (PM) demagnetization and dangerous winding

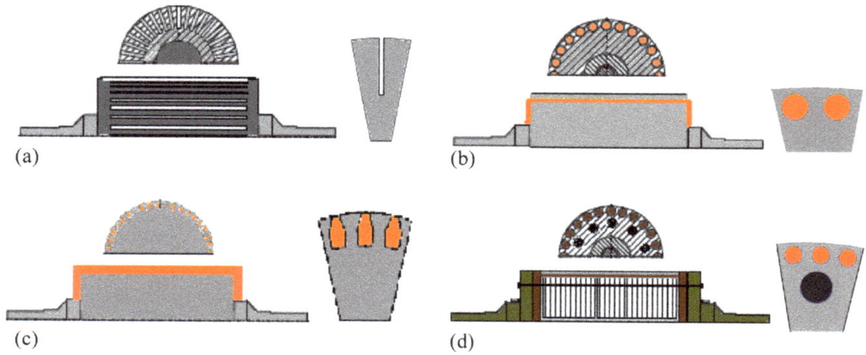

Figure 14.1 Different IM *rotor design for* HS *applications [126]: (a) slitted solid rotor, (b) solid rotor with embedded squirrel cage, (c) solid rotor with squirrel cage, (d) laminated rotor with embedded squirrel cage*

overheating [127,128]. In HS applications, the mechanical aspect represents an additional design key point to satisfy in order to ensure the machine structural integrity against the high centrifugal force that acts on the rotating parts. Moreover, the PMs' demagnetization risk is amplified by the higher PM eddy current that can arise on their surface. Also the thermal limit has to be reduced because of the higher harmonic core and joule losses [129–131]. The overall effect of these aspects worsen the PMSM performance in terms of efficiency and reliability so that the total losses limitation represent the most important challenge in the HS PM machine design. Typically, as shown in Figure 14.2(a), the PMs are glued on the rotor surface and the structural integrity is ensured by a retaining sleeve, which is often made of carbon fiber, for its high tensile strength. However, the low thermal conductivity of such a solution prevents the rotor heat exchange and increases the PMs thermal demagnetization risk [132]. Some HS PMSM design guidelines are presented in [133] where the uses of soft magnetic composite, as an alternative to the lamination core, and of a slotless stator are proposed as effective solution to reduce the total machine losses. In [134] a nonmagnetic conductive rotor sleeve is proposed as a valid solution capable of ensuring both structural integrity and reducing eddy-current PM losses. Alternatively, a nonconductive sleeve with a solid rotor core and a particular winding arrangement is proposed in [135] to reduce the rotor losses of a PM generator.

Instead of placing the PMs on the rotor surface, they can be *buried* in the rotor core so that the structural integrity is entirely ensured by the rotor iron, and no additional sleeve and cost are needed. This also involves a reduction of the total amount of PM materials [129] (lower air-gap length) but, on the other hand, higher Joule losses [140]. In [136], the rotor integrity is increased by means of H-shaped bars in the bottoms of the V-shaped poles as reported in Figure 14.2(b). The rotor sketches in Figure 14.2(c) and (d) show the *dovetail* shapes used to keep the PMs locked to the rotor [137,138]. Finally, Figure 14.2(e) and (f) reports the rotor sketches of two dovetail poles rotors [139]. Even though rare earth PMs involve high machine performance,

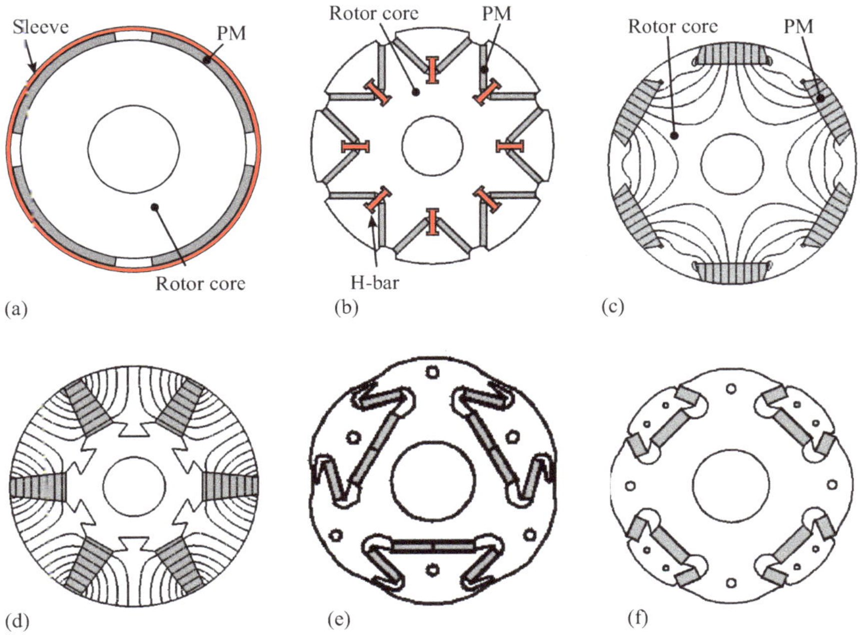

Figure 14.2 PM synchronous rotor for HS applications: (a) Surface-mounted Permanent Magnet (SPM), (b) H-shaped bars [136], (c) dovetail magnet 1 [137,138], (d) dovetail magnet 2 [137,138], (e) dovetail poles 1 [139], (f) dovetail poles 2 [139]

their high-cost fluctuations during the past years induced many researchers and PM machine manufacturers to find alternative solutions that yield a good compromise in terms of performance but less dependent on the PMs.

Among the PM-less machine scenarios, the SwR and SyR machines are the most promising suitable candidates in HS applications. The SwR machine's main advantages are intrinsic fault tolerance, low cost, robust design and low rotor inertia [141]. Unlike PMSM or IM, there are no additional components that add complexity to the mechanical or electrical design process but, conversely, it presents a simple rotor structure that allows one to reduce the manufacturing cost as well. Despite these advantages, the adoption of SwR machine in industry is still compromised by its high-torque ripple, low-torque density, high noise and high vibration [142]. Moreover, the SwR rotor structure contributes to a higher drag coefficient (i.e. higher air friction losses) even though some design solutions have been investigated to overcome such an aspect [143].

Similarly to SwR machine, the SyR one presents all the advantages due to the lack of rotor PMs, but it has been barely proposed for HS applications. However, thanks to its success in other industry applications, it has been receiving great attention in the recent years. The solid salient pole and transversely laminated rotor are the two

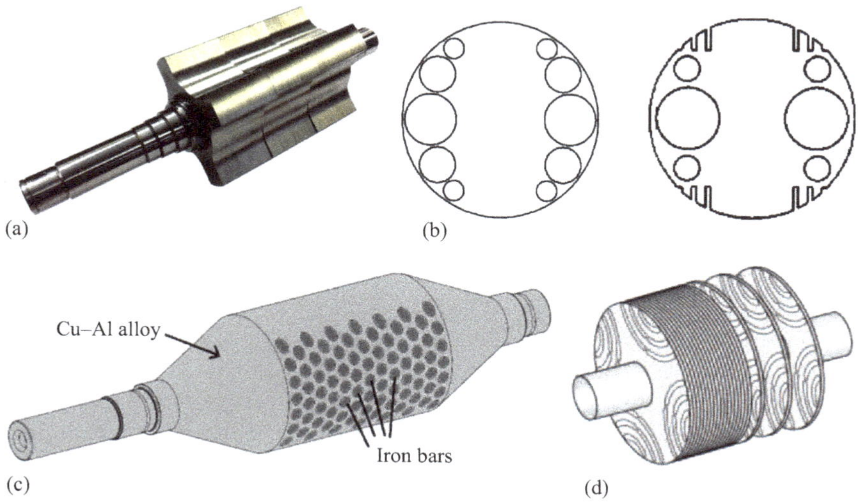

Figure 14.3 *HS-SYR solid rotor geometries: (a) salient pole [145], (b) drilled and drilled/slitted [146], (c) iron rods [147], (d) transversally laminated [26]*

main SYR rotor topologies. The HS SYR salient pole machine, shown in Figure 14.3(a), presents the simplest and cheapest rotor geometry, but it is well known for its intrinsic low rotor anisotropy [144] and high-torque ripple even though proper winding chording and rotor skewing allow to mitigate such a drawback [145]. As already pointed out for SWR machine, also the salient pole suffers from significant aerodynamic losses at HS so that a cylindrical rotor jacket could be required at the cost of increasing the air-gap length and reducing the torque density. The anisotropic solid rotor configurations in Figure 14.3(b) allow one to obtain rotor saliency higher than the salient pole with lower air-friction losses, thanks to the cylindrical rotor shape [146]. Exotic HS SYR rotor topology is presented in [147]. As shown in Figure 14.3(c), the rotor consists of a round iron rod array holding a number of rods evenly spaced from each other. The goal of such a solution is to achieve low magnetic reluctance along the d-axis by means of the iron rods and high q-axis reluctance with the nonmagnetic rods holder material (e.g. Cu–Al alloy). The quite simple rotor construction and robustness are the main advantages of this design; however, it can easily be manufactured for two poles only.

In normal applications, the transversally laminated SYR rotor, shown in Figure 14.3(d), is designed to maximize the rotor anisotropy and to minimize the torque ripple giving to the mechanical design aspect a secondary role. In particular, the structural integrity of such a machine is ensured by thin iron bridges, the size of which is kept small as much as possible (i.e. lamination width) to limit their negative effect on the rotor anisotropy. However, in HS applications, the iron bridges have to be accurately designed and their effects on the machine performance have to be taken

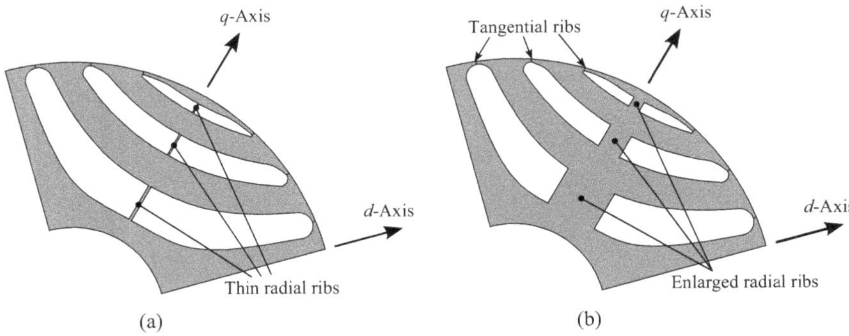

Figure 14.4 SyR rotor sketches with three flux-barriers per pole designed for different applications: (a) typical geometry, (b) HS geometry

into account. In fact, as the rated speed increases, their sizes have to be increased as well to limit the centrifugal mechanical stress but at the same time they have to limit the q-axis flux that compromises the rotor anisotropy.

In [148], this topic was introduced providing some design guidelines for a multi-flux-barrier SyR rotor without any retaining sleeve for HS aerospace application. In particular, the work mainly focuses on the lamination material selection and the analytical radial ribs design. A comparative study of two Finite Element (FE)-based optimization strategies for HS-SyR is carried out in [149,150] where it is found that a physically decoupled electromagnetic and mechanical optimization is more suitable than the coupled approach in which both electromagnetic and structural analyses are performed for objective evaluation. Figure 14.4 reports a comparison between two SyR rotor geometries designed for common and HS application where the larger radial rib sizes of the second one are evident.

14.2 Design methodology for high-speed synchronous reluctance machines

In this section, an analytical model is first adopted to derive a preliminary geometry [34]. The stator geometry is fixed and the focus is on the rotor structure. The effect of iron ribs and their influence on the increase of q-axis flux is taken into account. The flux-barrier-end angles are selected so as to minimize the torque oscillation [34,52,55,151]. Then, the purpose is to maximize the average electromagnetic torque, according to fixed speed. This is done through adjusting the quantity of air present in the rotor (Figure 14.5).

An FE analysis (FEA) is carried out to validate the results obtained by means of the analytical approach. Then, optimization is employed to optimize the position of the flux-barrier in the rotor, with the purpose of reducing the q-axis flux through the iron bridges (which causes the decrease of rotor anisotropy and average torque) and the

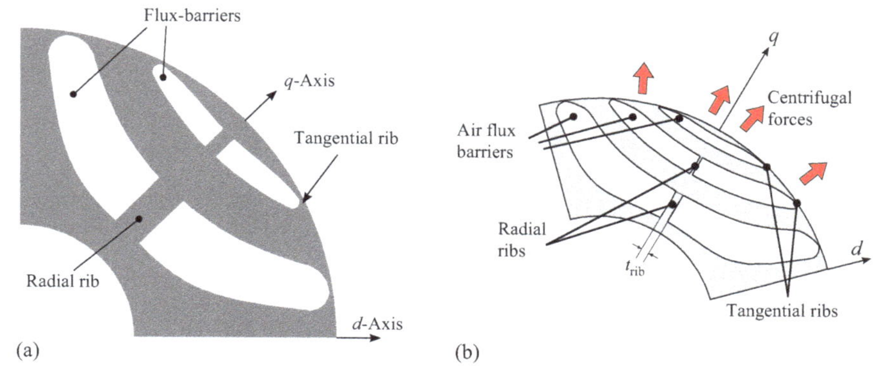

Figure 14.5 Sketches of synchronous reluctance machines for high-speed applications: (a) example of one rotor pole with two flux-barriers per pole; (b) sketch of one pole with three flux-barriers; distinction between the tangential and the radial ribs is made

torque ripple. The Pareto front is reported highlighting the potentials and the drawback of the optimal solutions. Finally, a thorough analysis of one optimal solution has been performed. It also includes a full electromagnetic analysis to determine accurate iron losses and a mechanical analysis under centrifugal stress to determine local stresses on the rotor and to verify the integrity of the structure.

14.2.1 Analytical approach

14.2.1.1 Choice of flux-barrier angles

Thanks to the model presented in Chapter 6, it is possible to easily predict the combinations of flux-barrier angles that lead to the lower torque ripples.

One flux-barrier rotor

Figure 14.6(a) reports the torque behavior due to the first two slot harmonics as a function of the electrical flux-barrier angle, ϑ_b^e. According to a four-pole machine ($2p = 4$), they are the 17th and 19th for the machine with $Q = 36$. These waveforms are obtained through the complete model of the SyR machine, shown in Chapter 5. This figure shows how much the choice of the flux-barrier angle affects the torque ripple resulting from the harmonic under study. In addition, the vertical lines identify the flux-barrier angles found by the simple formulation derived in Chapter 6. It can be noted that the prediction of the torque minima and maxima perfectly corresponds to the minima computed by means of the complete analytical magnetic model of the machine. As an example, a proper choice of the flux-barrier angle would be about 75° and not 70°.

Two flux-barriers rotor

Figure 14.6(b) reports the torque behaviors due to the first two slot harmonics in the plane of the two electrical flux-barrier angles ϑ_{b1}^e and ϑ_{b2}^e. These harmonics are of

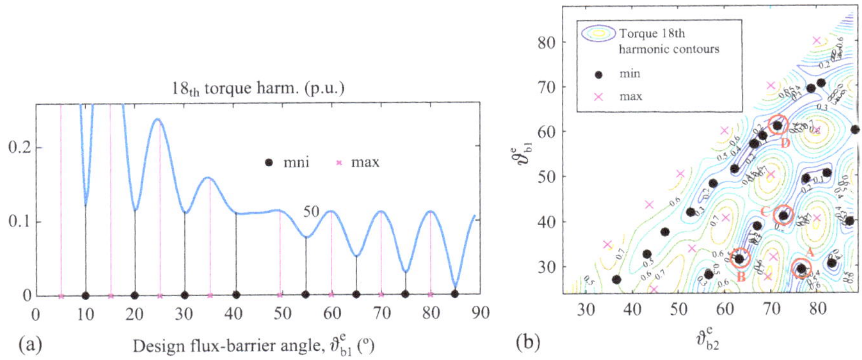

Figure 14.6 Analytical torque maps of the first slot harmonic as a function of the design flux-barrier angles for a 36-slot 4-pole machine. The continuous lines are computed through the complex analytical model presented in Chapter 5, while the lines and the markers are obtained with the model described in Chapter 6. (a) One flux-barrier machine. The vertical lines represent the positions of the torque ripple minima and maxima found by means of the model presented in Chapter 6. (b) Analytical torque maps of the first slot harmonic in a two flux-barriers machine. The black dots show the minima found through model presented in Chapter 6.

the same order of the previous ones. The two-dimensional maps are again obtained through a complete model of the two flux-barrier SyR machine, shown in Chapter 5. Then the points marked by black dots and magenta crosses are found by means of the model proposed in Chapter 6. They identify the flux-barrier angles that correspond to the torque ripple minima and maxima, respectively. Once again this model is able to predict these points with satisfactory precision, despite additional assumptions.

14.2.1.2 Flux-barriers and flux-carriers design

While the choice of the flux-barrier-ends mainly affects the torque ripple, the selection of the magnetic insulation quantity along the q-axis mainly affects the average torque capability of the machine. The thickness of the flux-barriers is an important design parameter in order to achieve good rotor saliency and, therefore, high performance. However, their thicknesses are limited by the level of saturation desired in the rotor iron. This is also the common guideline to choose a proper insulation ratio, k_{air}. In fact it is often desirable to let the rotor saturate more than the stator teeth, in order to reduce teeth iron losses. The coefficient k_{air} can be computed through the following equation:

$$k_{air} = 1 - \frac{B_g}{B_r} \cdot \frac{D_r}{p(D_r - D_{sh})} \tag{14.1}$$

where B_g is the peak air-gap flux density, B_r is the desired rotor flux density, D_r the rotor diameter, D_{sh} the shaft diameter and p the number of pole pairs. It represents the sum of the flux-barrier thicknesses along the q-axis, which means

$$k_{air} = \frac{l_{air}}{l_{tot}} \tag{14.2}$$

where $l_{air} = t_{b1} + t_{b2}$ is the amount of magnetic insulation along the q-axis, t_{b1}, t_{b2} the thickness of the first and second flux-barriers, respectively, and l_{tot} is the available space given by $(D_r - D_{sh})/2$. Its value typically ranges between 0.4 and 0.6.

Flux-barrier thicknesses

Once the total amount of air is chosen through the coefficient k_{air}, the individual thickness of the flux-barriers is chosen according to the stator Magnetomotive Force (MMF) in front of each flux-barrier: the higher the MMF step between two flux-barriers, the thicker the flux-barrier.

Figure 14.7(a) shows a sketch of a two flux-barrier rotor with a sinusoidal q-axis magnetization. $\mathscr{U}_s(\vartheta_r)$ and $U_r(\vartheta_r)$ represent the stator and the rotor scalar magnetic potentials, respectively. The mean value of $\mathscr{U}_s(\vartheta_r)$ computed between the flux-barrier end angle $\vartheta^e_{b(j-1)}$ and ϑ^e_{bj} is f_j. For a rotor with two flux-barriers, they are

$$f_1 = \frac{\sin\vartheta^e_{b1}}{\vartheta^e_{b1}} \qquad \Delta f_1 = f_1 - f_2$$
$$f_2 = \frac{\sin\vartheta^e_{b2} - \sin\vartheta^e_{b1}}{\vartheta^e_{b2} - \vartheta^e_{b1}} \qquad \Delta f_2 = f_2 \tag{14.3}$$

A common design rule adopted for the flux-barrier thickness ratio is [52]

$$\frac{t_{b2}}{t_{b1}} = \frac{\Delta f_2}{\Delta f_1}\sqrt{\frac{\vartheta^e_{b2}}{\vartheta^e_{b1}}} \tag{14.4}$$

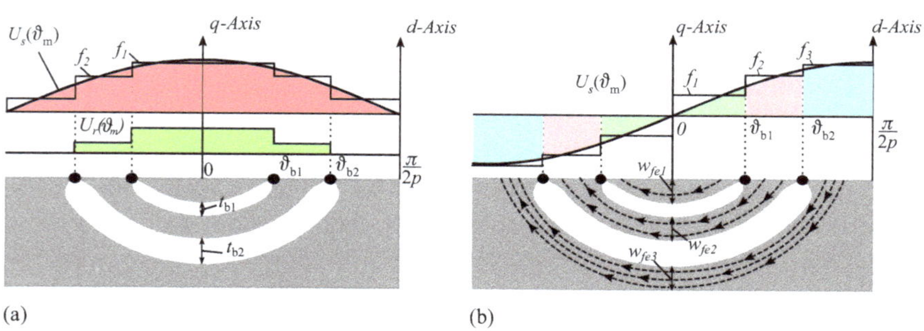

Figure 14.7 Sketches of a two flux-barriers per pole rotor with different stator magnetization: (a) q-axis magnetization—the flux-barrier thicknesses are also indicated; (b) d-axis magnetization—the flux-carrier widths are also indicated

and then [27]

$$t_{b1} = \frac{k_{air} l_{tot}}{1 + \frac{\Delta f_2}{\Delta f_1} \sqrt{\frac{\vartheta^e_{b2}}{\vartheta^e_{b1}}}} \quad (14.5)$$

so that t_{b2} can be easily obtained through (14.4). The thickness of each flux-barrier is computed with respect to the q-axis as shown in Figure 14.7(a).

Flux-carrier widths

The flux-carrier widths are designed assigning a width proportional to the MMF at the air gap. Figure 14.7(b) shows a sketch of a two flux-barrier rotor with a sinusoidal d-axis magnetization. The meaning of f_j is the same of the previous subsection and the flux-carrier widths are computed with respect the q-axis.

The flux-carrier width ratio is defined as

$$\frac{w_{cj}}{w_{ck}} = \frac{f_j}{f_k}, \quad j, k = 1, 2, 3 \quad (14.6)$$

where

$$\begin{aligned} f_1 &= \frac{1 - \cos \vartheta^e_{b1}}{\vartheta^e_{b1}} \\ f_2 &= \frac{\cos \vartheta^e_{b1} - \cos \vartheta^e_{b2}}{\vartheta^e_{b2} - \vartheta^e_{b1}} \\ f_3 &= \frac{\cos \vartheta^e_{b2}}{\frac{\pi}{2} - \vartheta^e_{b2}} \end{aligned} \quad (14.7)$$

Then

$$w_{c1} = \frac{(1 - k_{air}) l_{tot}}{1 + \frac{f_1}{f_2} + \frac{f_1}{f_3}} \quad (14.8)$$

and the remaining flux-carrier widths can be determined from (14.6).

14.2.1.3 Iron ribs computation

The thickness of the tangential iron bridges near the air gap is set equal to the minimum manufacturing length or to the lamination width. Their effect on the structural integrity is negligible.

The radial iron ribs, in the middle of the flux-barriers, are properly designed at the rated speed of the motor with the aim of containing the centrifugal stress due to the rotation. A simple analytical method to design the radial rib of each flux-barrier has been described in [117]. Once the flux-barrier-end angles are known, it is possible to compute the thickness of the jth flux-barrier as follows:

$$t_{rib,j} = \frac{\nu_\sigma (1 - k_{air}) \gamma_{Lam} \omega_m^2 D_r^3 \left(2\vartheta_{bj} - \sin 2\vartheta_{bj}\right) \cos \vartheta_{bj}}{8\sigma_r} \quad (14.9)$$

where ν_σ is a safety factor, γ_{Lam} is the lamination mass density, D_{r} is the rotor diameter, ω_{m} is the mechanical rotor speed, and σ_{r} is the tensile strength of the lamination. ω_{m} is often selected slightly higher than the nominal one, and ν_σ is set in the range between 2 and 3 [42].

14.2.2 Optimization approach

For SyR machines, an optimization of the rotor is almost always necessary to obtain a good design. This statement is reinforced when dealing with HS SyR machines. In fact the presence of iron ribs along the q-axis greatly affects the performance of the machine.

One of the most used optimization algorithms in electromagnetic designs is Differential Evolution (DE).* Typically it guarantees good performance and convergence in a reasonable amount of time.

Typical objectives for HS machines include the maximization of average torque (or power), of the saliency, of the power factor, the minimization of the torque ripple, of the losses or of the cost. The solutions found should also exhibit proper robustness against slight variations of the design parameters. It is evident that the optimization of such machines is multi-objective from the start. Therefore, there will not be a single optimum, but many optimal solutions that are called Pareto solutions. In fact, they have at least one of their objectives that is the best among all the solutions.

Figure 14.8 shows the flowchart of the evaluation procedure performed for each individual. The geometrical feasibility is evaluated considering the optimizer input parameters. The unfeasible individuals are discarded setting zero average torque and a high-torque ripple. For the feasible individuals, the algorithm computes the radial rib thicknesses, finds the Maximum Torque Per Ampere (MTPA) current angle and evaluates the average torque, its oscillation and the motor losses in a sixth of electrical period.

14.2.3 Sensitivity analysis

To estimate the robustness of good candidates (high average torque and low torque ripple), a sensitivity analysis on the results has been carried out. The dimensions of the parameter space are equal to the number of parameters, four in the considered optimization. Furthermore all the parameters are normalized within the interval 0 and 1.

It is useful to consider a three-dimensional space first. In the case study, the parameters can be the flux-barrier-end angles as shown qualitatively in Figure 14.9. Each dot in the space corresponds to an individual with an associated average torque and a torque ripple. For example, the point j in Figure 14.9 is characterized by the objectives ($T_{\mathrm{m}j}$, $\Delta T_{\mathrm{m}j}$).

*See Chapter 9.

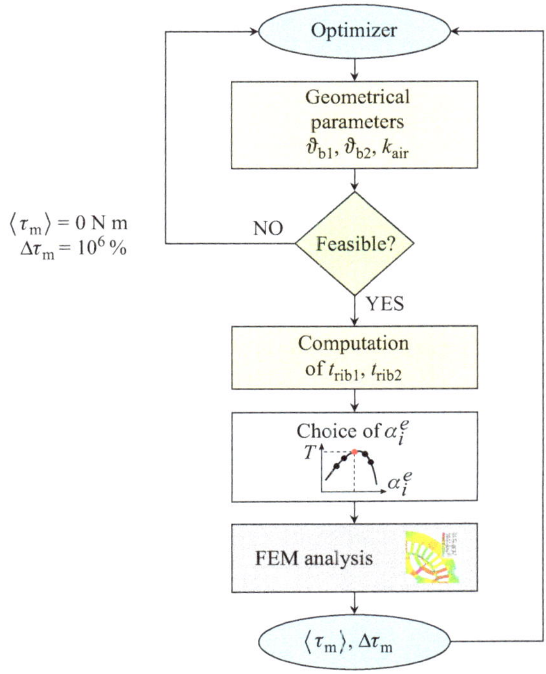

Figure 14.8 Flowchart of the objectives evaluation done by the optimizer

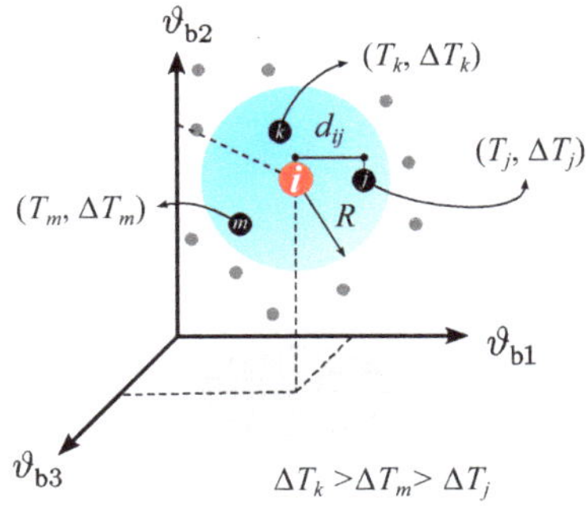

Figure 14.9 Space of parameters. Meaning of the hypersphere and distance d_{ij} between the points i and j. T_{mk} and ΔT_{mk} are the average torque and the torque ripple of the kth individual.

The sensitivity analysis is carried out for each point in the parameter space. The distance between two points is computed with the Euclidean norm. It is verified that $d_{ij} = d_{ji}$ since the distance is a symmetric property. The matrix of the distances, $\boldsymbol{D}$, is symmetric and its order is the number of the individuals N. The element in the ith row and in the jth column is the distance d_{ij} defined earlier. Obviously the matrix has all the diagonal elements equal to zero.

$$\boldsymbol{D} = \begin{bmatrix} 0 & d_{12} & \cdots & d_{1N} \\ & \ddots & & \\ d_{i1} & \cdots & 0 & d_{iN} \\ d_{N1} & \cdots & \cdots & 0 \end{bmatrix} \tag{14.10}$$

To the purpose of analyzing a restricted volume around each single point, it is useful to define a radius R of an hypersphere. The sensitivity can be computed locally in each point. In the normalized research space, the maximum normalized radius of the hypersphere that includes all the individuals from each one of them is equal to $\sqrt{N_{\mathrm{par}}}$, where N_{par} is the number of parameters.

Referring to Figure 14.9, the hypersphere around the individual i includes only the points j, k and m. The other ones are not considered. This means that a restricted number of elements of the ith row of the matrix $\boldsymbol{D}$ are useful. Among these remaining individuals, the maximum and the minimum torque ripples are researched. The figure reports that the maximum and the minimum torque ripples correspond to the individual k and j, respectively.

The sensitivity of the ith individual is finally given by

$$s_i = \frac{\max\left(\Delta T_{\mathrm{m}}\right)|_i - \min\left(\Delta T_{\mathrm{m}}\right)|_i}{\Delta T_{\mathrm{m}i}} \tag{14.11}$$

where the subscripts at the numerator indicates the maximum and the minimum torque ripples among the individuals inside the hypersphere centered on the individual i. The result of (14.11) is obviously a function of the hypersphere radius.

14.3 First application example

An example of HS SyR motor design is presented hereafter. Two different design methods are used: a Semi-Analytical Design Approach (SADA) and Finite Element Optimization (FEO). In the former, the analytical method is used to select the proper flux-barrier angles combination and to compute the radial rib thicknesses while the FEA is used to determine the best insulation ratio and the motor outputs (i.e. average torque, inductances) [106]. The latter needs an optimization algorithm and FE simulations only. The SADA procedure is faster since it does not require optimization. However, the iron saturation is not considered in the flux-barrier angle choice. On the contrary, the total FE procedure requires a higher computational time, but it is more accurate since it takes into account the iron saturation.

The design only deals with the rotor geometry. The stator, the winding and the materials are given and reported in Table 14.1. Through (14.9), it is possible to estimate if a design for such speed is feasible. In particular, an estimation of the required maximum rib thickness can be made based on the desired design speed. It is sufficient to select ϑ_{b} equal to the maximum allowable one, which is $\pi/2p$.

Table 14.1 Main motor data for the first HS *application*

Motor output			
Rated voltage (RMS)	V_N	400	V
Number of poles	$2p$	4	1
Mechanical speed	n	30 000	rpm
Geometrical data			
Outer diameter	D_e	152	mm
Inner diameter	D_s	90	mm
Stack length	L_{stk}	100	mm
Number of poles	$2p$	4	1
Number of slots	Q	36	1
Slot height	h_s	16.3	mm
Tooth width	w_t	4.14	mm
Mechanical air gap	g	0.7	mm
Winding			
Conductors in slot	n_c	6	1
Number of parallel paths	n_{pp}	2	1
Coil pitch	y_q	8	1
Materials			
Iron lamination			N020

Then, the rib thickness will be a fraction of the maximum space available, which means $t_{\text{rib}} = c\,D_r \sin\vartheta_b$, where c measures the fraction. So

$$c = \frac{\nu_\sigma (1 - k_{air})\gamma_{Lam}\omega_m^2 D_r^2 \left(\frac{\pi}{p} - \sin\frac{\pi}{p}\right)}{8\sigma_r \tan\frac{\pi}{2p}} \tag{14.12}$$

In our case, considering $k_{air} = 0.5$, $\sigma_r = 500\,\text{MPa}$, $\nu_\sigma = 2.5$, $\gamma_{Lam} = 7\,650\,\text{kg/m}^3$, it results in $c = 0.1$.

14.3.1 Analytical design

In Table 14.2, some flux-barrier angle combinations, which exhibit the lowest torque ripple, are reported. They are selected from the map in Figure 14.6.

The magnetic insulation ratio, k_{air}, can be selected through (14.1), which gives a value around 0.55 for this machine. However, in this case the optimal k_{air} has been selected by means of a parametric FEA for each considered solution. The results are shown in Figure 14.10. The figure also shows the rotor geometry variations with the insulation ratio. It is evident that k_{air} strongly affects the flux-barrier shape and the radial rib thicknesses. In fact, as the coefficient increases, the amount of rotor iron is lower and so the radial ribs become thinner.

Table 14.2 Flux-barrier-end angles combinations that exhibit the lowest torque ripple found through Figure 14.6

Rotor	ϑ_{b1}^{e} [°]	ϑ_{b2}^{e} [°]	ϑ_{b1} [°]	ϑ_{b2} [°]
A	29.25	76.63	14.63	38.32
B	31.49	63.25	15.75	31.63
C	40.96	72.73	20.48	36.37
D	61.02	71.61	30.51	35.81

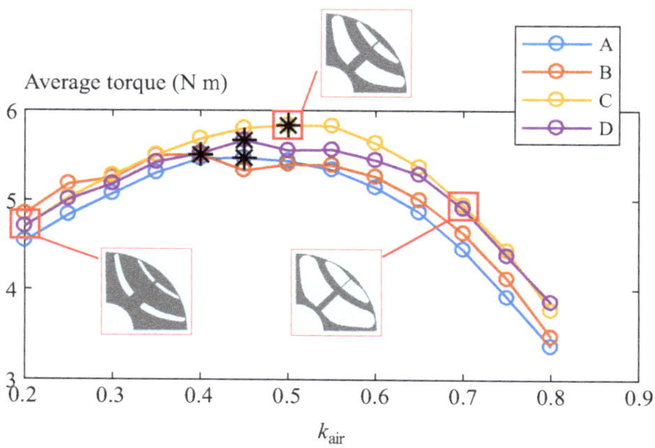

Figure 14.10 Average torque versus magnetic insulation ratio for different flux-barrier angles

It can be noted that the maximum average torque is achieved with k_{air} between 0.4 and 0.55. The motor C exhibits the highest average torque among the selected ones. The main motor outputs are reported in Table 14.3.

14.3.2 Finite-element optimization

The FE optimization is carried out with an RMS current density equal to 5.6 A/mm^2. The selection of the current density is based on an a priori thermal analysis of the machine.

In this case, the chosen objectives are the average torque, $\langle T_m \rangle$, and the torque ripple, ΔT_m. The geometrical parameters are the two flux-barrier angles, ϑ_{b1}, ϑ_{b2}, and the magnetic insulation ratio, k_{air}, while the other geometrical quantities are derived from these parameters [27].

The optimization has been carried out with a population size of 25 individuals and 80 generations and it took about 2 days. The final objectives plane is shown in Figure 14.11. The black diamonds represent the Pareto front. For HS applications it is

Table 14.3 Main electromagnetic results of the motors analyzed

(a) Results obtained through FEA of the machines designed with SADA and FEO

Design method	α_i^e [°]	$\langle T_m \rangle$ [Nm]	ΔT_m [%]	L_d [mH]	L_q [mH]	ξ 1	$\cos\varphi$ 1
SADA	52.5	5.83	17.6	0.896	0.344	2.61	0.43
FEO	47.5	5.91	9.16	0.910	0.367	2.48	0.40

(b) Motor losses in different parts

Part	Losses [W]		
	Fund. Harm.	Space Harm.	Total
Copper	194	0	194
Stator iron	479	100	579
Rotor iron	0	159	159
Total	673	259	932

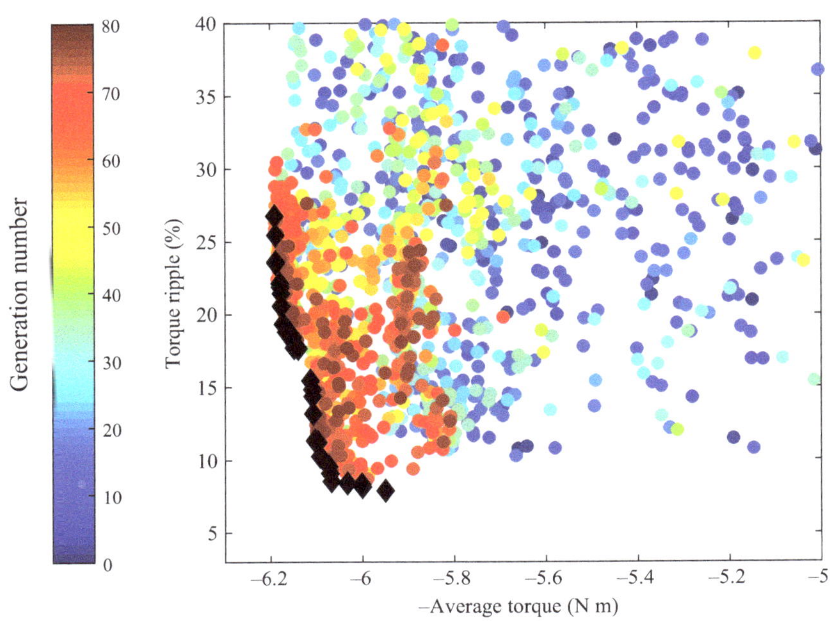

Figure 14.11 Simulated individuals in the objectives plane and Pareto front (black diamonds) at 30 000 rpm

mandatory to limit the torque oscillation that causes dangerous mechanical vibrations and acoustic noise. For this reason, the individual with the lowest torque ripple has been selected. The average torque is about 6 Nm while the torque ripple is about 9% with respect to the average torque. The relative flux-barrier angles are $\vartheta_{b1} = 20.70°$ and $\vartheta_{b2} = 31.81°$ while the insulation ratio is $k_{air} = 0.429$. With respect to the flux-barrier-end angles analytically predicted, the first one is practically the same while the second one is 9.12° (electrical) smaller. The different rotor configurations justify the mismatch of the torque ripple obtained with the two methods.

14.3.3 Electromagnetic analysis

After the selection of the most promising solution, an in-depth electromagnetic analysis is required to precisely evaluate the machine performance. In particular, the average torque and the torque ripple found through the optimization are confirmed and better estimated, as reported in Figure 14.13(a). The numerical results are also reported in Table 14.3.

The flux density plot of the FE optimal motor is shown in Figure 14.12(a). It can be noted that the iron saturation within the tangential ribs covers a larger area and the iron saturation of channels is not negligible. These aspects highly affect the analytical model hypothesis and explains the different torque ripple predicted by the approaches.

The HS of rotation involves high frequency for the current supply. Even though thin lamination sheets have been considered, iron losses may play an important role in the overall machine performance. So, to accurately estimate the losses, the flux density for every mesh element has been recorded in both x and y directions for a complete electrical rotation. This allows one to split the flux density fluctuations in Fourier series and, thus, to find the losses for all the harmonics involved. The harmonic losses are caused by the staircase MMF waveform and by the effect of stator slots. Current distortion may introduce additional losses, but it is not considered in this study.

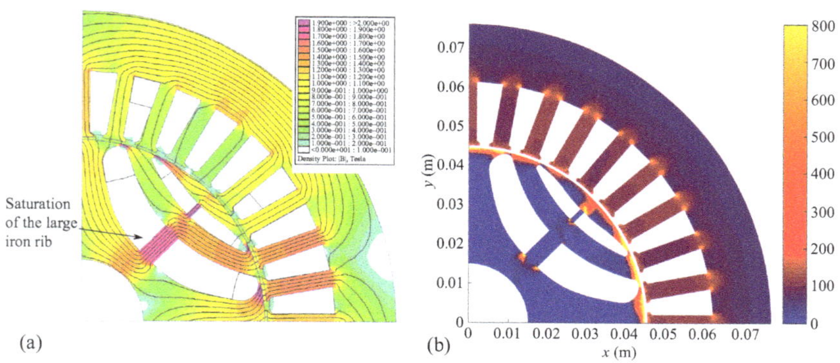

Figure 14.12 Plots of the rotor designed with the FEO approach: (a) flux density plot, (b) loss density map at the MTPA operating point

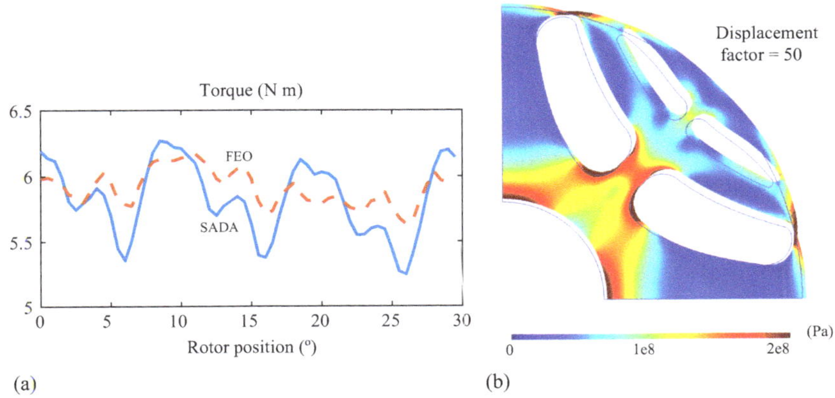

Figure 14.13 Plots resulting from the electromagnetic and mechanical FE analyses: (a) finite-element torque versus rotor position with different flux-barrier angles, (b) plot of von Mises stress of the optimized rotor for 30 krpm

The loss density map is reported Figure 14.12(b). It can be noted that the maximum loss density appears near the rotor periphery and near sharp corners. However, the volumes involved are rather small. On the other hand, stator teeth and back-iron see lower loss densities but they are diffused, filling almost the entire volume. In fact, the stator iron is also subjected to the fundamental harmonic of the flux density, which is obviously the highest harmonic.

The comparison among the losses is reported in Table 14.3. The majority of losses are due to the iron losses in the stator caused by the fundamental of the flux density. Joule losses in the stator winding are the second main cause. Then it can also be seen that the additional losses introduced by the harmonics are not negligible at this speed, as they represent almost 30% of the total electromagnetic losses.

14.3.4 Structural analysis

Lastly, a structural FEA has to be performed to verify whether the radial iron ribs are able to sustain the centrifugal load of the rotor iron paths [152]. In the present study, only the rotor design with the FE optimizations is considered since it achieves better performance. Furthermore, the two rotor geometries are similar from the mechanical point of view.

The rotor sharp corners have been smoothed out in order to avoid dangerous stress concentrations. The von Mises stress and the displacement plots are shown in Figure 14.13(b). The ultimate tensile strength of the selected material is 500 MPa, while the chosen safety factor is 2.5 [42]. The scale of the von Mises stress has been set to the desired stress value, which is 200 MPa for this study. It can be observed that the target stress value is never reached within the radial iron ribs. Stress concentration is present near corners, but it will lead only to some yielding.

14.3.5 Discussion

The aim of this section was to propose a design procedure for HS SyR motors. Only the rotor geometry has been properly designed while the stator and the winding have been kept fixed. The goals of the rotor design were to achieve the maximum torque density with the lowest torque ripple and to ensure the structural integrity against the high centrifugal force. To do that, the flux-barrier geometries (i.e. angles, thicknesses) and the radial ribs have been designed. Two design approaches have been proposed: one is semi-analytical while the other is based on FE optimization. The former is faster and finds suitable flux-barrier angles by means of a simple analytical model. The motor performance is then precisely evaluated through FEA. The latter is more accurate since it considers the iron saturation but requires more time because of the optimization algorithm. The comparison between the two methods revealed that the nonlocalized saturation of the tangential ribs and the radial rib thicknesses strongly affect the torque ripple estimation of the analytical model. For this reason, the torque oscillation obtained by means of the FE optimization is about 52% of that resulting from the semi-analytical approach.

The losses computation has shown that rotor losses have to be taken into account for accurate efficiency prediction, especially at such HSs. In fact, the rotor losses represent about 30% of the total. The structural analysis confirmed the validity of the radial ribs design method giving a suitable stress distribution within the rotor.

In conclusion, the semi-analytical method is a quick procedure that allows one to obtain a good initial design of an SyR rotor. However, to get more sound designs, a multi-objective optimization algorithm should be employed.

14.4 Second application example

The aim of this section is to investigate the potential of HS SyR machines. Optimization is carried out so as to maximize the machine performance (high power, proper power factor, low vibration) at given speed. The machine size is fixed and the focus is on the rotor geometry, with the purpose of maximizing the electromagnetic torque, according to the necessary thickness of the ribs. It is shown that, even if the ribs are thick, it is possible to reach a proper torque density. However, the optimal solutions are quite sensitive to the geometrical variations, so that a particular care is required in the manufacturing of the machine. Finally, the power limit of the SyR motor with a barrier rotor is found.

A sketch of a four-pole SyR rotor with three flux-barriers per pole is shown in Figure 14.5. It also distinguishes the radial iron ribs (in the middle of the barrier) from the tangential iron bridge (near the air gap). The radial bold arrows represent qualitatively the centrifugal force that acts on the rotor lamination.

To the purpose of investigating the potentials of HS SyR machines, some optimizations are carried out. The objective is to maximize the machine performance, expressed as electric power, power factor and low vibration at given HS, which is set to be equal to 20 000 rpm.

Table 14.4 Main data of the machine under optimization for the second HS application

Motor part	Symbol	Value
Outer diameter	D_e	240 mm
Inner diameter	D_s	140 mm
Stack length	L_{stk}	140 mm
Pole number	$2p$	4
Slot number	Q	48
Slot height	h_s	25 mm
Tooth width	w_t	5 mm
Air gap	g	1 mm
Iron lamination	NO20	

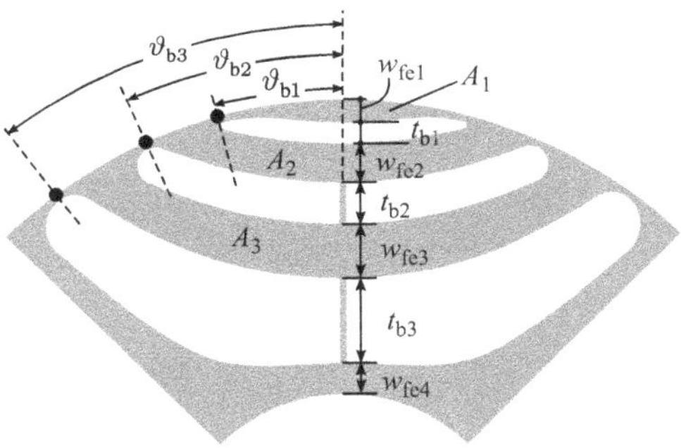

Figure 14.14 Sketch of a one pole REL rotor with three flux-barriers per pole. The flux-carrier widths and the flux-barrier thickness are also marked.

The dimensions of the stator are fixed and reported in Table 14.4. The optimizations are focused on the rotor geometry. The position of the flux-barriers is quite important to reduce the centrifugal forces, while the position of the ends of the flux-barriers is important to reduce the torque oscillation [34,151]. The purpose is to maximize the electromagnetic torque, according to the necessary thickness of the ribs.

14.4.1 Choice of the optimization parameters and iron ribs computation

14.4.1.1 Optimization parameters and objectives

The selected rotor parameters are the flux-barrier angles shown in Figure 14.14 and the magnetic insulation ratio k_{air} defined as

$$k_{air} = \frac{l_{air}}{l_{air} + l_{fe}} \tag{14.13}$$

where $l_{\mathrm{air}} = \sum_j t_{\mathrm{b}j}$ and $l_{\mathrm{fe}} = \sum_i w_{\mathrm{c}i}$ are the total central thickness of the flux-barriers and of the iron, respectively. Usually three flux-barriers per pole are employed because they lead to a higher and smoother torque. The shape of such flux-barriers derives from the magnetic flux lines that would flow if the rotor were solid (see Section 3.1). As reported in Figure 14.14 the flux-barrier angles are sorted in ascending order from the outer flux-barrier to the inner one. The objectives of the optimizations are the average torque and the torque ripple, computed as

$$\langle T_{\mathrm{m}} \rangle = \frac{1}{2\pi} \int_0^{2\pi} T_{\mathrm{m}}(\vartheta_{\mathrm{m}})\, \mathrm{d}\vartheta_{\mathrm{m}} \tag{14.14}$$

and

$$\Delta T_{\mathrm{m}} = \frac{\max\left[T_{\mathrm{m}}(\vartheta_{\mathrm{m}})\right] - \min\left[T_{\mathrm{m}}(\vartheta_{\mathrm{m}})\right]}{\langle T_{\mathrm{m}} \rangle} \tag{14.15}$$

respectively, where $T_{\mathrm{m}}(\vartheta_{\mathrm{m}})$ is the torque computed with the Maxwell stress tensor (see Appendix E). The agreement between this torque computation and experimental tests has been proven in numerous works, such as [34,50].

14.4.1.2 Computations of the radial iron bridges

At HS operation the rotor structure has to sustain high centrifugal stresses that may cause rotor deformations and mechanical issues. In order to improve the rotor integrity, the following two solutions are evaluated:

1. use of radial iron ribs;
2. addition of a containing sleeve with high tensile strength.

The design of the radial ribs is carried out with the following safety precautions. The ultimate tensile strength of the rotor lamination is set to $\sigma_{\mathrm{R}} = 500$ MPa while the safety factor is $\nu_\sigma = 2.5$ [45]. This is a proper choice to guarantee an infinite lifetime. Furthermore, the structural contribution of the tangential iron ribs is neglected.

The mass and the center of gravity of each island are directly computed from the actual geometry, and they are referred to as m_i and $R_{\mathrm{G},i}$, respectively, for the ith island. The jth rib has to sustain the centrifugal load due to the mass M_j given by

$$M_j = \sum_{i=1}^{j} m_i \qquad R_{\mathrm{G},j} = \frac{\sum_{i=1}^{j} m_i R_{\mathrm{G},i}}{\sum_{i=1}^{j} m_i} \tag{14.16}$$

It follows that the centrifugal force of the jth mass is

$$F_j = M_j\, \omega_{\mathrm{m}}^2 R_{\mathrm{G},j} \tag{14.17}$$

where $R_{\mathrm{G},j}$ is the radius of the center of gravity of the jth mass while ω_{m} is the mechanical angular speed.

The iron areas and centers of gravity can be estimated through some geometrical relationships [153], or it can be directly measured from the actual geometry in the FE program. The latter approach has been used in the following. Then the ith rib thickness

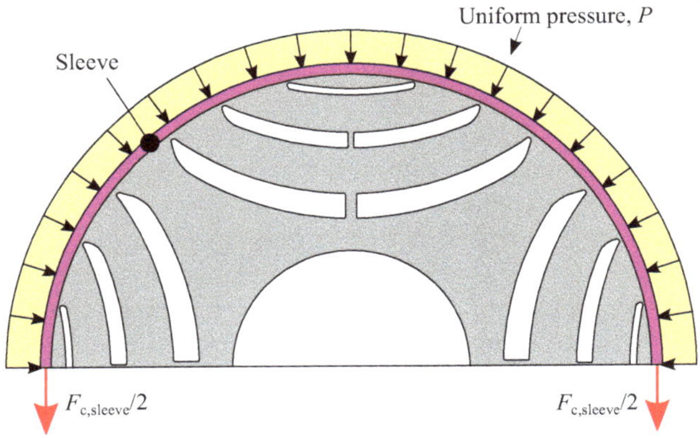

Figure 14.15 Scheme of the computation of the sleeve thickness

can be computed from the maximum iron stress taking into account the safety factor, ν_σ, typically set in the range between 2 and 3 [42,117]. The result is

$$t_{\text{rib},i} = \nu_\sigma \frac{F_i}{\sigma_R L_{\text{stk}}} \tag{14.18}$$

14.4.1.3 Computation of the sleeve thickness

The sleeve counteracts the centrifugal force through uniform radial pressure sketched in Figure 14.15. The chosen sleeve material is a carbon fiber composite with an ultimate tensile strength equal to $\sigma_{\text{CF}} = 2\,000$ MPa and safety factor $\nu_{\text{CF}} = 2$.

The formulation comes from the membrane theory, since the sleeve thickness, t_{sleeve}, is much smaller than the rotor radius. Hence the relationship between the centrifugal force and the radial pressure is the following:

$$F_{\text{c,sleeve}} = \int_{-\pi/2}^{\pi/2} P \cos\vartheta \, \frac{D_{\text{r}} L_{\text{stk}}}{2} \, \mathrm{d}\vartheta = P D_{\text{r}} L_{\text{stk}} \tag{14.19}$$

where P is the pressure at the air gap and ϑ is the angular coordinate. The pressure is assumed to be uniformly distributed along the rotor periphery so as to link the centrifugal force and the sleeve thickness through

$$t_{\text{sleeve}} = \nu_{\text{CF}} \frac{F_{\text{c,sleeve}}}{2\sigma_{\text{CF}} L_{\text{stk}}} \tag{14.20}$$

The sleeve allows one to reduce the stress on the radial ribs, hence their thickness. However, the air-gap length is kept constant whilst the rotor diameter is decreased to include the sleeve. So the equivalent gap becomes $g_{\text{eq}} = g + t_{\text{sleeve}}$.

Both solutions affect negatively the average torque because the iron ribs increase the q-axis flux while the sleeve increases the effective air gap. Thus a comprise between the two solutions could exploit better performances and it will be evaluated later on.

14.4.2 Optimization

The optimization procedure used in the present section implements a multi-objective DE algorithm.† The optimization is carried out with an electrical loading of about 47 000 A/m, which corresponds to an RMS current density equal to 7.15 A/mm^2.

The chosen objectives have been defined earlier and they are the average torque, $\langle T_{\mathrm{m}} \rangle$, and the torque ripple, ΔT_{m}. The geometrical parameters are the flux-barrier angles, $\vartheta_{\mathrm{b1}}, \vartheta_{\mathrm{b2}}, \vartheta_{\mathrm{b3}}$, and the magnetic insulation ratio, k_{air}. The remaining geometrical quantities are derived from these four parameters. The choice of such parameters greatly affects the final performance of the machine. The FE-aided design has been validated in previous works [34,50] for lower speeds. The experimental results have confirmed the ones found by means of simulations, in particular, for the objectives of interest.

The optimization flowchart, without computation of the sleeve, is shown in Figure 14.8. At first, the input parameters defined earlier are selected and the feasibility of the rotor geometry is evaluated. If it is not realizable, the individual is discarded setting zero average torque (i.e. $\langle T_{\mathrm{m}} \rangle = 0$) and a high-torque ripple (i.e. $10^6\%$). Otherwise, the algorithm computes the radial rib thicknesses according to (14.17) and the FE motor model is built.

For each individual, the MTPA point at rated current is found varying the current angle α_i^{e}. This requires about four additional simulations. The objectives are thus evaluated using the actual MTPA angle. Finally, the optimizer processes the outputs and proceeds with the next individual.

14.4.3 First optimization

14.4.3.1 Results

The first optimization is performed with 1-mm air gap and no sleeve, at the speed of 20 000 rpm. The optimization is carried out with 80 generations constituted by 40 individuals. The results of the whole optimization is reported in the objectives plane in Figure 14.16, together with the Pareto front. On the x-axis, there is the opposite of the average torque, $\langle T_{\mathrm{m}} \rangle$, while on the y-axis the torque ripple, ΔT_{m}, is reported.

It can be observed that the solutions converge to the optimal ones, located near the bottom left corner of the plot. The resulting Pareto front is steep, since the torque ripple has high variations around similar average torque values. Nonetheless, the results show that the rotor with three flux-barriers is able to reach a proper average torque with many combinations of the input parameters.

In order to highlight the dependence of the torque ripple on the flux-barrier angles, Figure 14.17(a) shows the barrier angles sorted by the relative torque ripple. The points lying on the Pareto front are highlighted by black dots. Many of them are

†See Chapter 9.

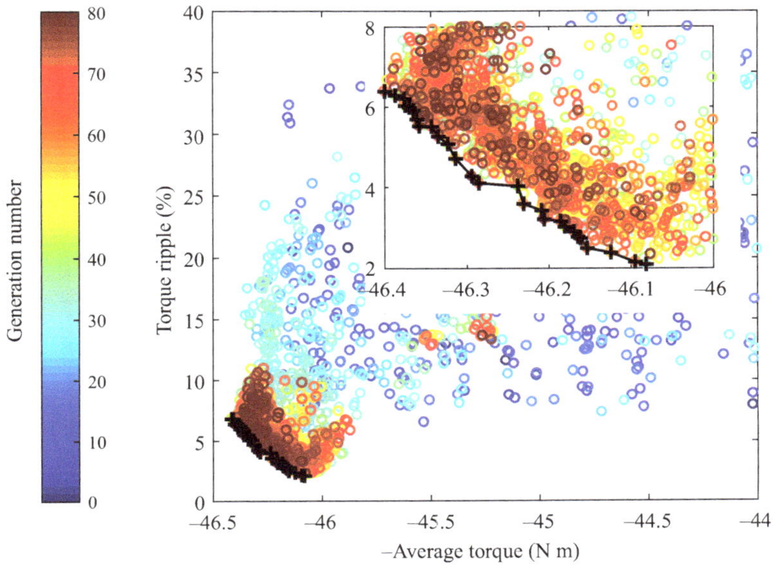

Figure 14.16 Simulated motors in the objectives plane and Pareto front at 20 krpm

near the angles $(\vartheta_{b1}, \vartheta_{b2}, \vartheta_{b3}) = (17^\circ, 29^\circ, 34^\circ)$. The value of the fourth parameter, k_{air}, is close to 0.42. The Pareto points are all close to the aforementioned angles, and slight deviations from these values cause an increase in the torque ripple and/or the leaving of the front.

Moreover, Figure 14.17(a) shows that the flux-barrier angles are not evenly spaced since the third flux-barrier angle is close to the second one. In fact, the third angle is directly related to the average torque so, generally, the greater the third flux-barrier angle the higher the torque. However, the decrease of such an angle permits to reduce the respective island mass and, as a consequence, the rib thickness. It is clear that a compromise solution has been found by the optimizer.

14.4.3.2 Sensitivity analysis of the results

Figure 14.17(b) shows the sensitivity of the optimization results for two different radii, $R_1 = 0.1$ and $R_2 = 0.3$. The radius has to be sufficiently small to correctly evaluate the sensitivity of each point, but also sufficiently large to find other points around each one. Figure 14.17(b) also reports the tendency lines of the sensitivity, which would otherwise be very scattered. These lines clearly point out how the lower torque ripple values are more sensible to geometric variations of the chosen parameters. Additionally the larger the variation—so the larger R—the higher the torque ripple change. This fact highlights the importance of good manufacture of the rotor, which therefore needs small tolerances.

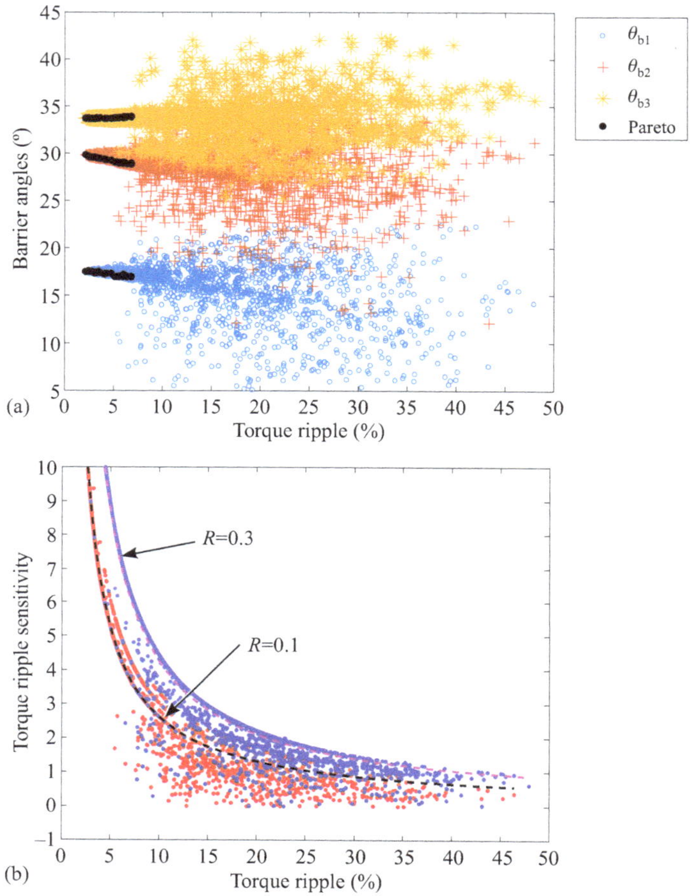

Figure 14.17 Plots resulting from the multi-objective optimization procedure: (a) angles of flux-barrier-ends versus torque ripple at 20 000 rpm, (b) torque ripple sensitivity versus the torque ripple percentage for different radii at speed 20 000 rpm

14.4.3.3 Structural analysis of the optimized rotor

The structural integrity of one optimal individual has been verified with FE using GetDP [154]. The von Mises stress map is shown in Figure 14.18. The acceptable rib stress limit is considered to be 200 MPa. It can be noted that the average iron rib stress is always lower than the maximum limit according to the preliminary design. Higher stress values are concentrated in the proximity of the edges.

The tangential iron ribs of the first flux-barrier (the outer one) are not stressed while the other ones present values higher than 200 MPa. This is due to the local deformation of these ribs that do not participate in counteracting the centrifugal force though.

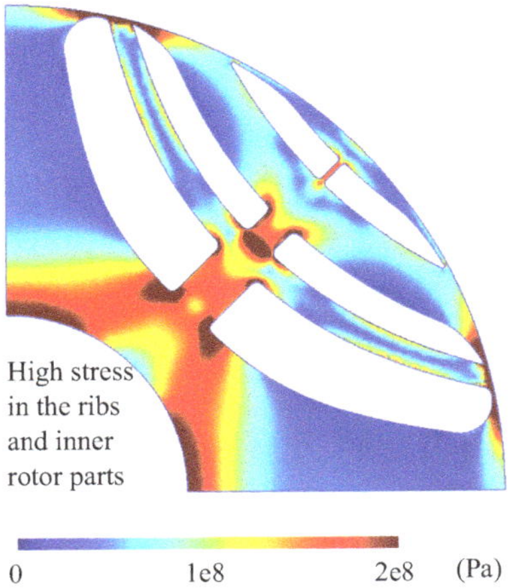

Figure 14.18 Map of the von Mises stress of the optimized rotor for 20 000 rpm

Table 14.5 Average torque for different thicknesses of the sleeve at 20 krpm

t_{sleeve}	[mm]	0	0.25	0.5
$\langle T_{\mathrm{m}} \rangle$	[Nm]	46.3	43.9	42.1

The maximum displacement is about 83 μm, and it has been observed in the third island, while close to the air gap the value is about 70 μm. This results in a 7% variation of the air-gap thickness, which is quite negligible.

14.4.3.4 Optimizations with sleeve

The same optimization has been carried out with the following sleeve thicknesses: 0.25 mm and 0.5 mm. Hence the effective gaps are 1.25 mm and 1.5 mm, respectively.

The corresponding sleeve force is obtained through (14.20). The force that the jth radial rib has to sustain is decreased by $F_{\mathrm{c,sleeve}}$. If such difference is negative, the thickness of the jth rib is set equal to the lamination width. Finally, the radial iron ribs are computed from (14.18), neglecting the tangential rib contribution as well.

The results, reported in Table 14.5, show that the adoption of the sleeve greatly reduces the available torque. This aspect has been verified with the optimistic assumption of no sleeve precompression. In fact, the rotor iron ribs should be designed according to the precompression stress at zero speed; otherwise lateral deflection of radial ribs might occur. This requires thicker radial ribs, decreasing the machine anisotropy, hence the average torque. Therefore, a rotor with a precompressed sleeve would have an worst reduced performance.

14.4.3.5 Salient pole comparison

The performance of a synchronous salient-pole rotor with the same stator has also been investigated. This structure allows one to avoid the mechanical constraints of the flux-barrier geometry. The pole shoe spans 50% of the pole pitch. The average torque results about 42 N m with a torque ripple higher than 78% so that a rotor skewing is mandatory. With a skewed rotor, the torque oscillation has been reduced to 4.1% with an average torque of about 40 N m.

Even if the salient pole is mechanically robust, for the rated speed, it does not represent a valid alternative to the flux-barrier rotor since the average torque is lower.

14.4.4 Optimization at higher speeds

Further optimizations at different speeds are presented hereafter. The aim of these investigations is to determine the influence of the speed on the overall performance and on the rotor flux-barriers shape. The stator geometry is kept fixed with and the same slot current is imposed. As in the previous case, the optimization is focused on the rotor geometry.

In Figure 14.19(a) the objectives plane of the optimization at 30 000 rpm is reported. It can be noted that the average torque is lower than before and it is around 36 N m. In addition, the minimum torque ripple achievable is slightly higher than the one at 20 000 rpm. The Pareto front is very steep and different from the front obtained in Figure 14.16.

The optimization has been repeated for a speed of 40 000 rpm. The objective plane is reported in Figure 14.19(b). As expected, the average torque results in about 26 N m and is lower than the previous optimizations since the leakage flux of the radial iron ribs is higher.

14.4.5 Comparison of the optimization results

Figure 14.20 shows the flux density plots of the selected individuals at the speeds 20 krpm, 30 krpm and 40 krpm. It is worth noticing that the radial iron ribs increase as the speed increases because of the higher centrifugal load. Furthermore, the magnetic insulation ratio, k_{air}, increases too. This effect is because the optimizer finds the compromise between wider ribs and lighter islands. It can also be noted that the saturation of the radial iron ribs decreases as the speed increases. On the contrary, the iron paths between the flux-barriers become more and more saturated. The behaviors of the torque versus the rotor position of the optimal individuals are reported in Figure 14.21 for the different speeds. For each plot, the average torque and the ripple are also reported. It can be noted that the average torque at 20 000 rpm is higher than the other ones. In particular, at 30 000 rpm the average torque is about 21% lower while at 40 000 rpm the decrease is about 43%. Furthermore, it can be observed that the torque ripple tends to increase with the speed. In fact, at the lower speed, it is 4.77% while at 40 000 rpm it is 12.88%. In Table 14.6 the d- and q-axis inductances, the saliency ratio (ξ), the difference between the inductances and the power factor are reported for motors designed for different speeds. All values are expressed in per unit, referring to the direct axis inductance obtained with the lowest radial rib thicknesses,

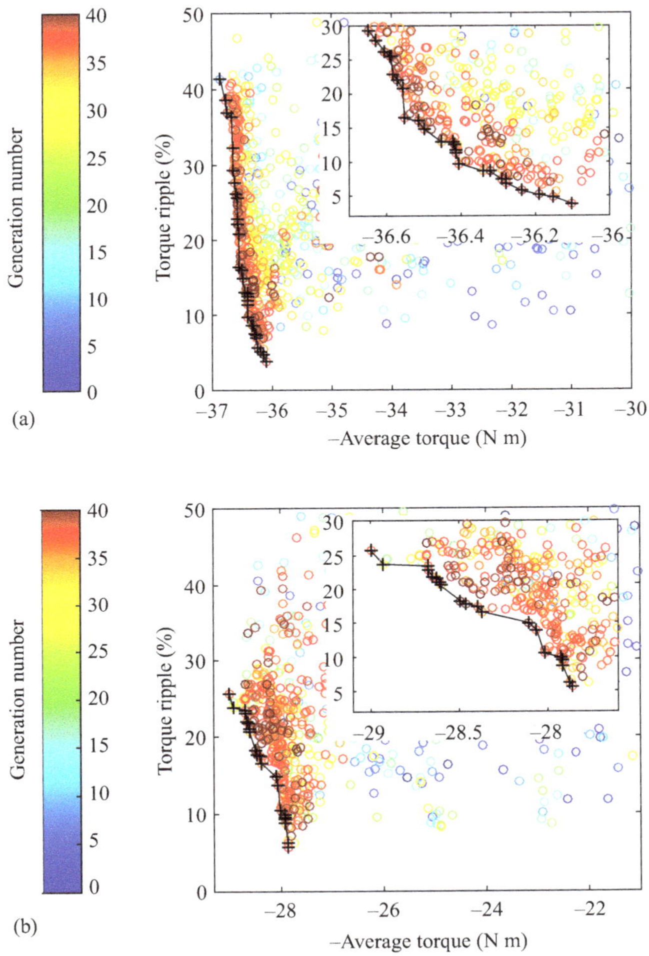

Figure 14.19 Simulated individuals in the objectives plane (average torque and torque ripple) and Pareto front. Two different speeds have been considered: (a) 30 000 rpm and (b) 40 000 rpm

i.e. for motor running at 1 500 rpm. It is possible to observe that at the highest speed the d-axis inductance is 30% lower than the reference one. The q-axis inductance increases of about 120%, which is essentially due to the larger radial ribs' thicknesses. The reduction of rotor anisotropy related to the speed is evident considering both the saliency ratio and, mainly, the difference between d- and q-axis inductances. With respect to the reference one, the former parameter decreases about 68% while the latter 57%. The power factor worsens with the speed decreasing from 0.652 to 0.339.

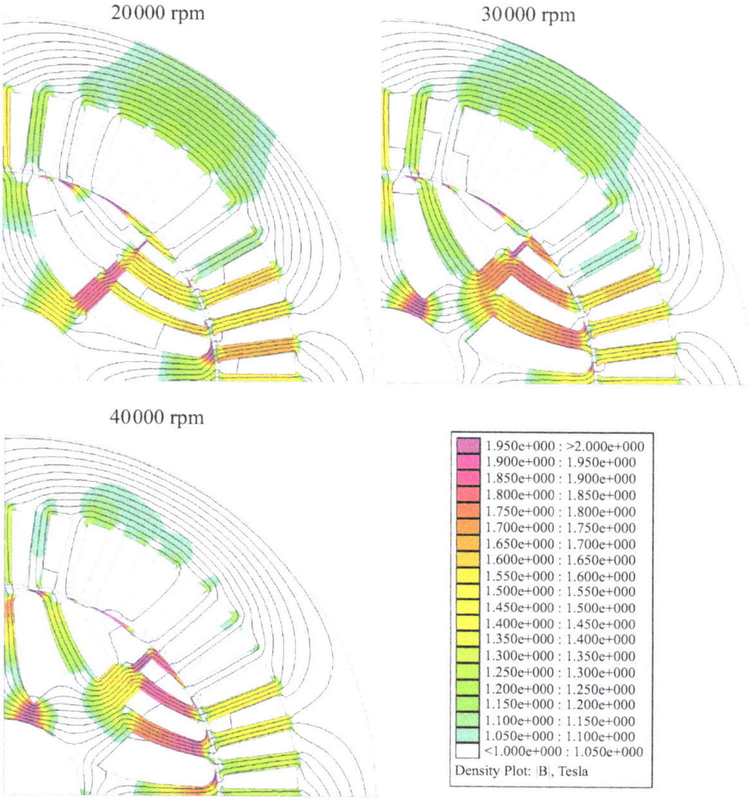

Figure 14.20 Comparison of the flux-density plots of the optimized motors for different speeds

This aspect can be explained considering the flux density plots at different speeds reported in Figure 14.20. It is evident that, at low speed, the radial ribs are saturated and so the iron islands can be considered magnetically isolated from each other. Conversely, as the speed increases, the radial ribs are no more saturated and so the islands are far from being magnetically isolated.

The results of all the optimizations at different speeds are summarized in Figure 14.22. It can be noted that the maximum output power is a function of the speed.

It increases up to the speed 37 500 rpm, then it decreases for higher speeds. This aspect represents a limit of the transverse-laminated reluctance machines for the geometry under investigation.

Another interesting aspect that can be deduced from Figure 14.22 is that the magnetic insulation, measured by k_{air}, increases as the speed increases. Clearly, such increase leads to a decrease of the overall iron mass and, consequently, a decrease of the rib thicknesses.

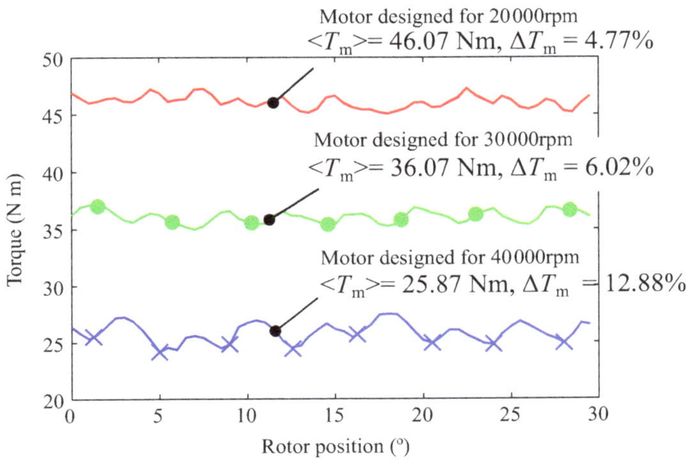

Figure 14.21 Torque versus rotor position waveforms for the motors optimized at different speeds

Table 14.6 Per-unit d- and q-axis inductances and saliency ratios for different rated speeds. The reference value is the d-axis inductance at 1 500 rpm.

n [rpm]	L_d [p.u.]	L_q [p.u.]	ξ [1]	$L_d - L_q$ [p.u.]	PF [1]
1 500	1	0.1568	6.3785	0.8432	0.65152
20 000	0.9371	0.2673	3.5	0.6699	0.54174
30 000	0.8097	0.3125	2.5913	0.4972	0.42722
40 000	0.7075	0.3483	2.03	0.3592	0.33874

14.4.6 Discussion

This section dealt with the investigation of the potential and the optimization of the geometry of HS SyR motors. The optimization was focused on the rotor geometrical parameters, such as flux-barrier angles and insulation ratio. The goals were high-power and low-torque ripple.

It has been verified that, at rated speed of 20 000 rpm, the flux-barrier geometry is the most suitable solution. Alternative solutions, such as the use of a retaining sleeve and a salient pole rotor, do not meet the requirements. A sensitivity analysis on the optimization results has also been carried out. It has been observed that low-ripple solutions have a high sensitivity and thus require a high manufacturing precision.

To the aim of investigating the potential of the flux-barrier SyR motor, the analysis has been carried out for HSs. Because of the increase of the rib thicknesses with the

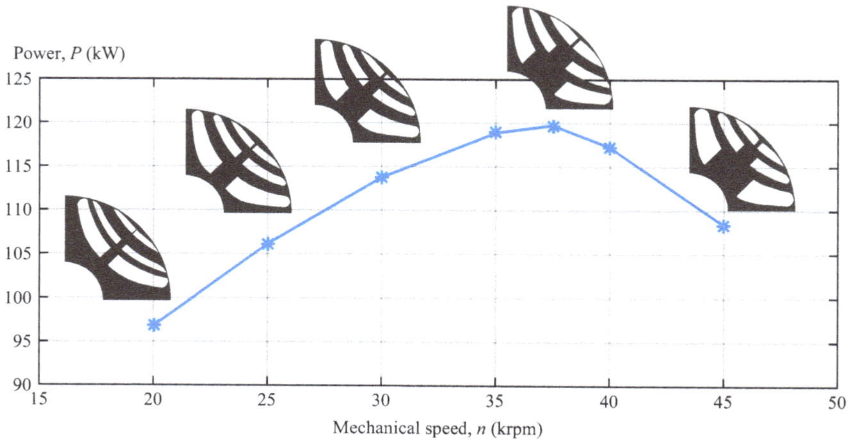

Figure 14.22 Maximum output power versus mechanical speed. For each speed, the rotor geometry is different, as obtained by specific optimization.

speed, the motor with flux-barrier rotor presents a power limit, corresponding to a speed close to 37 500 rpm. Furthermore the higher rib thickness worsens the rotor anisotropy and consequently the power factor.

14.5 Comparison of high-speed synchronous reluctance and PM machines

In this section, a performance comparison between an SyR machine and an SPM machine designed for HS application is carried out. The considered output power is 120 kW at the nominal speed of 20 000 rpm. The stator geometry and winding are fixed for both machines, while the rotors are properly designed. The main motor data are reported in Table 14.7.

A motor drive system, consisting of an electric motor directly coupled to an HS load, is shown in Figure 14.23. A three-phase inverter controls the power flow from the DC bus to the electric motor, operating at high variable frequency corresponding to the HS application. The current time harmonics are also qualitatively reported in the inverter output.

14.5.1 Surface-mounted PM motor

This subsection deals with the design of the SPM motor. Its geometry is sketched in Figure 14.24 and the main data are reported in Table 14.8.

Table 14.7 Main data of both machines

Motor output			
Rated power	P_N	120	kW
Frequency	f	667	Hz
Mechanical speed	n	20 000	rpm
Geometrical data			
Outer diameter	D_e	240	mm
Inner diameter	D_s	140	mm
Stack length	L_{stk}	140	mm
Number of poles	$2p$	4	–
Number of slots	Q	48	–
Slot height	h_s	25	mm
Tooth width	w_t	5	mm
Mechanical air gap	g	1	mm
Winding			
Conductors in slot	n_c	6	–
Number of parallel paths	n_{pp}	4	–
Number of conductors per phase	N_s	24	–
Coil pitch	y_q	10	–
Materials			
Iron lamination			N020
Solid steel (SPM rotor)			C45
Permanent magnet			N38UH
Sleeve			Carbon fiber

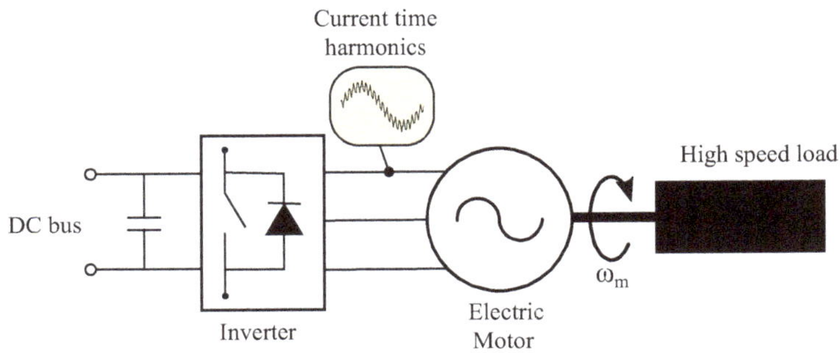

Figure 14.23 High-speed motor system

14.5.1.1 Lamination characteristics

Due to high rated frequency of the machine, an NO20 lamination, characterized by 0.2 mm thickness and 0.52 μΩ m resistivity, has been chosen. In this way, it is possible

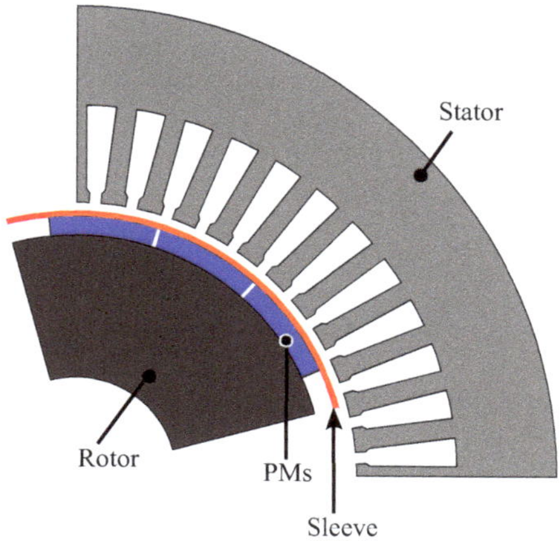

Figure 14.24 Sketch of SPM motor with discretized magnets and retaining sleeve

Table 14.8 Characteristic data of SPM *motor*

SPM motor data			
Rated line to line voltage	$\hat{V}$	520	V
Rated phase current (peak)	$\hat{I}$	230	A
Carbon fiber Young's modulus	E_{sleeve}	132	GPa
PMs coefficient of thermal expansion	$\lambda_{T,PM}$	10e−6	1/K
Iron coefficient of thermal expansion	$\lambda_{T,fe}$	12e−6	1/K
Carbon fiber bandage thickness	t_{sleeve}	2.5	mm
Magnet thickness	t_{PM}	8	mm

to reduce eddy-current losses. The lamination is characterized by 12.2 W/kg specific loss at $B = 1\ T$ and 29.7 W/kg specific loss at $B = 1.5\ T$.

14.5.1.2 Permanent magnet characteristics

PMs have to resist to high temperature since there is no rotor cooling system. For this reason, an NdFeB magnet UH grade has been chosen. The residual flux density at estimated operating temperature of 120°C is B_{res}=1.12 T. However, the magnetic field at the knee of *BH* curve decreases down to 800 kA/m. PM thickness has been chosen equal to $t_{pm} = 8$ mm, to avoid the risk of irreversible demagnetization. The PMs cover 80% of the pole pitch. To reduce eddy-current rotor losses, they have been segmented in both radial and axial directions and insulated one from the others.

14.5.1.3 Sleeve design

A retaining sleeve is necessary to oppose the centrifugal force and keep PMs on the rotor since glue is not enough due to the high operating temperature and the HS of

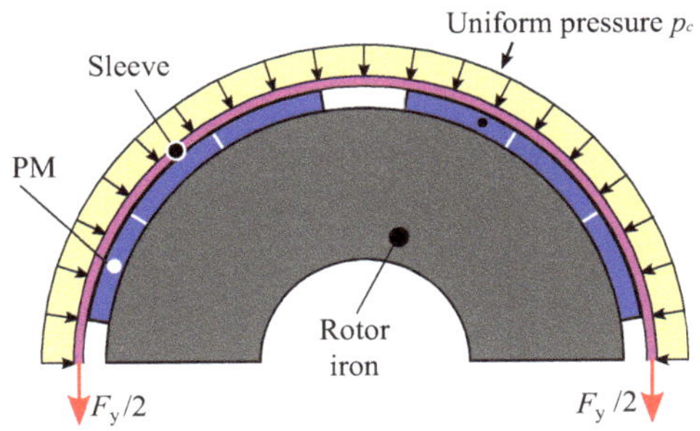

Figure 14.25 Computation of the SPM sleeve

rotation. The chosen sleeve material is carbon fiber. It is a nonconductive material characterized by a low mass density and high traction strength ($\sigma_{t,max} \approx 2\,500$ MPa). However, it exhibits some drawbacks. First of all its thermal conductivity is quite low, so that it can be considered as a thermal insulator. As a consequence, the rotor cooling is quite difficult since the heat mainly flows in the axial direction. Then it is an anisotropic material and its Coefficient of Thermal Expansion (CTE) is negative, so that an additional stress occurs in the sleeve.

The thickness of this sleeve is designed on the basis of total stress at the highest speed and maximum temperature with the following procedure (Figure 14.25). According to the centrifugal force of the sleeve and PMs, the corresponding pressures are

$$p_{\text{sleeve}} = \rho_{\text{sleeve}}\omega_{\text{m}}^2 t_{\text{sleeve}} R_{\text{sleeve}} \tag{14.21}$$

$$p_{\text{PM}} = \rho_{\text{PM}}\omega_{\text{m}}^2 t_{\text{PM}} R_{\text{PM}} \tag{14.22}$$

where ω_{m} is the rotor speed, ρ_{sleeve}, ρ_{PM}, R_{PM} and R_{sleeve} are the mass density and the radius of the sleeve and PM, respectively.

As a design choice, the sleeve has to ensure a zero residual contact pressure p_{c} between magnets and rotor iron at 20% over-speed, without exceeding the maximum allowable tangential stress. This is possible imposing a prestress pressure p_{pre} during manufacture; thus the total contact pressure is

$$p_{\text{c}} = p_{\text{pre}} - p_{\text{PM}} - p_{\text{sleeve}} \tag{14.23}$$

the corresponding prestress results to be

$$\sigma_{\text{pre}} = \frac{p_{\text{pre}} R_{\text{sleeve}}}{t_{\text{sleeve}}} \tag{14.24}$$

An additional stress has to be considered in order to take into account the radial expansion of the rotor iron and PMs due to the temperature rise. In fact, their CTE is higher than the carbon fiber one.

The radial expansion of the rotor iron and PMs is

$$\Delta R_{\text{th.exp}} = R_{20}\,\lambda_T\,\Delta T \tag{14.25}$$

where R_{20} is the rotor radius with no sleeve at 20°C, λ_T is the material rotor CTE (reported in Table 14.8) and ΔT is the temperature increase. The corresponding stress can be computed as follows:

$$\sigma_{\text{th.exp}} = E_{\text{sleeve}} \frac{\Delta R_{\text{th.exp}}}{R_{\text{sleeve}}} \tag{14.26}$$

where E_{sleeve} is the Young modulus of the sleeve material. Finally, the total stress in the sleeve is

$$\sigma_{\text{tot}} = \sigma_{\text{pre}} + \sigma_{\text{th.exp}} \tag{14.27}$$

For the rotor under analysis the sleeve thickness results in $t_{\text{sleeve}} = 2.5$ mm. Thus, assuming a mechanical air gap of 1 mm, the total magnetic air gap results in 3.5 mm.

14.5.1.4 PWM current harmonics

The current harmonics due to the inverter are computed considering the electric circuits shown in Figure 14.26. The inverter is represented by the voltage generator $v_{\text{An}}(t)$ while the motor is modeled with an inductance, L, and a sinusoidal voltage generator $e_{\text{A}}(t)$. The voltage generated by the PWM inverter is the sum of the fundamental-frequency component and its ripple:

$$v_{\text{An}}(t) = v_{\text{An},1}(t) + v_{\text{ripple}}(t) \tag{14.28}$$

therefore, the current ripple is computed as follows:

$$i_{\text{ripple}}(t) = \frac{1}{L}\int_0^t v_{\text{ripple}}(\zeta)\,d\zeta + i_0 \tag{14.29}$$

The resulting waveform and the ripple spectrum of the SPM motor current are reported in Figure 14.27(a) and (b), respectively. It can be noted that the highest current harmonics are located close to the modulation ratio, which is $m_{\text{f}} = f_{\text{sw}}/f = 15$, and its multiples (i.e. 30, 45, etc.). The inverter switching frequency, f_{sw}, is set to 10 kHz.

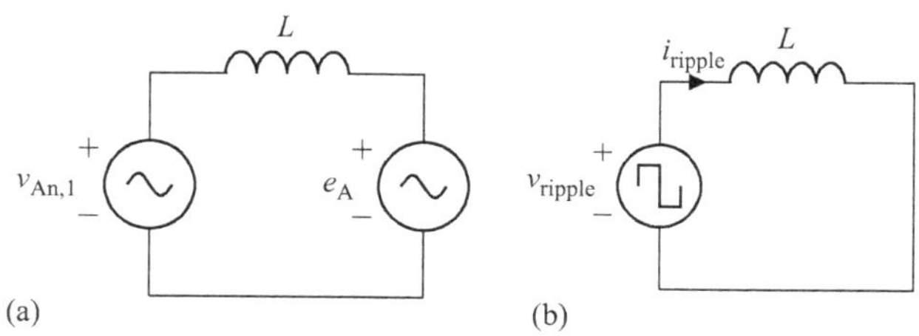

Figure 14.26 Single-phase equivalent circuit of the PWM motor supply: (a) fundamental-frequency component, (b) ripple-frequency component

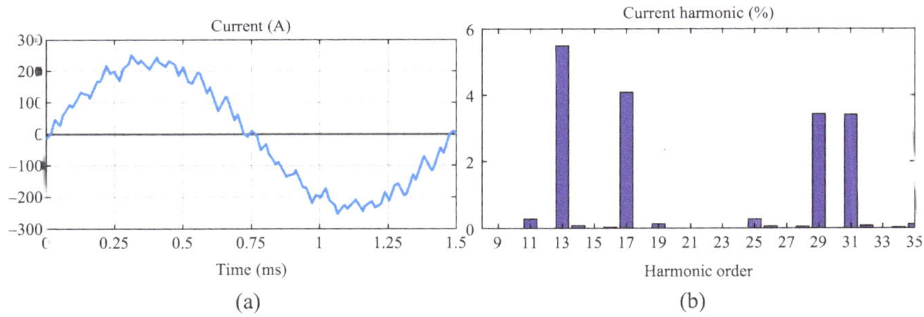

Figure 14.27 (a) Current waveform and (b) current ripple spectrum of an SPM motor, according to the switching frequency $f_{sw} = 10\,kHz$

Table 14.9 Time harmonic rotor losses of and SPM motor

ν –	f_r [kHz]	I [A]	P_{Fe} [W]	P_{PM} [W]	P_{tot} [W]
13	9.33	12.32	155	57	212
17	12	9.24	110	35	145
29	20	7.7	93	31	124
31	20	7.5	88	29	117

14.5.1.5 Computation of the rotor losses

The rotor losses are due to eddy currents induced in the rotor conductive parts. The three main electromagnetic sources of such losses are:

1. current time harmonics;
2. MMF space harmonics;
3. slot openings.

Time harmonics

A time harmonic simulation has been carried out for each current component computed in Section 14.5.1.4. PMs segmentation has been considered imposing the total net current flowing through each magnet to be zero. The resulting losses in the solid iron and PMs are reported in Table 14.9. Most of the losses are in the rotor iron rather than in the PMs.

Space harmonics

The effect of MMF harmonics is taken into account replacing the stator with a uniform current sheet. The stator electric loading is split into its Fourier series. The amplitude of each harmonic and the corresponding frequency are reported in Table 14.10. The total losses in the rotor iron and PMs are also reported. It can be seen that the higher losses occur for multiples of slot harmonics, which are of order $Q/p \pm 1$. Anyway,

Table 14.10 Losses produced by the MMF *space harmonics in the SPM motor*

h –	K_s [A/m]	f_h [Hz]	Losses [W]
1	50	0	0
5	2.86	4 000	5.00
7	2.18	4 000	0.88
11	6.42	8 000	3.37
13	6.35	8 000	1.40
17	2.07	12 000	0.07
23	44.57	16 000	9.38
25	43.63	16 000	6.13
29	2.39	20 000	0.01
31	1.78	20 000	0.01
35	5	24 000	0.05
37	4.83	24 000	0.04
41	1.5	28 000	0.00
43	1.87	28 000	0.01
47	29.62	32 000	1.80
49	28.12	32 000	1.62

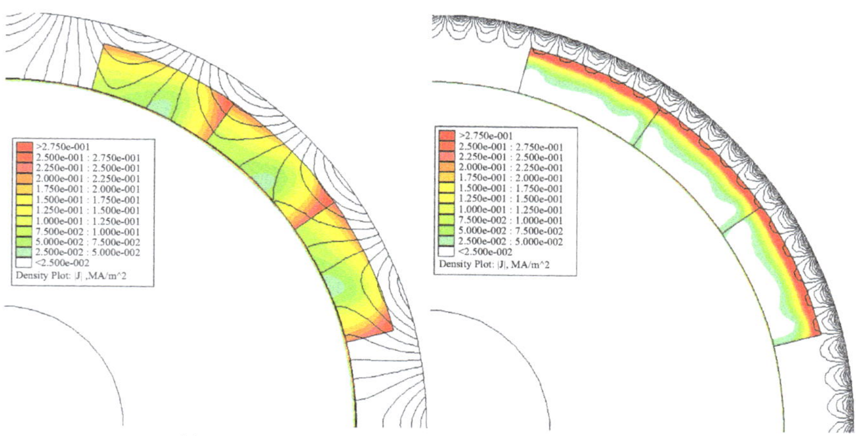

(a) (b)

Figure 14.28 Time harmonic simulations of MMF *space harmonics with an SPM rotor: (a)* $\nu = 5$*, (b)* $\nu = 23$

comparing the losses in Tables 14.9 and 14.10, it is noticed that space harmonic losses are quite lower than time harmonic losses.

The eddy-current density plots due to 5th and 23rd current harmonic are reported in Figure 14.28. As expected the higher the harmonic order, the lower the skin depth.

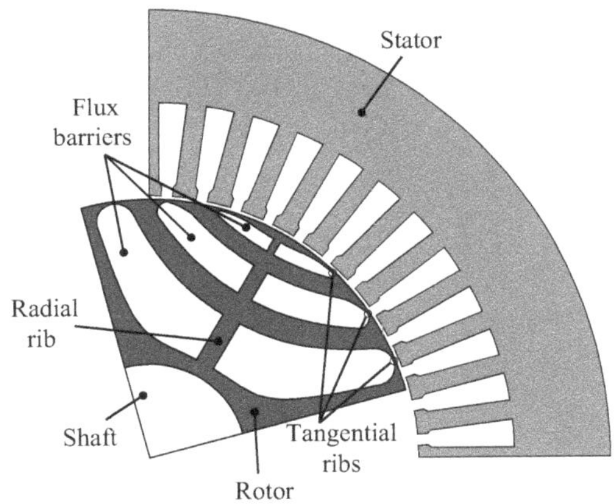

Figure 14.29 Synchronous reluctance machine structure

Slot opening harmonics

The variation of air-gap flux density due to the slot openings causes rotor eddy currents. This variation can be analyzed adopting a slotless stator and imposing a linear current density along the inner stator circumference. Due to the high air gap, these losses are also negligible. The resulting rotor losses computed with time harmonic simulations are 50 W.

14.5.2 Synchronous reluctance motor

A sketch of a four-pole SyR rotor with three flux-barriers per pole is shown in Figure 14.29. It is possible to identify the radial iron ribs (in the middle of the barriers) and the tangential iron ribs (close to the air gap).

Due to HS, the rotor has to be mechanically robust. This aspect can be achieved (a) increasing the thickness of the rotor iron ribs and (b) adopting a retaining sleeve, similarly to the SPM machine.

Both methods present drawbacks. The thick iron ribs represent magnetic short circuits until saturation is reached, so that a magnetic flux flows through such iron paths reducing the available machine torque [153]. The retaining sleeve increases the air-gap width reducing the inductances and, consequently, the torque.

14.5.2.1 Optimization without sleeve

Due to the complex rotor geometry design, the optimization procedure implements a multi-objective DE algorithm. Furthermore, the magnetic analysis of the motors is performed by means of the FE method. For the SyR machine, the optimization is often needed in order to reach a proper average torque and torque ripple. The optimization is carried out with the same stator geometry of the SPM but with a higher electric loading (53.5 kA/m). The objectives of the optimization are to maximize the average torque and to minimize its oscillation (ripple).

The chosen rotor parameters are the barrier-end angles ($\vartheta_{b1}, \vartheta_{b2}, \vartheta_{b3}$) and the magnetic insulation ratio k_{air}, defined as follows:

$$k_{air} = \frac{2\sum_i t_{bi}}{D_r - D_{sh}} \tag{14.30}$$

where t_{bi} is the thickness of the ith flux-barrier and D_{sh} is the shaft diameter.

The optimization has been carried out with 60 generations consisting of 30 individuals. For each individual the radial ribs are designed according to (14.18).

The optimization results are reported in Figure 14.30(a), where the objective plane is shown together with the Pareto front (black markers). The average torque with negative sign is reported on the x-axis while the torque ripple is reported on the y-axis. It can be observed that the Pareto front is steep, since the torque ripple has high variations around similar average torque values. In fact, the torque ripple varies from

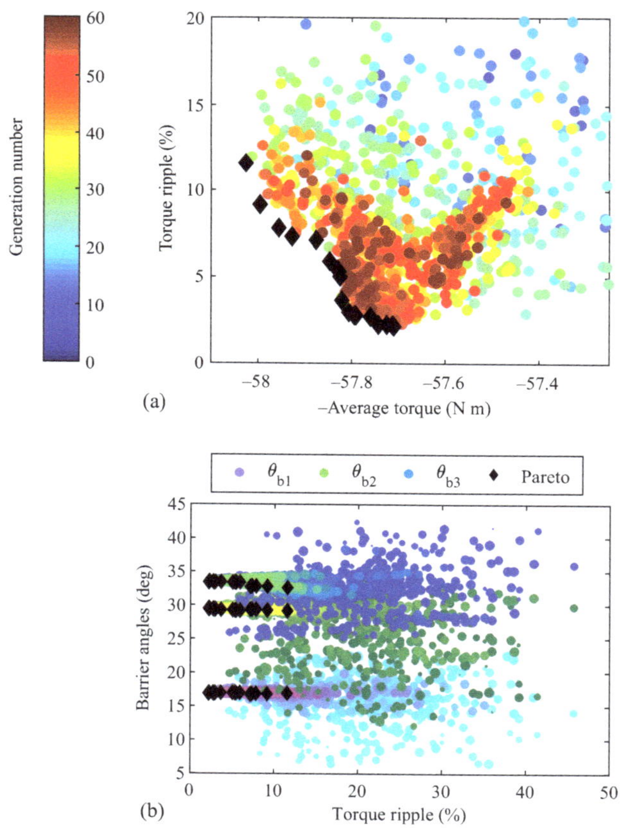

Figure 14.30 Results of the SyR motor multi-objective optimization: (a) objective plane and Pareto front and (b) end-barriers angle versus torque ripple

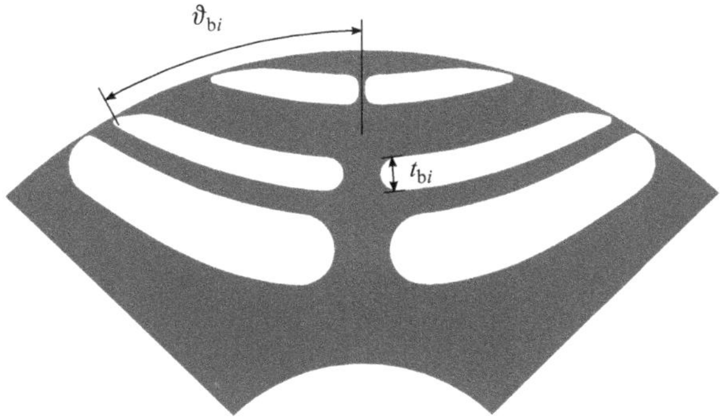

Figure 14.31 Optimized REL rotor geometry without sleeve

2% to 12% while the average torque is always between 57.7 Nm and 58 Nm. Thanks to this fact, the motor geometry with the lowest torque ripple has been selected.

Figure 14.30(b) shows the barrier-end angles versus the torque ripple. It also shows the values of k_{air} through the marker size (the larger the point, the smaller k_{air}) and the age of the individual through the color. The black markers belong to the Pareto front. The barrier-end angles corresponding to the lowest torque ripple are quite defined close to the values $(\vartheta_{b1}, \vartheta_{b2}, \vartheta_{b3}) = (16.9°, 29.4°, 33.5°)$, and the magnetic insulation ratio, k_{air}, approaches 0.41.

The optimized geometry is shown in Figure 14.31. It can be noted that the barrier-end angles are not evenly distributed along the rotor periphery unlike the common design rule used in [95,155]. In this case, the second and the third flux-barriers are quite close to each other and so the second iron island is wider than the third island thickness. The DE algorithm tends to decrease the rib thickness, so as to reduce the flux flowing along the q-axis that reduces the average torque. To do that, the optimizer converges to geometries with smaller values of the third barrier angle in order to reduce the iron mass. In general, the thickness of radial ribs is quite large due to the HS of the rotor.

Optimization with sleeve

The same optimization has been carried out with the presence of the carbon fiber sleeve. Its thickness has been set to 1 mm, and the iron ribs have been designed accordingly. Although the thickness of the iron ribs result in quite thin, the torque reduces strongly due to the higher air gap. Figure 14.32 shows that this solution is not suitable to reach the target torque. Therefore, this solution is not considered in the following analysis. The torque behavior versus the rotor position of the optimized SyR rotor with and without sleeve is shown in Figure 14.33.

14.5.2.2 Structural analysis

The von Mises stress and the displacement plots are reported in Figure 14.34. The full-scale value is set to 200 MPa, representing the acceptable rib limit, as said earlier.

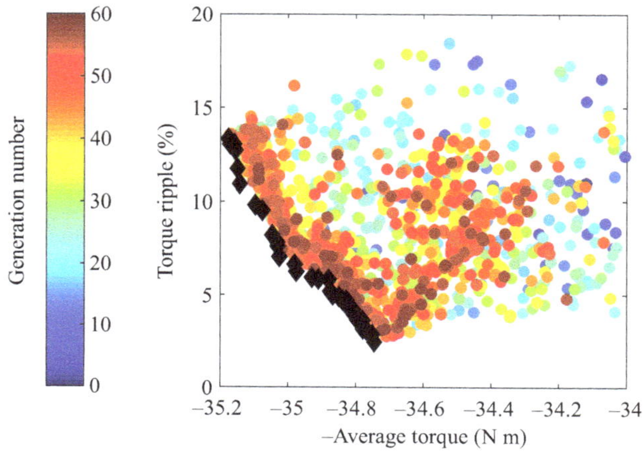

Figure 14.32 Results of the SyR motor optimization with sleeve

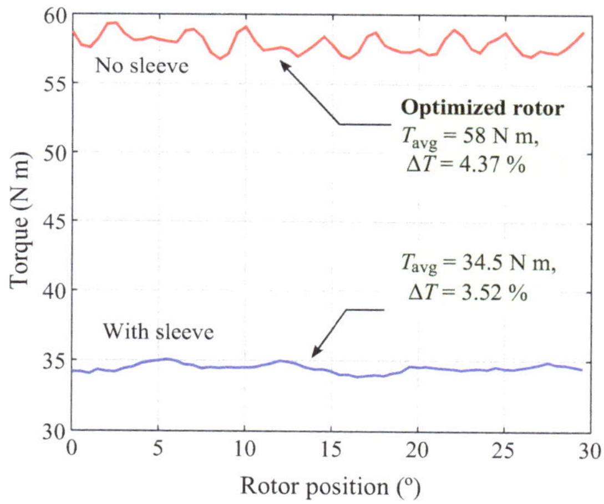

Figure 14.33 Torque versus rotor position with different SyR rotors

The lamination stress is concentrated close to the tangential ribs, the radial rib sides and the shaft diameter. The radial ribs have been rounded in order to avoid the concentration of stress. In general, the stress levels are always lower than the maximum limit and this proves a good accuracy of the proposed method described in Section 14.4.1.2.

14.5.2.3 Computation of the rotor losses

Time harmonic

The current harmonics have been computed with the procedure discussed in Section 14.5.1.4. The current ripple spectrum is reported in Figure 14.35. It should be noted

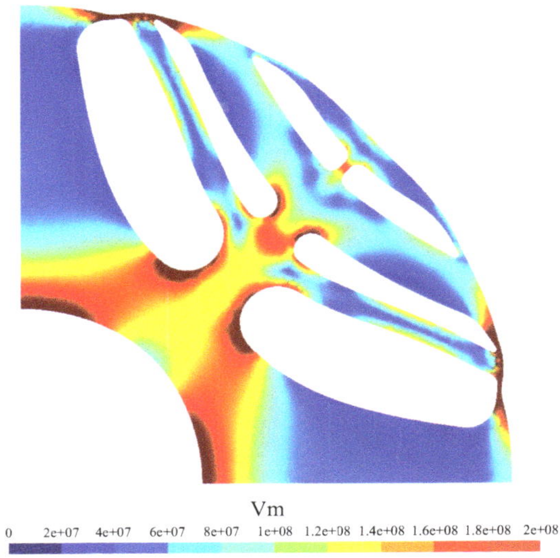

Figure 14.34 The von Mises and displacement plots of the REL rotor at 20 krpm

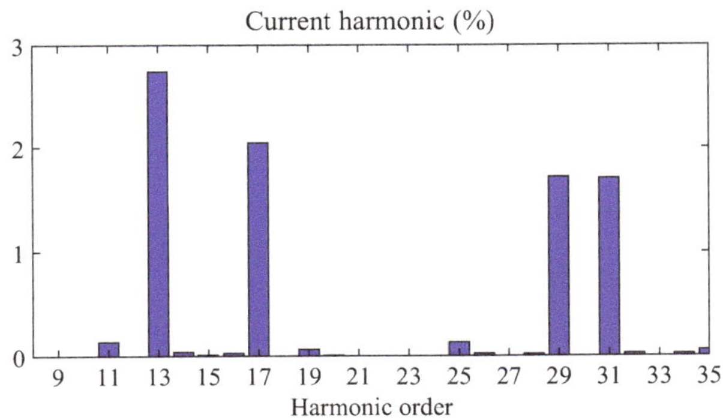

Figure 14.35 Current ripple spectrum of SyR motor

that the current ripple of the SyR motor is about 50% of the SPM one since the SyR inductances are higher. This is due to the lower air-gap length. The iron losses produced by the highest current harmonics are reported in Table 14.11.

Space and slot opening harmonics

The computation of the losses produced by the space harmonics and the slotting effect have been carried out as reported in Section 14.5.1.5. The numerical results are not reported since their magnitude is negligible.

Figure 14.36 shows the results of time harmonic simulations for $\nu = 5$ and $\nu = 23$. The flux lines in Figure 14.35 flow through the inner part of the rotor while with the higher harmonic, the flux lines are shielded near the rotor periphery.

Table 14.11 Time harmonic rotor losses of SyR motor

ν –	f_r [kHz]	I [A]	P_{tot} [W]
13	9.33	9.1	183
17	12	6.6	185
29	20	5.6	276
31	20	5.6	276

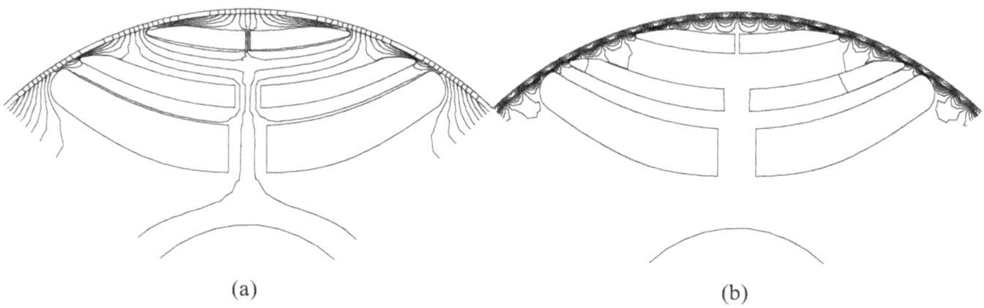

Figure 14.36 Time harmonic simulations of space MMF harmonics with an SyR rotor: (a) $\nu = 5$ and (b) $\nu = 23$

14.5.3 Final comparison

A final comparison between the SPM and SyR motors is reported in Table 14.12. The comparison is made with the same average torque and rotor speed (so the same output power). As expected, the SyR motor requires more current to reach the same performance: the stator currents have to produce the magnetic field and to produce torque as well. This is also highlighted by the different current phase angle. Obviously, higher currents cause higher Joule losses in the winding that result in 140% higher for the SyR motor. Then the higher current leads to a higher magnetization of the machine. The peak flux densities in the stator back-iron and teeth are higher as shown in Figure 14.37. This also causes increased stator iron losses, which are about 26% higher for the SyR machine.

The higher SyR d-axis inductance is beneficial for the current ripple, which results to be quite lower. Nonetheless, the overall SyR rotor losses, essentially due to current time harmonics, are higher compared to the SPM ones due to the lower air gap that does not limit the magnetic field caused by the current harmonics.

Even though the SyR motor losses result in higher than the SPM ones, the electric efficiency is not heavily compromised, being 96.5% for the SyR and 97.6% for the SPM motor. Furthermore, the SPM machine requires particular care due to the

Table 14.12 Comparison between the SPM *and* SyR *motors*

		SPM	SyR	
Conductor current density	J_c	5.2	7.5	A/mm^2
MTPA angle	$\alpha^e_{i,\text{MTPA}}$	90	56	°
Torque	T_m	58	58	N m
Torque ripple	ΔT_m	2.99	4.37	%
Flux linkage	Λ	0.087	0.094	Vs
Inductance	L_d	230	500	μH
Inductance	L_q	230	150	μH
Tooth flux density	$B_{t,\max}$	1.49	1.8	T
Back-iron flux density	$B_{bi,\max}$	1.16	1.3	T
Losses				
Stator Joule losses	P_{Cu}	445	1 070	W
Stator iron losses	P_{Fe}	1 875	2 360	W
Rotor losses	$P_{r,\text{loss}}$	630	920	W
Total losses	P_{tot}	2 950	4 350	W
Material cost				
Lamination	C_{NO20}	73	73	€
Copper	C_{Cu}	47	47	€
Permanent magnets	C_{PM}	140	0	€
Rotor steel	C_{C45}	7	0	€
Carbon fiber	C_{CF}	73	0	€
Total cost	C_{tot}	340	120	€
Material weights				
Lamination	G_{NO20}	48.5	48.5	kg
Copper	G_{Cu}	6.7	6.7	kg
Permanent magnets	G_{PM}	1.76	0	kg
Rotor steel	G_{C45}	11.8	0	kg
Carbon fiber	G_{CF}	0.27	0	kg

temperature rise in the magnets that could be damaged. This is especially exacerbated by the presence of the carbon fiber retaining sleeve that is a bad thermal conductor.

The material weights are also reported in Table 14.12. With the nowadays specific material costs, it results in the fact that the SPM is about three times more expensive than the SyR. This is justified by the presence of the PMs and carbon fiber retaining sleeve.

14.5.4 Discussion

This section dealt with the comparison of synchronous motors for HS applications. A surface-mounted PM motor and an SyR motor have been analyzed taking into

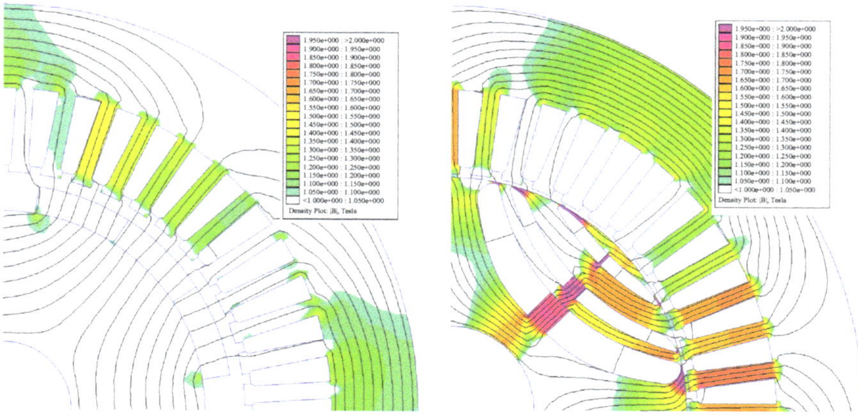

(a) (b)

Figure 14.37 Flux density plot of the analyzed motors: (a) SPM and (b) REL

account both magnetic and mechanical aspects. The effects of the current harmonics (produced by the inverter), the space harmonics and the slot openings are considered in the rotor losses computation.

The SPM motor has segmented PMs and solid rotor to limit the rotor losses and has a retaining sleeve to counteract the centrifugal force of the PMs. From the FE results, it has been demonstrated that the current harmonics have the highest contribution on the total rotor losses.

The SyR rotor has been optimized by means of a multi-objective algorithm setting the average torque and the torque ripple as objectives. The flux-barrier-end angles and the magnetic insulation ratio have been selected as optimization parameters. The thicknesses of the radial ribs have been designed according to the rotor geometry with the purpose of sustaining the centrifugal stress in the rotor iron. Since larger ribs are needed for HS applications, the machine rotor anisotropy decreases. A retaining sleeve has been considered to limit the iron rib thickness, too. However, it has been proven that such retaining sleeve greatly affects the average torque giving a reduction of about 41%.

Eddy-current losses in the rotor have been computed for both the SPM and SyR motors. From a quantitative point of view, it has been proven that the SPM total losses are about 70%, the one of the SyR one. The absences of PMs and sleeve in the SyR rotor imply a reduction of the active part costs of about 65% with respect to the SPM one.

Chapter 15
Overview of fault-tolerant SyR machines

The Fault-Tolerant (FT) reliability of an electric drive system is a mandatory requirement in those applications where the failure could cause economic costs or, more importantly, injure the human safety as in the automotive, aerospace and military applications [156–158].

The fault-tolerance of a drive system is defined as the ability to ensure proper speed or torque reference tracking in the event of fault. In other words, more the fault-tolerance is high, more the drive system performance in fault condition approaches the rated one. Faults can be first classified on the basis of their nature: mechanical and electrical. The most common mechanical faults are due to eccentricity problems [159] and misalignment of bearings [160]. From the electric point of view, the faults mainly occur in the power converter and electric machine [157,159,161]. In AC drives, a failure of the inverter switches can cause Short-Circuit (SC) of the DC bus or a disconnection of one or more phases. In electrical machines, the winding is typically subject to faults due to insulation failures (caused by thermal stress, abrasion or voltage surge), unbalanced supply conditions, a bad connection at the motor terminal, or high resistance contact.

Several solutions have been adopted to improve the FT reliability of different drive systems focusing on advanced control strategies and dedicated machine design aimed to increase their redundancy [162].

The Induction Motor (IM) is definitely the most used electric machine thanks to its robustness, reliability and mature technology. Most of the proposed solutions to increase the FT of IM drive are based on the principle of systems redundancy. Some examples are to adopt multi-phase (greater than three) windings [163–165], to make the winding neutral connection available, or to use inverters with a redundant number of controlled power switches. On the other hand, a very poor solution deals with the IM design itself. A typical IM drive system is shown in Figure 15.1(a). In the event of one-phase disconnection, the IM can be controlled in single-phase operation mode obtaining as result a very high-torque ripple component. To overcome such a problem, a two-phase operation mode has been proposed in [166,167] where the IM neutral point is connected to the middle point of the DC source capacitor bank, as in Figure 15.1(b), and the shifting angle between the healthy phases is set to $\pi/3$. Alternatively, in [168], the IM neutral point is connected to a converter fourth leg as reported in Figure 15.1(c). In this way, the FT reliability is increased; however additional components, i.e. costs, are necessary. In addition, in faulty condition the

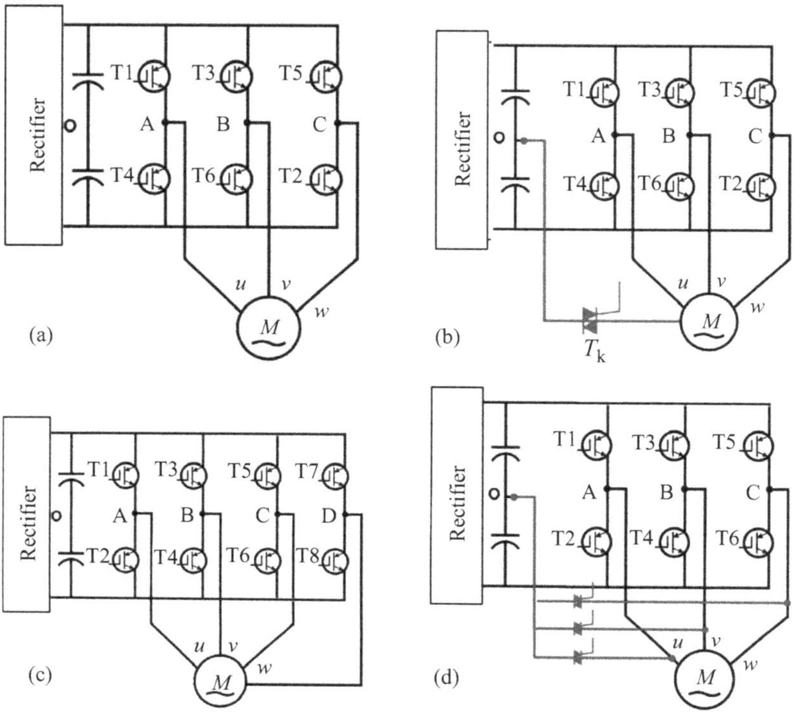

Figure 15.1 IM drive system arrangements: typical and custom solutions for maintaining the machine under balanced operation in the event of one inverter leg fault: (a) healthy condition, (b) Liu et al. solution [167], (c) Correa et al. solution [168], (d) Mendes' and Marques Cardoso's solution [169]

current amplitudes are $\sqrt{3}$ times higher than the healthy ones so that a high inverter overrating is required. A further solution proposed in [169] is to connect the IM phase terminals to the middle point of the DC bus link as in Figure 15.1(d). With this reconfiguration mode, the available voltages at the IM terminals are $\sqrt{3}$ times lower than the healthy ones. To overcome such a problem it is possible to increases DC bus voltage or the stator winding has to be changed from star connection to delta connection. This involves higher faulty current amplitudes; hence, an inverter overrating is still required. Otherwise, the IM may continue to operate in faulty condition yielding only a 50% of the rated torque. Instead of changing the IM drive system connections or adding additional components, it is possible to mitigate the characteristic low-frequency pulsating torque by injecting odd harmonic voltages at appropriate phase angles [170].

A more advanced solution is the adoption of a dual three-phase system as reported in [171]. A sketch of such a drive system is illustrated in Figure 15.2(a). It consists of two separate standard three-phase inverters and an IM with two independent

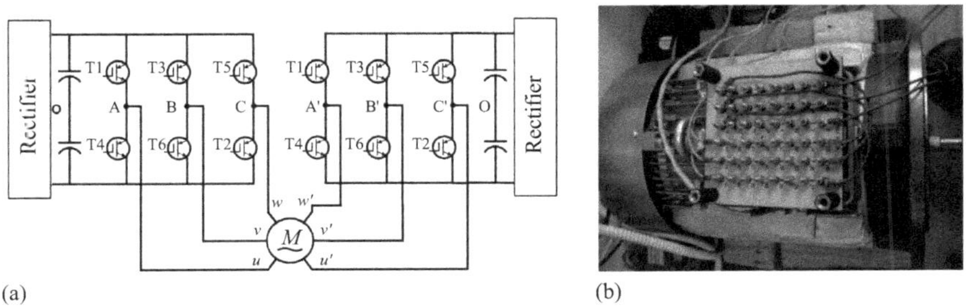

Figure 15.2 IM dual-three phase: (a) drive system [171] and (b) prototype [175]

three-phase windings. Each inverter supplies one of the two IM windings. In healthy conditions, both windings are fed and the IM produces the rated performance. In the event of fault of one of the two three-phase windings, only the healthy one is supplied while the faulty one is disconnected by means of a dedicated power switch. In this condition, the IM performance is reduced unless a proper power converter overrating is adopted. Despite the high FT reliability of this drive system, it still represents a cheaper solution with respect to the fully redundant systems, composed of two inverters and two motors, or the multi-phase machines that need dedicated power converters and switches [172–174].

Thanks to its working principle and robust structure, the Switched Reluctance (SwR) machine received great attention in FT applications. In addition to the wide variety of control techniques, some solutions that focus on the machine design can be found in literature. The simplest solution to improve its torque ripple is to increase the number of teeth [158] as shows the sketch in Figure 15.3(a). In [176], the modular SwR stator design, illustrated in Figure 15.3(b), is proposed. It consists of a certain number of modules isolated from each other by nonmagnetic spacers [176], and the rotor has a classical geometry instead. The magnetic insulators between the stator modules force the magnetic flux lines to flow in the energized module and the corresponding rotor poles only, as highlighted in Figure 15.3(b). Since the coils of each phase are fed independently and one stator slot is not shared by coils from different phases, such a solution exhibits high reliability in the case of SCs or overheating. An other interesting solution, patented by Mansir as shown in Figure 15.3(c), consists of adding an additional winding in a typical SwR salient pole rotor to compensate the unbalanced magnetic radial forces that arise in the event of stator winding fault [177]. In healthy condition, the rotor flux paths are symmetrically distributed in all rotor poles so that the resultant electromotive force and induced current in the compensating winding are zero. In faulty condition, the flux is nonsymmetrical and an electromotive force different from zero is induced in the rotor winding. Hence, a rotor current arises and its reaction flux tends to compensate the unbalanced radial force.

Although all the proposed solutions allow one to effectively increase the SwR machine FT reliability, they have no effect on the well-known low-power factor and torque density. To overcome such a drawbacks the segmental rotor topology, in Figure 15.3(d), can be adopted. In fact, it has been proven that it yields about

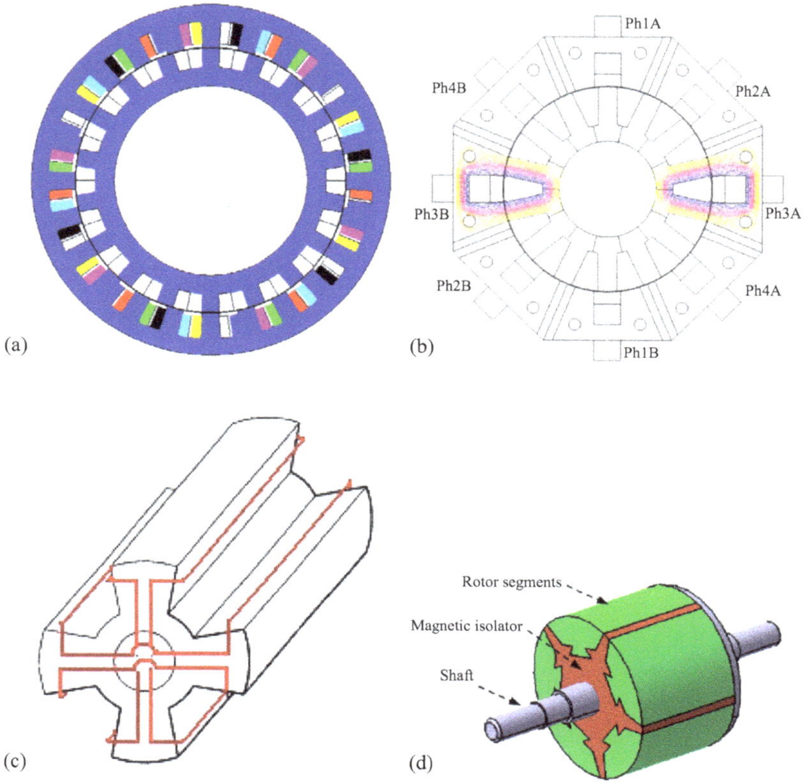

Figure 15.3 SwR *machine design solutions to improve the FT reliability: (a) a high number of teeth, (b) modular stator, (c) active rotor, (d) segmental rotor*

40% increase in torque for the same copper loss with respect to the usual toothed rotor topology [178,179]. Its rotor is constructed from a series of discrete magnetic segments insulated from each other by nonmagnetic materials.

The Permanent Magnet Synchronous Machine (PMSM) is usually the first choice in the applications where high-torque density and efficiency are required even though it presents the highest cost among the electric machines scenario. However, unlike IM and SwR machine, it does not show an intrinsic FT capability. For example, in the event of a winding SC, the faulty current is continuously fed by the rotor bemf and its amplitude may exceed of several times the rated one with a consequent risk for the Permanent Magnets (PMs) magnetization and winding insulation [180,181]. In the case of an Open-Circuit (OC) fault, dangerous high voltages are induced in the faulty phases. Several research activities have been proposed in literature to solve or mitigate these aspects. The risk of SC fault can be reduced with a complete electrical insulation between the phases driving each of them with a dedicated single-phase bridge [182]. Concerning the SC fault reliability, two alternative design solutions can be adopted.

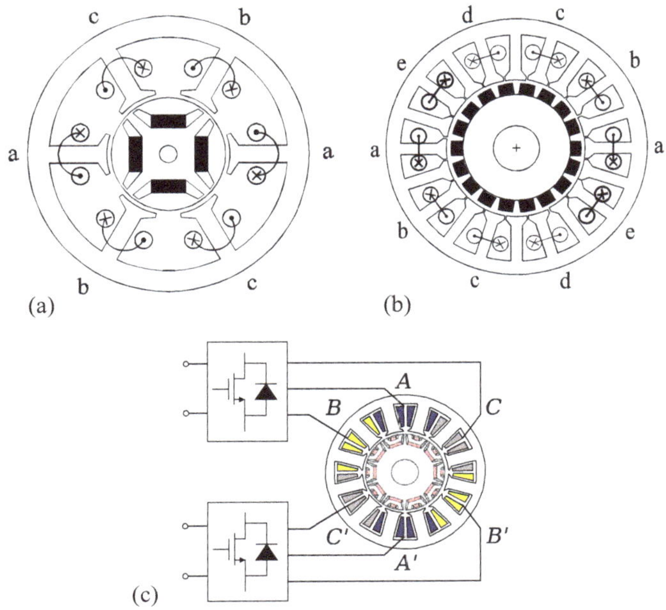

Figure 15.4 PMSM for FT applications: utilization of the fractional slot windings and multi-phase systems—(a) six-slot four-pole machine, (b) five-phase machine, (c) dual three-phase machine

The aim of the first one is to minimize the per unit inductance in order to have SC current amplitudes high enough to determine a rapid deterioration of the failure parts and, consequently, to lead to an OC condition. Conversely, in the second and more FT reliable solution, the d-axis inductance is maximized to limit the SC current amplitude below the rated one and therefore to increase the tolerance of the drive system to such a fault even for a long time. Because of its large air-gap length, the PMSM exhibits a very low inductance so that fractional slot windings are usually adopted for their high slot leakage inductance [183]. In such winding arrangements, the phase coils are wound around single teeth, as shown in Figure 15.4(a), so that the contact among the phases is reduced as well as the SC risk and overheating propagation. In addition, the fractional slot winding could be arranged to minimize the mutual magnetic coupling too [184]. On the other hand, the main drawbacks of the fractional slot configurations are a slightly lower winding factor, a high harmonic contents in the MMF distribution and, consequently, higher core and rotor losses [161,183,184].

As usual, more than three phases can be adopted to increase the drive system redundancy. The good FT reliability of a five-phase PMSM has been demonstrated in [174,185] while current control strategies and detailed thermal analyses for an analogous motor, sketched in Figure 15.4(b), have been presented in [186,187] obtaining good results in terms of FT reliability. In [188,189], the impact of different dual

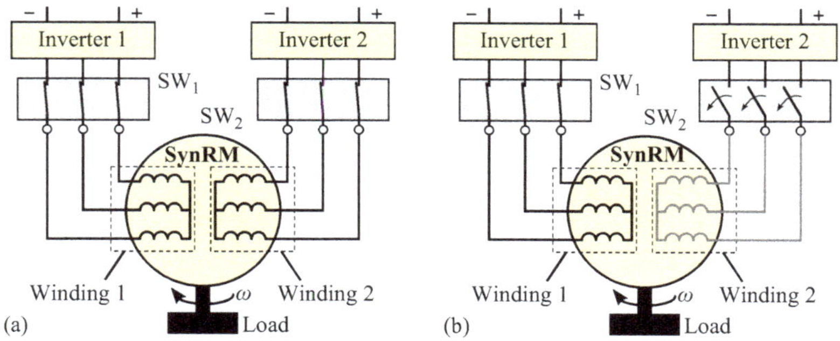

Figure 15.5 Dual three-phase drive system layout in (a) full winding mode (FWM) and (b) open-circuit half winding mode (OC-HWM)

three-phase winding arrangements in a 12-slot 10-pole PMSM machine, shown in Figure 15.4(c), is investigated in several operating conditions focusing on many aspects such as average torque, torque ripple, unbalanced radial force and thermal limits.

Despite the well-known peculiarities of the Synchronous Reluctance (SyR) machine and its potential FT capability due to the lack of rotor PMs, its use in FT applications has been barely proposed during the last years in literature. As for the IM, there are appreciable electrical, magnetic and thermal interactions among the winding phases. The adoption of fractional slot windings with non-overlapped coil to improve the physical and magnetic coupling is not suitable for SyR machine due to the high harmonic content of the stator MMF. In [190] a control strategy for an SyR machine supplied by a three-phase full bridge inverter with neutral point accessible is proposed to mitigate the dangerous effect of one-phase winding SC failure. Likewise, the behavior of an SyR machine supplied by a multilevel inverter with neutral point clamped is analyzed in [191] for different single-power switches SC faults. No study noteworthy investigated the SyR machine design for FT application and focused on the impact of different winding arrangements on its FT capability.

The following part of the work deals with the design of SyR machines with dual three-phase winding with the aim of improving its FT reliability. The dual three-phase winding solution is selected on the basis of the good performance exhibited in the IM and PMSM applications mentioned before. The main operating conditions investigated to evaluate the SyR machine FT reliability are the *full winding mode* and *OC half winding mode*. In the first one, reported in Figure 15.5(a), both the three-phase windings are normally supplied by two independent inverters and the SyR machine produces the rated outputs. In the second case, reported in Figure 15.5(b), one of the two windings (i.e. winding 2) is disconnected while the other one is normally supplied.

Chapter 16
Impact of winding arrangement in dual three-phase synchronous reluctance machine

16.1 Overview

This chapter introduces the study of the FT-Synchronous Reluctance (SʏR) machine investigating the impact of different dual three-phase winding arrangements on the machine performance by means of Finite Element (FE) analysis [192].

The machine performances under investigation are average torque, torque ripple, unbalanced radial force on the rotor and mutual magnetic coupling between the two three-winding phases. The first three items are evaluated Full Winding Mode (FWM) and Open-Circuit Half Winding Mode (OC-HWM).

The FE analyses are performed on the SʏR motor sketched in Figure 16.1 whose main data are reported in Table 16.1. The stator geometry is standard while the rotor one has been optimized accordingly. A single-layer three-phase winding was initially adopted to have only one-phase coils per slot. The parameters used for the optimization were the magnetic insulation ratio and the three flux-barrier angles, and the other rotor parameters have been computed as reported in Section 3.3. The optimization cost functions were the maximum average torque and the minimum torque ripple in maximum torque per ampere condition. The optimal rotor parameters are reported in

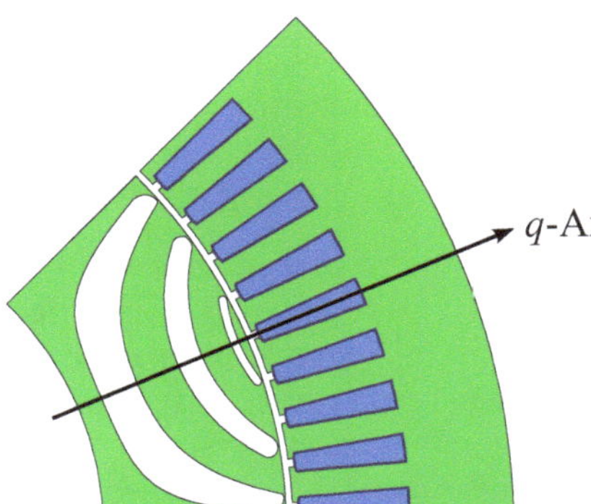

Figure 16 1 Sketch of portion of an eight-pole synchronous reluctance motor with three flux-barriers per pole

Table 16.1 Main data of the motor

Motor part	Symbol	Value	Unit
Outer diameter	D_e	535	mm
Inner diameter	D_s	375	mm
Stack length	L_{stk}	100	mm
Pole number	$2p$	8	–
Slot number	Q_s	72	–
Slot height	h_s	42.5	mm
Tooth width	w_t	9.6	mm
Air gap	g	0.5	mm
Flux-barrier angles	$\vartheta_{b1}, \vartheta_{b2}, \vartheta_{b3}$	$\{20.6, 51.9, 75.3\}$	°
Magnetic insulation ratio	k_{air}	0.4	–
Winding layers	N_{Lay}	1	–
Series conductor per slot	n_{cs}	1	–
Phase current density	J	5 rms	A/mm^2
Iron lamination			M530-50A

Table 16.1. The objective evaluation procedure reported in Chapter 9 has been adopted in the optimization. A rated current density of 5 A/mm^2 has been selected according to the typical operating condition.

The first part of the chapter describes the dual three-phase winding arrangements under analysis. Then, the motor performance, in FWM and OC-HWM, are evaluated for each winding configuration. Finally, a comparison between the different winding arrangements is carried out highlighting their pros and cons.

16.2 Dual three-phase winding arrangements

In the following, the phases of the two windings are labeled as A_1, B_1, C_1 and A_2, B_2, C_2, respectively. The latter phases are open-circuited in OC-HWM condition. In all the analyses, the root-mean-square current density is kept fixed to 5 A/mm^2.

16.2.1 Configurations with three winding axes

The distribution of the phase coils and relative phase axes, A, B, C (indicated with arrows) in a conventional single layer three-phase winding with 72-slot and 8-pole, labeled W-3ph, is reported in Figure 16.2. It presents three slots per pole per phase, a coil throw equal to nine slots, a winding factor $k_w \approx 0.96$ and a periodicity equal to four (i.e. the winding can be split in four equal parts).

From such coil distribution, it is possible to arrange two different dual three-phase windings arrangements. In the first configuration, referred to as W-11-22, the stator is split into two separate parts, each of them containing one of the two three-phase windings. In the second one, referred to as W-12-12, the stator is split into four parts and the dual winding coils are alternated for each pole pair. The two configurations

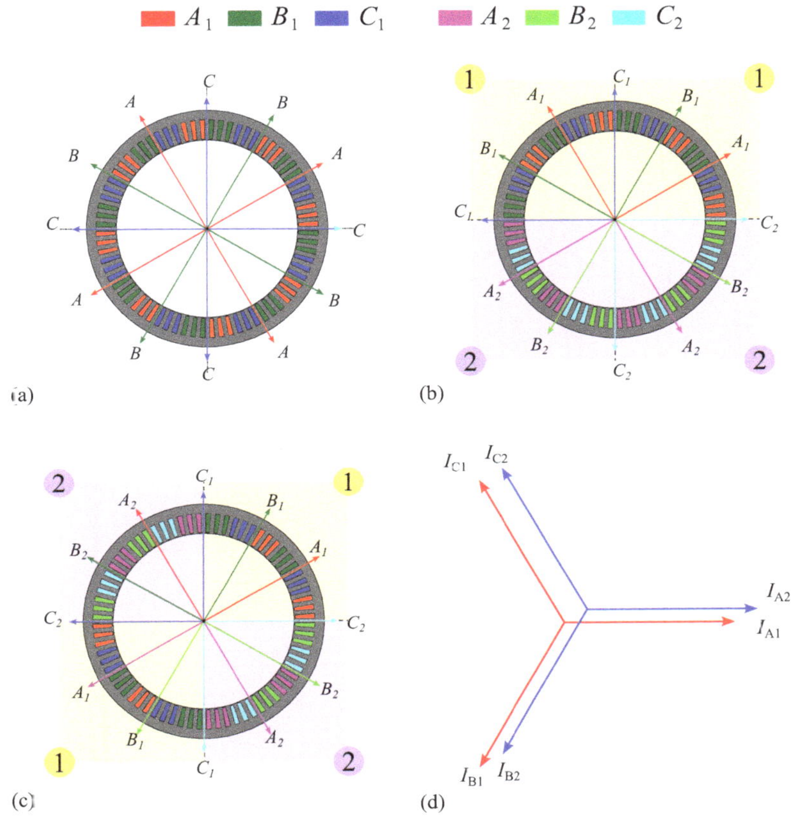

Figure 16.2 Phase coils distributions: (a) standard three-phase, (b) dual three-phase W-11-22, (c) dual three-phase W-12-12, (d) in phase *currents supply condition*

are shown in Figure 16.2(b) and (c), respectively. The background colors highlight the regions that belong to the three-phase winding 1 and 2. Even though the number of electrical phases is six, the actual number of winding axes is still three. For example, the A_1- and A_2-axis coincide with the A-axis of the standard three-phase winding, the same for the left phases. This implies that, in both cases, the dual three-phase windings has to be supplied with an equal number of three-phase current systems *in phase* as reported in Figure 16.2(d). Moreover, it is worth noticing that, in FWM, all the windings in Figure 16.2 are equivalent by definition while, as will be explained in Sections 16.3 and 16.5, they exhibit different characteristics in OC-HWM.

16.2.2 Configuration with six winding axes

With the adopted slot-pole configuration, the unique coil distribution that allows one to obtain six different winding axes is reported in Figure 16.3(a) and labeled as W-6ph.

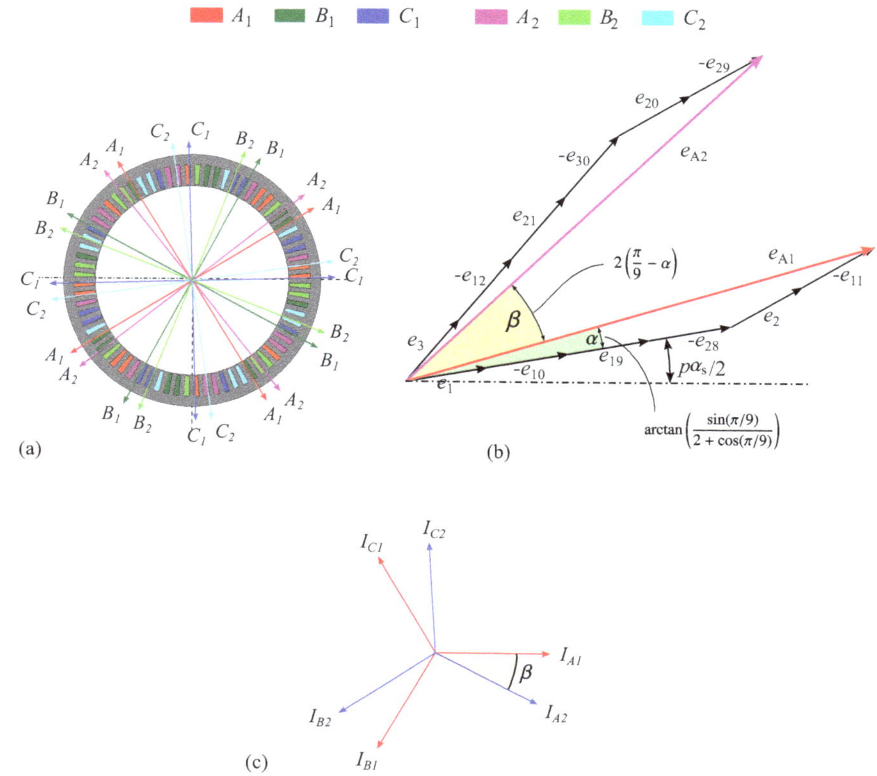

Figure 16.3 (a) Phase coils distribution in the dual three-phase winding arrangement with six different winding axes, (b) phases displacement computation, (c) relative out of phase *current supply condition*

It can be noted that the phases of the two three-phase windings are distributed along the whole stator periphery so that it is not possible to distinguish areas that belong to only one winding as in Figure 16.2(b) or (c). The winding is characterized by a periodicity of 2 and a winding factor equal to 0.986. From the vector diagram in Figure 16.3(b), the electrical displacement between the phases A_1 and A_2 [105,193] results to be

$$\beta = 2\left[\frac{\pi}{9} - \arctan\left(\frac{\sin(\pi/9)}{2 + \cos(\pi/9)}\right)\right] = 0.466\,\text{rad} = 26.7^\circ \tag{16.1}$$

that defers from the classical 30° of the six-phase windings.

In addition to the *in phase* current supply condition, in Figure 16.2(d), this winding can be also supplied by currents *out of phase*, as reported in Figure 16.3(c). In the first case, the winding FWM behavior in the same of the standard three-phase one in Figure 16.2(a). In the second case, the winding behaves as an actual six-phase

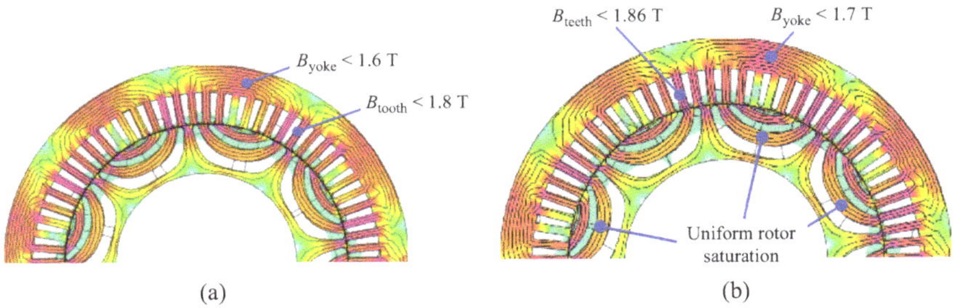

Figure 16.4 FWM flux density plots with in phase (W-3ph) (a) and out of phase (W-6ph) (b) current supply conditions. The first one refers to the winding arrangements in Figure 16.2 while the second one to Figure 16.3(a).

winding yielding higher winding factor (i.e. 0.986) and lower Magnetomotive Force (MMF) harmonics. Hereafter, only this current supply condition will be considered for the considered winding. In OC-HWM, the winding exhibits completely different characteristics from the other ones as will be described later on.

16.3 Torque behaviors

This section shows that the FE results about the torque behavior of the different winding configurations in FWM and OC-HWM. Table 16.2 summarizes the numerical results in terms of average torque, torque ripple and maximum unbalanced force on the rotor.

16.3.1 FWM torque

From Table 16.2, it can be noted that the average torque of the winding W-3ph is slightly lower than the W-6ph one while the torque ripple is about 3% higher. The better performances of the winding W-6ph are mainly due to the higher winding factor and the lower MMF harmonic content. The flux density plots obtained with the two different windings are reported in Figure 16.4(a) and (b), respectively. From the comparison, it can be observed that the maximum flux density in teeth is about the same while in the stator yoke the flux density increases of about 6.2% from W-3ph to W-6ph.

16.3.2 Open-circuit half winding mode

16.3.2.1 Winding W-11-22

Figure 16.5(a) reports the torque behavior obtained with the winding configuration W-11-22 in OC-HWM. It is possible to see that the average torque is about 44% of the FWM one so that the torque ripple results to be more than double (i.e. 37.5%).

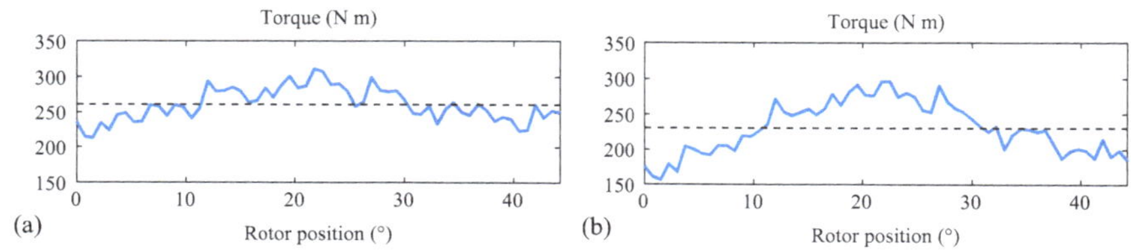

Figure 16.5 Torque versus rotor position of winding (a) W-11-22 and (b) W-12-12 in OC-HWM

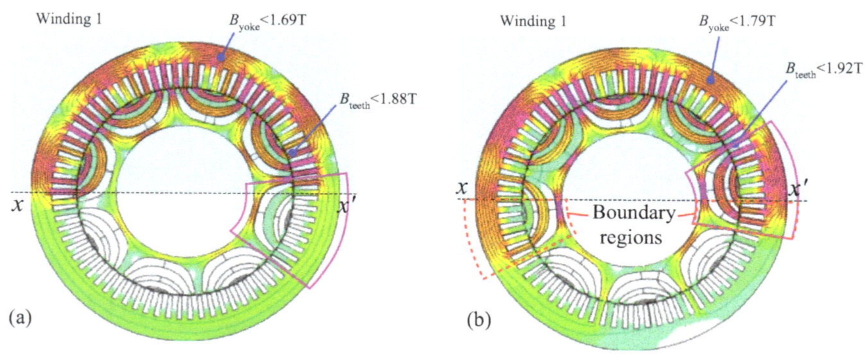

Figure 16.6 Flux density plots of the winding W-11-22 in OC-HWM for different rotor positions: (a) $\vartheta_m = 22.5°$, maximum torque and (b) $\vartheta_m = 0°$, minimum torque

The torque behavior has a period of 180 electrical degrees. Figure 16.6(a) and (b) shows the flux density plots obtained when the torque is minimum, $\vartheta_{\mathrm{m}} = 0°$, and maximum, $\vartheta_{\mathrm{m}} = 22.5°$, respectively. The shaded area highlights the region of the supplied winding (i.e. winding 1). The horizontal dashed line that splits the two winding regions is called boundary axis $\overline{xx'}$. The continuous line that surround a rotor pole indicates its position during the rotation.

It is possible to see that the flux-barriers position, with respect to the axis $\overline{xx'}$, greatly affects the flux density distribution. In particular, when the flux-barriers do not cross the axis $\overline{xx'}$, as in Figure 16.6(a), most of the flux lines flow in the region of the winding 1. The flux lines flowing in the stator yoke of the winding 2 (open-circuited) are due to the subharmonic of the stator MMF. With respect to the FWM, the flux density in the stator yoke and in the teeth increase up to 1.69 T and 1.88 T, respectively. As the flux-barriers cross the axis $\overline{xx'}$, as in Figure 16.6(b), a large number of flux lines, surrounded by red dashed line, flow in the stator area of the winding 2 causing a reduction of useful flux, a higher saturation near the boundary axis and, consecutively, the torque reduction.

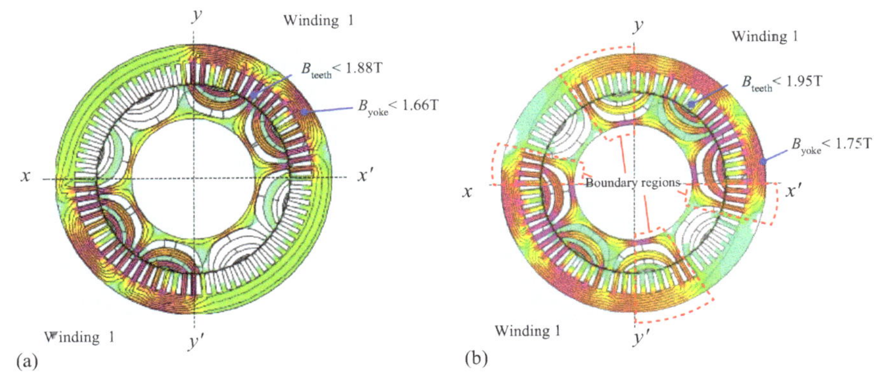

Figure 16.7 Flux density plots of the winding W-12-12 in OC-HWM for different rotor positions: (a) $\vartheta_m = 22.5°$, maximum torque and (b) $\vartheta_m = 0°$, minimum torque

16.3.2.2 Winding W-12-12

Figure 16.5(b) shows the torque behavior of the winding configuration W-12-12 in OC-HWM. The average torque is 62% lower than the FWM one while the absolute torque ripple is 50% higher. The two boundary axes $\overline{xx'}$ and $\overline{yy'}$ (dashed lines) delimit the winding 1 and 2 regions. The torque period is 180 electrical degrees but in this case the difference between maximum and minimum torque is higher than the previous one.

The flux density plots, corresponding to the minimum and the maximum torque of the winding W-12-12, are reported in Figure 16.7(a) and (b), respectively. Similarly to the winding configuration W-11-22, the maximum torque is achieved when the flux-barriers are within the winding 1 regions (dashed areas in Figure 16.7(a)). Conversely, when the flux-barriers cross the axes $\overline{xx'}$ and $\overline{yy'}$, as in Figure 16.7(b), a part of the main flux flows in the winding 2 regions. The wider boundary regions of such a winding arrangement with respect to W-11-22 explain the lower average torque and higher oscillation.

16.3.2.3 Winding W-6ph

The torque behavior of the winding W-6ph in OC-HWM is shown in Figure 16.8(a). The motor achieves an average torque that is 41% of the FWM one while the torque ripple is about four times higher. The lower average torque with respect to the dual three-phase arrangements (W-11-22 or W-12-12) is due to the fact that, in the current case, the whole air-gap motor has to be magnetized. The higher torque ripple is justified by the worst harmonic content of the stator MMF due to the lower number of effective slots.

From the flux density plot in Figure 16.8(b), it can be noted that the magnetic loading is not uniform in the motor. In particular, the stator yoke flux density in the

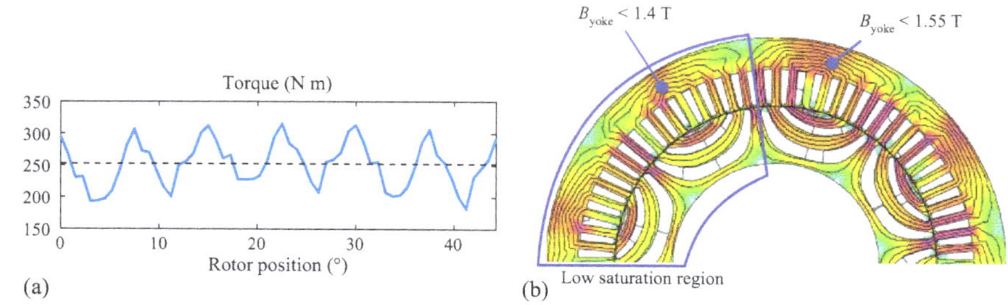

Figure 16.8 Results obtained with the winding W-6ph in OC-HWM: (a) torque behavior and (b) flux density plot

left area is higher than the right one since the number of slots occupied by the winding 1 is higher in the right machine quarter.

16.4 Radial force

The radial force on the rotor has been computed for each winding configuration. From Table 16.2, it results that in OC-HWM only the winding configuration W-11-22 exhibits nonnegligible radial force (i.e. 12 kN). This is expected since, in such configuration, the winding 1 covers only a part of the stator periphery. On the contrary, in all other arrangements, a balanced force remains even in OC-HWM since a mechanical winding symmetry remains.

16.5 Magnetic coupling between winding 1 and 2

This section deals with the evaluation of the magnetic coupling between winding 1 and winding 2 phases for different winding configurations. The self- and mutual inductances between the phases are computed setting a constant current in the phase A_1 and considering the other phases disconnected. The current density is still 5 A/mm^2.

Figure 16.9 shows the behavior of self- and mutual inductances versus rotor position for different winding configurations. The self-inductance of the phase A_1 is labeled as L_{A1} while the mutual inductance with phase B_1 and C_1 are labeled as M_{A1B1} and M_{A1C1}, respectively. The mutual inductances between phase A_1 and the winding 2 phases are labeled as M_{A1A2}, M_{A1B2} and M_{A1C2}.

All self- and mutual inductances of the winding 1 phases exhibit about the same maximum value (i.e. 0.077 mH). Only W-6ph inductance curves, shown in Figure 16.9(c), are symmetric with respect to the A_1-axis for symmetry reason. The magnetic coupling of each winding configuration is determined by the amplitude of the mutual inductance between phase A_1 and the winding 2 phases. In particular, the maximum value of the ratio between L_{A1} and the mutual inductances M_{A1A2}, M_{A1B2} and

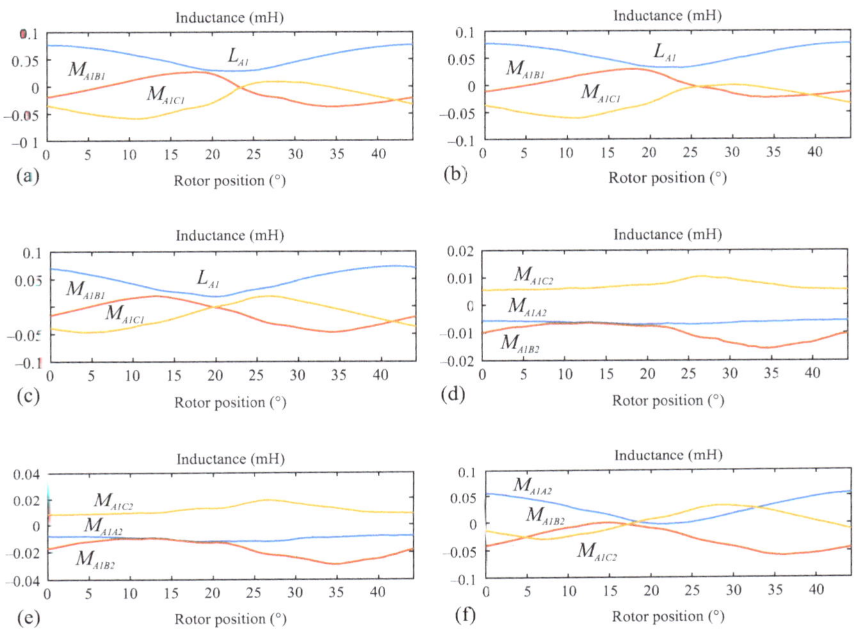

Figure 16.9 Self- and mutual inductances versus rotor position for different winding configurations. The columns refer to the winding W-11-22: (a) winding 1 phases and (d) winding 2 phases; W-12-12: (b) winding 1 phases and (e) winding 2 phases; and W-6ph ((c) winding 1 phases and (f) winding 2 phases), respectively.

M_{A1C2}, indicated as r_{mc}, gives the grade of coupling and allows one to compare the different winding configurations. The behaviors of the mutual inductance are reported in Figure 16.9(d)–(f). It is possible to observe that the winding W-11-22 achieves the lowest magnetic coupling ($r_{\mathrm{mc}} = 0.37$) while the six-phase winding achieves the highest one ($r_{\mathrm{mc}} = 0.95$). The magnetic coupling ratio of the winding W-12-12 is 60% higher than the W-11-22 one ($r_{\mathrm{mc}} = 0.59$). The high magnetic coupling of the winding W-6ph is due to the fact that the coils of different phases are distributed along the whole stator diameter. Conversely, the windings W-11-22 and W-12-12 exhibit a lower magnetic coupling since, as can be observed in Figures 16.6(b) and 16.7(b), only a small part of the flux flows through the winding 2 regions.

16.6 Discussion

In this chapter, the open-circuit fault suitability of three different dual three-phase winding arrangements has been investigated by means of FE analysis on an SyR

Table 16.2 Comparison of simulated motor performance with various winding arrangements in different operating conditions

Operating condition	Winding arrangement	Average torque (N m)	Torque ripple (%)	Unbalanced radial force (kN)
FWM	W-3ph	595	16.4	≈ 0
	W-6ph	611	13.3	≈ 0
OC-HWM	W-11-22	261	37.5	$\approx$ 12 kN
	W-12-12	231	60.7	≈ 0
	W-6ph	253	53.2	≈ 0

machine. Various aspects are taken into account: average torque, torque ripple, unbalanced radial force on the rotor and magnetic coupling. The analyses have shown that when one of the two three-phase systems is disconnected (OC-HWM) all the winding arrangements exhibit an average torque that ranges between 35% and 44% of the rated one (FWM). The torque ripple in OC-HWM is always higher than the FWM one and ranges between 37% and 61%. These aspects suggest that, during the design of this SyR motor, the reduction of torque ripple in OC-HWM has to be an objective.

Furthermore, concerning the radial force, the configuration W-11-22 is the worst among the analyzed solutions achieving 12 kN. This is due to the asymmetric distribution of the stator winding. For this reason, the considered configuration cannot be applied in OC-HWM even if it exhibits the highest average torque and minimum ripple. The radial force of the other solutions is negligible for symmetry reasons.

The magnetic coupling of the different windings has been evaluated to estimate the amplitude of the maximum induced voltages in OC-HWM. The analyses showed that configurations W-11-22 and W-12-12 present the lowest magnetic coupling (i.e. $r_{mc} < 0.6$) while winding W-6ph reaches the maximum one (i.e. $r_{mc} \approx 0.95$). Hence, the latter winding is not reliable in OC-HWM since the induced voltages in the disconnected phases reach about the nominal value.

In conclusion, the winding W-12-12 represents a good compromise in terms of radial force and magnetic coupling but further investigations has to be carried out to limit its high-torque ripple in OC-HWM.

Chapter 17

Optimization of a synchronous reluctance machine for fault-tolerant applications

17.1 Introduction

In Chapter 16, it has been shown that the Synchronous Reluctance (SyR) machine with the dual three-phase winding arrangement W-12-12 exhibits, in general, good performance in both Full Winding Mode (FWM) and OC-FWM except for the torque ripple that is quite high in the second operating condition.

The aim of this study is to improve the Fault-Tolerant (FT) reliability of such a drive system focusing on the SyR rotor design. In particular, a standard geometry, whose main data are reported in Table 17.1, is selected for the stator lamination, while two alternative Finite Element (FE)-based optimization approaches are proposed for the rotor design. On the basis of the optimization results, the most promising individual is selected and analyzed in detail by means of FE analysis in FWM and Open Circuit Half Winding Mode (OC-HWM).

Table 17.1 Main motor data

Dimension	Value	Unit
Stator outer diameter	170	mm
Stator inner diameter	115	mm
Number of slots	36	–
Pole pairs	4	–
Slot height	17.5	mm
Tooth width	5.3	mm
Stack length	85	mm
Winding layers	2	–
Current density RMS	6	A/mm^2
Rated current	8	A
Voltage DC bus	350	V

17.2 Rotor optimization strategies

Two different optimization strategies are proposed in the following. In the first one, referred to as FWM optimization (FWM-OPT), the objectives are to maximize the average torque and to minimize the ripple in FWM only (FWM objectives). In the second one, referred to as Full-Half Winding Mode Optimization (FHWM-OPT), the average torque and ripple in OC-HWM are additional objectives of the optimization (OC-HWM objectives).

The rotor parameters are the flux-barrier angles, ϑ_{b1}, ϑ_{b2}, ϑ_{b3}, and magnetic insulation ratio k_{air}. The flux-barrier and iron channel thicknesses are computed as described in Section 3.3. A rated current density of 6 A/mm^2 has been set for the performance evaluation in FWM and OC-HWM. The torque ripple is computed as the relative standard deviation percentage of the FE torque waveform.

17.2.1 FWM-OPT strategy

Following the procedure described in Chapter 9, this optimization has been carried out in 80 generations with a population of 30 individuals each. The resulting objectives plane is reported in Figure 17.1. The horizontal and the vertical axes report the average torque and torque ripple, respectively. The circle markers indicate the Pareto front individuals. It can be noted that the average torque of the Pareto front ranges between 13.4 N m and 13.7 N m while the ripple presents a high variation between 1% and 10%. Some rotor sketches are reported in the plot to highlight how the geometry varies as the ripple decreases. It is possible to see that the shape of the flux-barriers near the air-gap tends to be sharper as the ripple decreases. This is because the shape of the flux-barrier strongly affects the flux density distribution within the tangential iron ribs. The individual with the lowest torque ripple (1%) and an average torque of

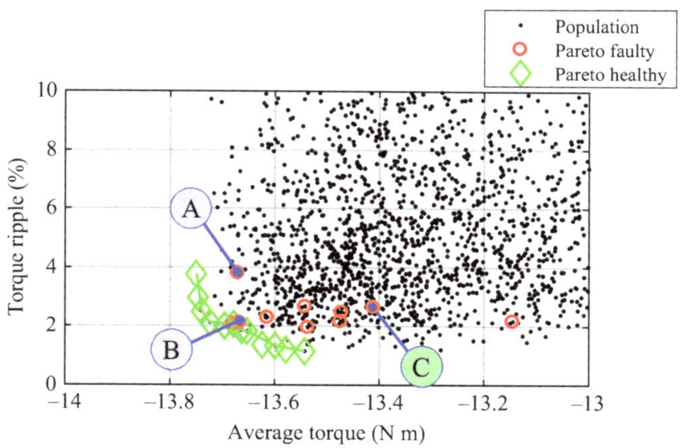

Figure 17.1 Objectives planes of the optimization strategy FWM-OPT: torque ripple versus average torque

13.5 N m has been selected. The corresponding flux-barrier angles are $\vartheta_{b1} = 34.8°$, $\vartheta_{b2} = 55.1°$ and $\vartheta_{b3} = 71°$ while the magnetic insulation ratio is $k_{air} = 0.44$. The same rotor geometry has been analyzed in OC-HWM yielding an average torque of 6.5 N m (50% of the FWM value) and a torque ripple of 11.7%, 11 times higher than the FWM one. The results are summarized in Table 17.2.

17.2.2 FHWM-OPT strategy

Figure 17.2 reports the flowchart of the objectives evaluation procedure performed for each individual of the optimization strategy FHWM-OPT. At first, the geometry feasibility of the motor is evaluated. The unfeasible individuals are rejected and their average torque and ripple values are set to 0 and 10^6, respectively. Otherwise, the radial ribs of the feasible individuals are computed according to the maximum speed.

Table 17.2 Average torque and ripple of the optimal individuals in FWM and OC-HWM

Optimization strategy	Operating condition	Torque [N m]	Ripple [%]
FWM-OPT	FWM	13.5	1.1
	OC-HWM	6.5	11.7
FHWM-OPT	FWM	13.3	2.6
	OC-HWM	5.6	5

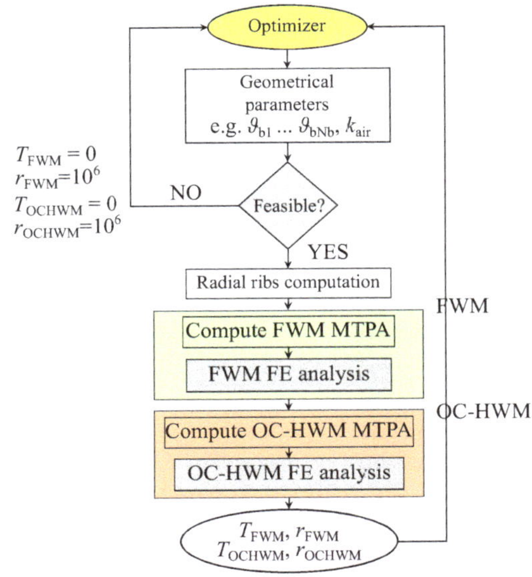

Figure 17.2 Flowchart of the FHWM-OPT strategy

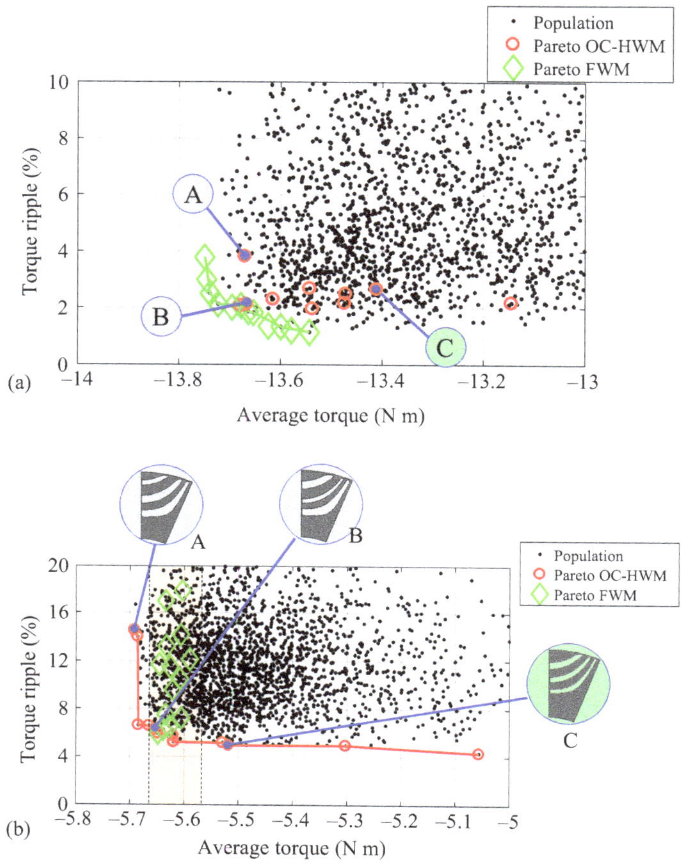

Figure 17.3 Objectives planes of the optimization strategy FHWM-OPT: torque ripple versus average torque—(a) FWM objective plane and (b) OC-HWM objective plane

Next, the Maximum Torque Per Ampere (MTPA) current angle in FWM is estimated in an iterative way by means of single-step FE simulations. The resulting current vector is then used to evaluate the average torque and torque ripple in FWM. After that, the MTPA current angle in OC-HWM is found like before and then used to get the OC-HWM cost functions.

The objectives planes of the FWM and OC-HWM cost functions are reported in Figure 17.3(a) and (b), respectively. The green diamond markers in the two figures are the non-dominated individuals in the FWM objectives plane (FWM Pareto front). Similarly, the red circle markers represent the non-dominated individuals in the OC-HWM objectives plane (OC-HWM Pareto front). It can be noted from the first plot that the average torque of the FWM Pareto front ranges between 13.5 N m and 13.8 N m while the ripple is always lower than 4%. Conversely, the OC-HWM Pareto

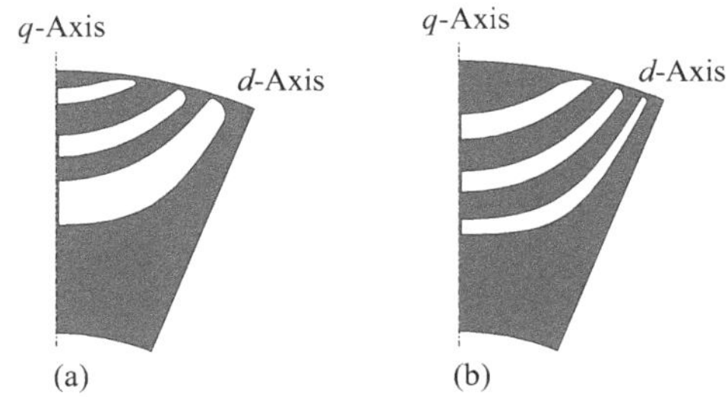

Figure 17.4 Rotor sketches of the optimal individuals obtained from (a) FWM-OPT and (b) FHWM-OPT

individuals, in the same plot, are distributed in a wide average torque range (from 13 N m to 13.7 N m) with limited torque ripple variation (between 2% and 4%). The OC-HWM objectives plane, in Figure 17.3(b), shows that the average torque of the OC-HWM Pareto individuals ranges between 5 N m and 5.7 N m with a limited torque ripple variation (i.e. between 4.3% and 6.6%). On the contrary, the FWM Pareto individuals, in the OC-HWM plane, are characterized by high-torque ripple variation with limited average torque variation (shaded area in Figure 17.3(b)). The rotor sketches reported in the plot show the rotor geometry variation along the OC-HWM Pareto front. It should be noted that the outer flux-barrier angle tends to increase as the torque ripple decreases while the inner flux-barrier angle moves away from the d-axis.

In this case, due to the higher number of cost functions, the selection of the optimal individual is less oblivious than in the previous case. For the present case study, the individual C, sketched in Figure 17.4(b), has been selected having chosen the OC-HWM torque ripple as the main objective to minimize. Such an individual has also been selected on the basis of the flux-barriers geometry feasibility. In fact, many geometries among the alternative solutions were actually unfeasible because of the very small iron channel thickness or the sharp flux-barrier fillets near the air gap.

17.3 Winding W-12-12 alternative configurations

The dual three-phase winding W-12-12 can be further arranged with two different phase coils distributions: the *aligned* one, in Figure 17.5(a), and the *shifted* one, in Figure 17.5(b).

The difference between the two configurations is the position of the A_1 magnetic axis with respect to the winding region boundary axes $\overline{xx'}$ and $\overline{yy'}$. In the winding W-12-12 aligned, the angular position of the A_1-axis, represented by the arrow in Figure 17.5(a), is 35° while in the shifted version, the axis is rotated of one slot pitch angle (i.e. 10°) in the clockwise direction. It is worth noticing that this aspect only matters in OC-HWM when the electric loading is not distributed along the whole stator periphery.

In FWM, the two configurations are equivalent by definition; however, this is not true in OC-HWM. Such an aspect will be investigated in the following FE analyses.

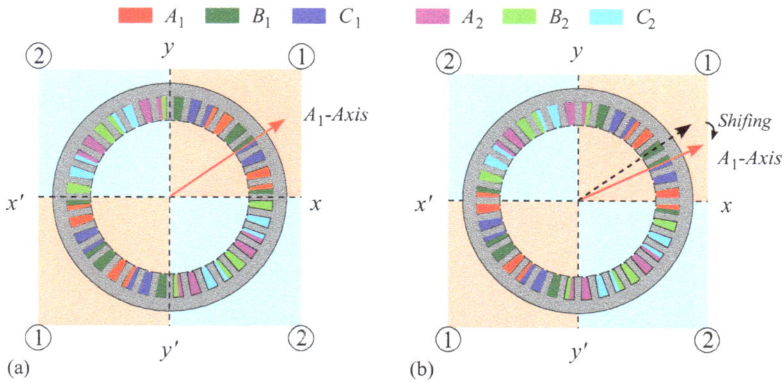

Figure 17.5 Dual three-phase winding arrangements: (a) W-12-12 aligned and (b) W-12-12 shifted. The shaded areas indicate the stator regions occupied by the two three-phase windings. The arrows show the position of the A_1 winding magnetic axis

17.4 Mutual magnetic coupling

This section deals with the analysis of the mutual inductances among the two three-phase windings. The simulations are carried out imposing constant currents in the winding 1 and considering the other one open-circuited. The winding 1 phases are fed with the following constant currents: $I_{A1} = 8$ A and $I_{B1} = I_{C1} = -4$ A.

Figure 17.6(a) shows the flux linkage behavior of the supplied phases. The flux linkage of each phase follows the behavior of the rotor permenance. In fact, it is maximum (minimum) when the rotor d-axis (q-axis) is aligned to the phase axis. Figure 17.6(b) and (c) report the flux linkage amplitude (i.e. without sign) behaviors of the winding 2 phases in the aligned and shifted configuration, respectively. The different magnetic coupling between the two configurations is evident. In the first configuration, the maximum induced flux linkage is about 76 mV s while in the shifted one is 40% lower. This is also confirmed by the flux density plots in Figure 17.7(a) and (b) obtained when the flux linked by the winding 2 phases is minimum. It is evident that the flux lines in the region of the disconnected phases are lower for the shifted arrangement. This observation suggests that the shifted solution should be preferred to reduce the magnetic coupling between the two three-phase windings.

17.5 Magnetic model of the machine

From the control point of view, it is particularly important to know the magnetic characteristics of the machine and its behavior in FWM and HWM. This section reports the most important curves used to determine the control of a machine. In

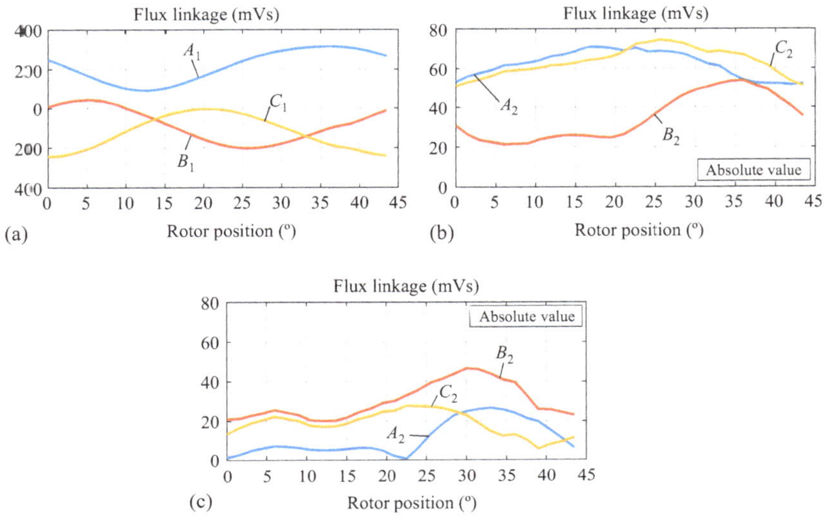

Figure 17.6 Flux linkages versus rotor position of the FWM and disconnected phases: (a) winding 1 phases, (b) winding 2 phases (W-12-12 aligned), (c) winding 2 phases (W-12-12 shifted). The curves are obtained feeding only one of the two windings with constant currents whose amplitudes are $I_{A1} = 8A$ and $I_{B1} = I_{C1} = -4A$

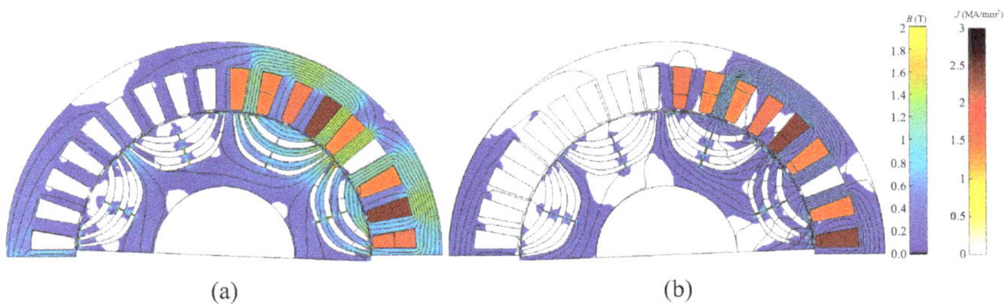

Figure 17.7 Flux density plots of the winding arrangements W-12-12 (a) aligned and (b) shifted in OC-HWM: rotor position that achieves the lowest mutual coupling between windings 1 and 2

particular they are the constant torque loci and the magnetizing curves along the d- and q-axis described by

$$\begin{cases} \lambda_d = \lambda_d(I_d, I_q) \\ \lambda_q = \lambda_q(I_d, I_q) \end{cases} \tag{17.1}$$

where λ_d (λ_q) and I_d (I_q) are the flux linkage and current along the d-axis (q-axis), respectively.

17.5.1 Magnetizing characteristics

17.5.1.1 *d*-Axis

The d-axis magnetizing curves obtained in FWM and OC-HWM are reported in Figure 17.8(a) and (b). The thicker black lines are obtained without cross-saturation, i.e. $I_q = 0$. It is possible to see that, in general, the flux linkage obtained in OC-HWM tends to be higher than in FWM. In particular, the maximum FWM flux linkage is about 0.36 Vs while in OC-HWM is 10% higher. Likely, for the same d-axis current $I_d = 8$ A, the maximum flux linkage in the two cases is 0.31 Vs (point D_{F}) and 0.35 Vs (point D_{H}), respectively. This aspect can be explained with the flux density plots in Figure 17.8(c) and (d) obtained in the point D_{F} and D_{H}, respectively. From the comparison, it is evident that in FWM the magnetic loading is higher than in the other

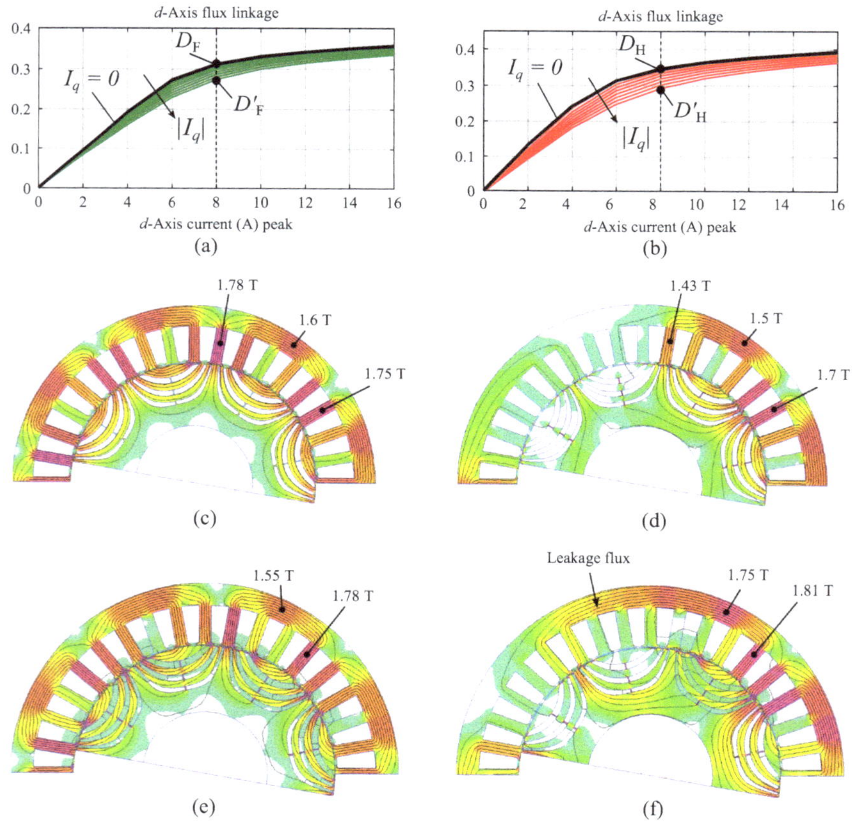

Figure 17.8 d-Axis magnetizing curves in FWM and OC-HWM and relatives flux density plots: (a) and (b) d-axis current (A) peak, and (c) point D_F, (d) point D_H, (e) point D'_F and (f) point D'_H

operating condition. In particular, the maximum flux density in the teeth is about 1.8 T in the first case while is 1.7 T in the second one. The cross-saturation impact is also different in the two operating conditions. With respect to a fixed d-axis current, i.e. 8 A, in FWM the flux linkage reduces of 12% (point D'_{F}) while in OC-HWM the reduction is about 15.5% (point D'_{H}). The higher flux linkage drop in the latter case can be explained with the flux density plots in Figure 17.8(e) and (f). In fact, in the second one, it is possible to note a higher stator yoke flux density and a higher number of flux leakage lines that flow in the winding 2 region (left machine portion). This "deviation" of the flux lines is more evident for low current levels in which the stator yoke saturation does not limit such a flux leakage.

17.5.1.2 *q*-Axis

The q-axis magnetizing curves in FWM and OC-HWM are reported in Figure 17.9(a) and (b), respectively. The thicker black lines are obtained with $I_d = 0$ that means without cross-saturation. In such a supply condition, the difference between FWM and OC-HWM is evident only for very high currents, i.e. $I_q = 16$ A, while is not meaningful for $I_q = 8$ A. As I_d increases, the variation of the q-axis flux is more evident than along the d-axis. More precisely, the flux linkage in the point Q'_{H} is 0.092 Vs (increment of 42% with respect Q_{F}) while in the point Q'_{F} is 0.112 Vs (increment of 72% with respect Q_{H}). From the flux density plots in Figure 17.9(c) and (d), corresponding to the points Q'_{F} and Q'_{H}, it can be noted that the teeth flux density in the

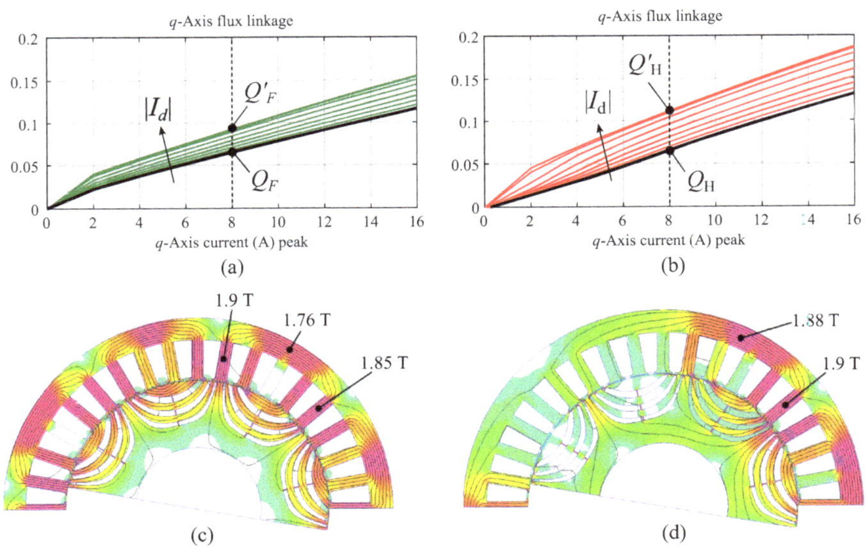

Figure 17.9 q-Axis magnetizing curves in FWM and OC-HWM and relatives flux density plots: (a) and (b) q-axis current (A) peak, and (c) point Q'_F, (d) point Q'_H

winding 1 region is similar in the two cases while the yoke flux density is higher in OC-HWM. The more evident effect of the cross-saturation on the q-axis flux linkage is due to the fact that the flux lines can easily bypass the flux-barriers and flow through the winding 2 region.

17.5.2 Constant-torque loci map

Figure 17.10 sketches the constant torque loci of the machine in FWM and OC-HWM. The shaded circle represents the current limit. The MTPA and maximum torque per voltage loci are also plotted. It can be observed that the same trajectories are valid both in FWM and OC-HWM. The working points A, B and C have been selected for the computation of torque ripple and power factor. In particular, A is the base point while B and C are flux-weakening points. The speed of each point is computed according to the machine flux linkage and voltage supply. The results are reported in Table 17.3. It can be noted that in FWM the torque ripple is always lower than 4% and is maximum at the highest speed because of the lower average torque. Moreover, the base power factor is quite low, i.e. 0.6, and increases up to 0.66 at the maximum speed. It can be noted that the average torque reduction in OC-HWM is higher than 50% in all cases. This because, as point out before, the anisotropy is mainly compromised by the higher q-axis flux that bypasses the flux-barriers. Consequently, also the OC-HWM torque ripple is higher than 50% but in any case its value is not higher than 10%.

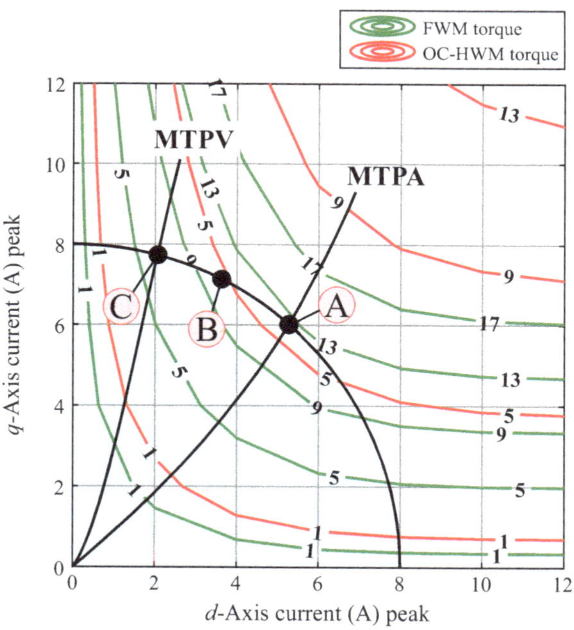

Figure 17.10 Iso-torque locus in the I_d–I_q plane in FWM and OC-HWM

Table 17.3 Simulated working points in FWM and OC-HWM

Working point				FWM			OC-HWM		
	I [A]	α_i^e [°]	n [rpm]	T [N m]	$r_{\%}$ [%]	PF –	T [N m]	$r_{\%}$ [%]	PF –
A	8	138	800	12.7	2.8	0.60	5.7	5.5	0.45
B	8	148	900	11.9	3.2	0.64	5.3	7.4	0.47
C	8	158	1 000	9.3	4.0	0.66	4.2	9.6	0.49

The reduction of rotor anisotropy also involves a very poor OC-HWM power factor that ranges between 0.4, in the base point, and 0.45, at the highest speed.

17.6 Conclusion

In this chapter, two alternative optimization procedures have been investigated to improve the FT reliability of a dual three-phase SyR motor drive. In the first one, the FWM average torque and torque ripple were optimized only. In the second one, both the FWM and OC-HWM average torque and ripple have been set as cost functions. The results have shown the better effectiveness of the second approach yielding good average torque and torque ripple both in FWM and HWM. On the basis of the optimization results, an optimal motor geometry has been selected and then analyzed in detail, by means of FE analysis.

Two alternative winding arrangements called W-12-12 aligned and W-12-12 shifted have been investigated. The analyses have shown that the two arrangements yield similar performance in terms of average torque and torque ripple, both in FWM and HWM, but the shifted version exhibits a lower mutual magnetic coupling between the two three-phase windings.

A high sensitivity of the magnetic characteristics to the operating conditions and to cross-coupling phenomenon has been observed. In particular, in OC-HWM the rotor saliency is lower than in FWM essentially because of the higher q-axis flux that can bypass the flux-barriers flowing through the region of the disconnected winding. This aspect also compromises the power factor. In fact, despite the rotor optimization, the power factor is always lower than 0.7 in FWM and reaches even lower values in OC-HWM, e.g. 0.45.

Chapter 18
Experimental validation of a synchronous reluctance machine for fault-tolerant applications

18.1 Introduction

This chapter continues the investigation of the Synchronous Reluctance (SyR) machine design for Fault-Tolerant (FT) applications optimizing both stator and rotor geometry [194]. On the basis of the knowledge gained in Chapters 16 and 17, the dual three-phases arrangement W-12-12 shifted is selected. Moreover, in order to permit a single-layer winding, a 48-slot 8-pole configuration is adopted. Two alternative stator geometries are proposed. The first one is a classical stator with slots all equal and evenly spaced along the stator periphery. The second one still presents all the slots of the same shape, but their displacement is not uniformly distributed along the stator periphery. The rotor presents three fluid flux-barriers per pole. In both the optimization, the strategy full-half winding mode optimization described in Chapter 17 is adopted. On the basis of the optimization results, an optimal individual is selected and manufactured. Such a prototype is tested in laboratory to verify the reliability of the proposed design procedure and the machine suitability for FT applications.

18.2 Dual three-phase winding arrangement

The coils distribution of the adopted winding W-12-12 shifted is shown in Figure 18.1(a). The winding connection scheme of an half-machine is reported in Figure 18.1(b). As can be observed, the frontal connections of phases B_1 and C_1 (or B_2 and C_2) have the same length and are 20% shorter with respect to the one of phase A_1 (or A_2). This has also been verified by the winding phase resistance measurement that has provided the following results:

$$R_{A1} = R_{A2} = 1.1\,\Omega \qquad R_{B1} = R_{C1} = R_{B2} = R_{C2} = 1.2\,\Omega$$

18.3 Motor geometry

Two stator geometries are proposed: *standard stator optimization* (SS-OPT) and *custom stator optimization* (CS-OPT). The geometry parameters selected for the stator

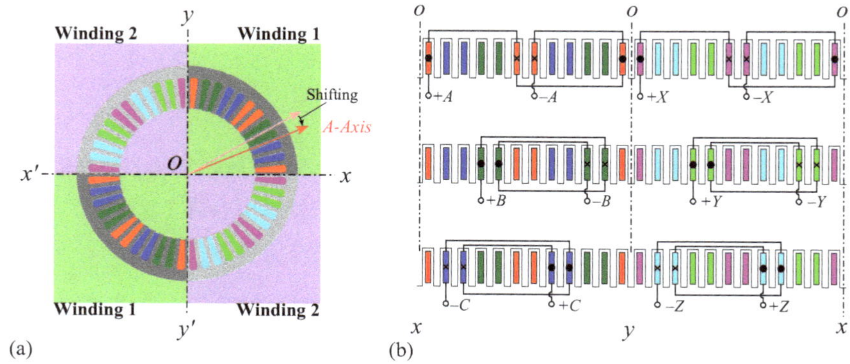

Figure 18.1 Dual three-phase winding arrangement W-12-12 shifted: (a) phase coils distribution and (b) connections scheme of two pole pairs

Table 18.1 Main motor data

Dimension	Value	Unit
Voltage DC bus	350	V
Stator outer diameter	170	mm
Rotor shaft diameter	38	mm
Air-gap length	0.4	mm
Number of slots	48	–
Pole pairs	4	–
Stack length	85	mm
Winding layers	1	–
Current density RMS	6	A/mm^2
Materials		
Lamination		M330-35A
Winding		Copper

and rotor geometry optimization are described in the following. In both optimizations, the outer stator diameter is fixed to 170 mm while the stack length is 85 mm. Other motor data are reported in Table 18.1.

18.3.1 Stator parameters

18.3.1.1 Standard stator

The SS-OPT refers to a stator in which the slots are all equal and evenly spaced along the stator circumference of a mechanical angle equal to $360/Q$, where Q is the slot number. This involves that all the teeth present the same rectangular cross-sectional area as in the typical stator geometries. In such optimization the stator parameters

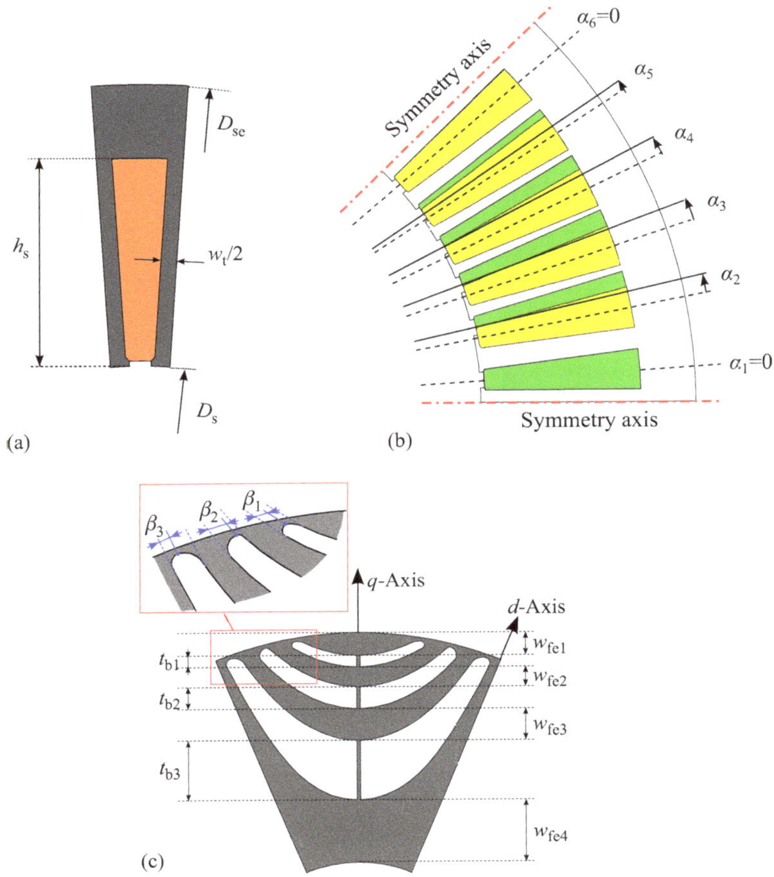

Figure 18.2 Geometrical parameters used for the stator and rotor optimization: (a) standard stator, (b) custom stator and (c) rotor

are the inner diameter, D_s, the tooth width, w_t, and the slot height h_s, as indicated in Figure 18.2(a). The total number of stator parameters is 3.

18.3.1.2 Custom stator

In the CS-OPT, all the slots have the same shape and are symmetric (to easy the winding insertion), but the angle between the slot axes is not constant. This means that the slot distribution is not uniform along the stator periphery and the tooth cross section is trapezoidal instead of rectangular. Figure 18.2(b) shows the parameters used for the stator optimization. The dashed lines represent the angular positions of the fictitious evenly spaced slots (the yellow ones). The continuous lines are the axes of the shifted slots whose angles are computed with respect to the original axes. For example, the shifting angle of the first slot is $\alpha_1 = 0$ which means that there is no angular displacement between the original position. The second slot is shifted of an angle α_2 and so on. In order to reduce the total number of parameters, the slot distribution is assumed to be symmetric with respect to the symmetry axis (dash dotted lines in Figure 18.2(b)) and all the pole pairs present the same slot distribution. Considering, D_s, h_s, w_t and the six slot angles, the total number of stator parameters is 9.

18.3.2 Rotor parameters

The fluid flux-barriers geometry, described in Section 3.1, is selected for the rotor design. The parameters used for the rotor optimization are reported in Figure 18.2(c). In particular, they are the iron channel widths, $w_{c1}, \ldots, w_{c4}$, flux-barrier thicknesses, $t_{b1}, \ldots, t_{b3}$, and flux-barrier angles, $\beta_1, \ldots, \beta_3$. Note that the *j*th angle, β_j, is actually computed as a certain percentage of the angle delimited by the outer and inner edge of the *j*th flux-barrier. The inner rotor diameter is fixed an equal to 38 mm according to the test bench requirement. The total number of rotor parameters is 10.

18.4 Optimization results

This section reports the optimization results obtained from the two approaches. In both the optimizations, a fixed current density of 6 A/mm^2 has been selected. The adopted objectives evaluation is described in Section 17.2.2. For the individual selection, a maximum torque ripple of 16.5% has been preliminary set. In this section, the Open Circuit Half Winding Mode (OC-HWM) will be indicated shortly as HWM for more clarity since no other operating conditions are evaluated during the optimization.

18.4.1 Standard stator optimization results

The optimization has been carried out with 300 generations of 65 individuals each. The two objective planes of the SS-OPT are reported in Figure 18.3(a). The first one, labeled as Full Winding Mode (FWM)-plane, plots the torque ripple versus average torque in FWM while the second one, labeled as HWM-plane, plots the same cost functions in HWM. The continuous line in the FWM-plane represents the Pareto front in FWM. The red circle markers indicate the individual of the Pareto front who exhibits an HWM torque ripple lower than 16.5%. The HWM performances of these survived individuals are also plotted in the HWM-plane with the same markers style. The dot markers are the whole individuals of the optimization.

Likely, the continuous line in the HWM plane is the Pareto front and the diamond markers represent the front individuals that present a torque ripple lower than 16.5%, which are plotted in the FWM plane as well. The markers size of the Pareto front individuals increases as the torque ripple decreases. The distribution of the two Pareto fronts in the FWM plane is quite different. In particular, the individuals of the FWM Pareto front exhibit an average torque that ranges between 19.5 N m and 21.6 N m with a torque ripple between 3% and 11%. The average torque of the survived HWM Pareto front individuals ranges between 12 N m and 20.6 N m, but their torque ripple

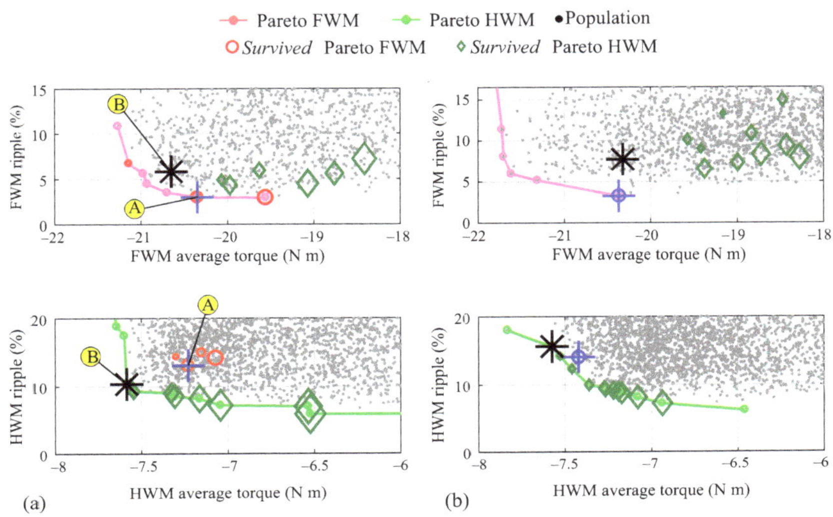

Figure 18.3 Objective planes of the (a) SS-OPT and (b) CS-OPT

is always lower than 10%. In the HWM plane, the individuals of the HWM Pareto front exhibit an average torque between 5.6 N m and 7.7 N m and, except for two individuals, all of them yield a torque ripple lower than 10%. Conversely, the average torque of the FWM Pareto front individuals is bounded between 7 N m and 7.5 N m with a torque ripple that ranges between 14.5% and 16.5%.

The individuals marked with cross (labeled as A) and star (labeled as B) yield good performance both in FMW and HWM. In particular, the individual A exhibits the lowest torque ripple in FWM, but its average torque is 1.5% lower than B. In HWM, the average torque of the individual B is 4.8% higher than B while the torque ripple is 2% lower.

18.4.2 Custom stator optimization results

The optimization has been carried out with 500 generations of 95 individuals each. Computationally speaking such an optimization is mo. The objectives planes of the CS-OPT are reported in Figure 18.3(b). The same graphical convention described before is valid. As can be seen in the figure, only one individual of the FWM Pareto front yields an HWM average torque lower than 16.5%. In HWM, the same individual exhibits an average torque of about 7.4 N m with a ripple of 14%. Conversely, many individuals of the HWM Pareto are present but their torque ripple is higher than 6.2%. In the HWM plane, the survived individuals of the HWM Pareto front yield an average torque between 6.5 N m and 7.8 N m while the torque ripple is between 6.2% and 18%.

It can be noted that the individual of the HWM Pareto front that yields the highest FWM average torque (marked with a star) has an HWM torque ripple of about 16.5%

Table 18.2 Prototype main data

Definition	Value	Unit
Rated current (peak)	8	A
Inner diameter	104	mm
Slot height	24.3	mm
Tooth width	3.6	mm
Conductors in slot	46	–
Flux-barrier angles	{9.86, 14.51, 20.28}	°
Flux-barrier thicknesses	{1.61, 3.17, 8.34}	mm
Iron channel width	{3.22, 2.81, 4.43, 8.98}	mm
Rotor skewing (3 steps)	7.5	°

that is higher than A and B in the SS-OPT. On the other hand, the FWM Pareto individual who exhibits the lowest FWM torque ripple (marked with a cross) achieves an average torque similar to B but with an HWM torque ripple of 14.5% that is higher than the one achieved by A and B.

In conclusion, the higher time consumption of such an optimization, due to the higher number of parameters, does not provide any design solution better than the SS-OPT. For this reason, it is not suitable for this application.

18.5 Prototype

The individual A of the SS-OPT has been selected for the prototype manufacturing. The main machine data are reported in Table 18.2. A view of the prototype is shown in Figure 18.4(a). The 24 terminals of the dual three-phase winding are all accessible by means of the terminal boards in Figure 18.4(b). The terminals on the left and right belong to the winding 1 and 2, respectively. In the considered picture, the winding 2 is short circuited by means of the short-circuit bridges to perform the test discussed in Section 18.8. The rotor prototype is visible in Figure 18.4(c). A three-step-skew rotor is adopted to further reduce the torque ripple. The slot pitch angle (i.e. 7.5°) is selected as skewing angle, so that the rotor steps are displaced of 2.5° each other. In Figure 18.2(c), it is possible to see a rotor pole and the three rotor steps used for the skewing.

18.6 Torque behaviors

Figure 18.5(a) reports the Finite Element (FE) and experimental torque behaviors obtained with different current amplitudes in Maximum Torque Per Ampere (MTPA) condition. It can be noted that for low currents (i.e. 4 A and 8 A), the FE average torque is higher than the experimental one. This is mainly due the manufacturing tolerance

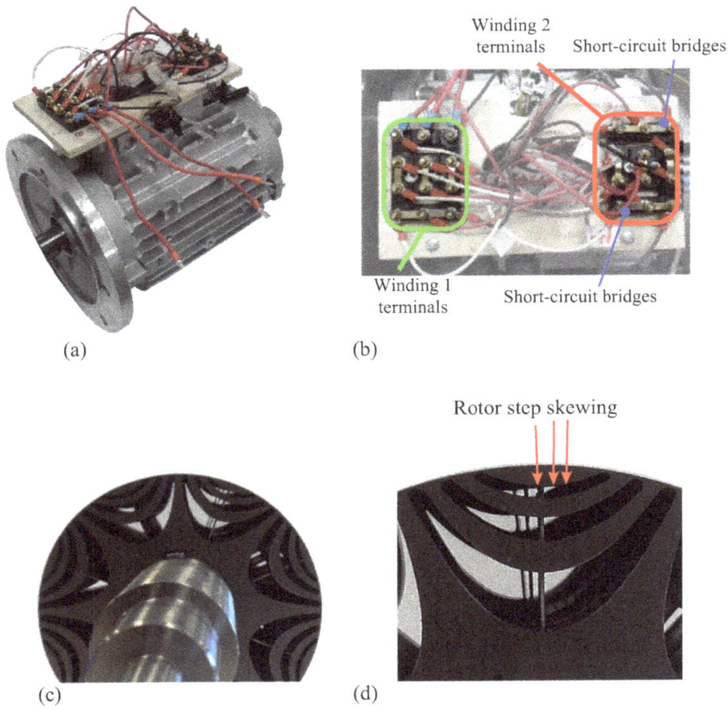

Figure 18.4 Views of the manufactured prototype parts: (a) prototype view, (b) terminals board and connections, (c) rotor prototype, (d) rotor skewing zoom

of the air gap and the different magnetic properties of the lamination with respect to the FE model. Conversely, for higher current amplitudes, there is a good agreement between the FE and experimental curves. It can be observed that the torque ripple in FWM condition is always lower than 3%.

Figure 18.5(b) reports the torque behaviors obtained in HWM for different current amplitudes. In this case, there is a good agreement in terms of average torque for current amplitudes higher than 8 A. The FE torque ripple is always higher than the experimental one and ranges between 8% and 20% except for the lowest current amplitude.

18.7 Magnetic coupling

In this section, the mutual coupling between the winding 1 and 2 is evaluated by means of FE analyses and experimental tests.

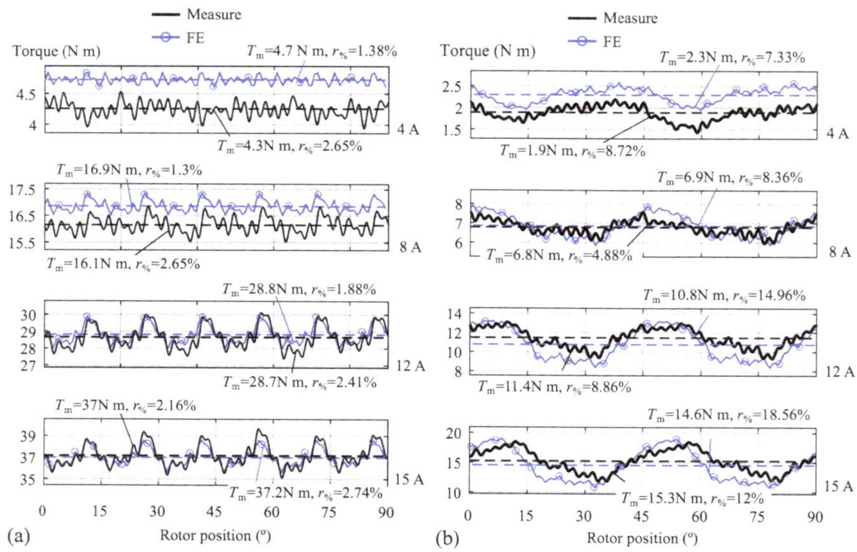

Figure 18.5 Torque versus rotor position: comparison between Finite Element (FE) and experimental results for different current amplitudes and operating conditions: (a) FWM and (b) HWM

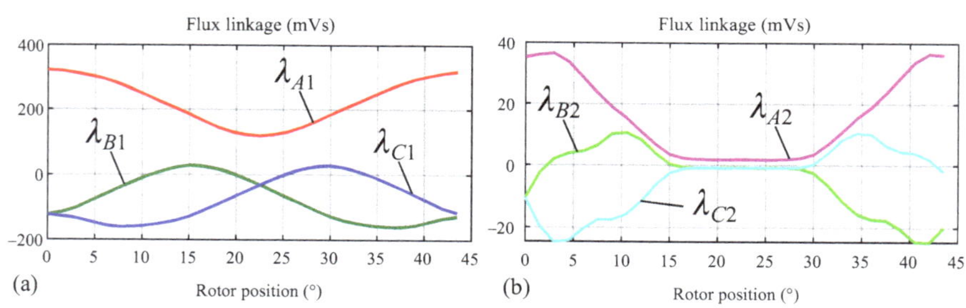

Figure 18.6 (a) Winding 1 phases and (b) winding 2 phases flux linkages versus rotor position when only A_1-phase is supplied

18.7.1 FE analysis results

The simulations are carried out feeding only the A_1-phase with the rated current (i.e. 8 A) and with all the left phases open circuited. The flux linkage of each phase is evaluated in an half electric period. Figure 18.6(a) and (b) shows the winding 1 and winding 2 flux linkages behaviors versus the rotor position. The maximum of the A_1-phase flux linkage is reached when the rotor d-axis is aligned to the A_1-phase magnetic axis. The flux-density plot corresponding to such a condition is shown in Figure 18.7(a). It can be noted that the stator region occupied by the winding 2 (left region) is interested by a higher magnetic loading. Conversely, when the rotor q-axis

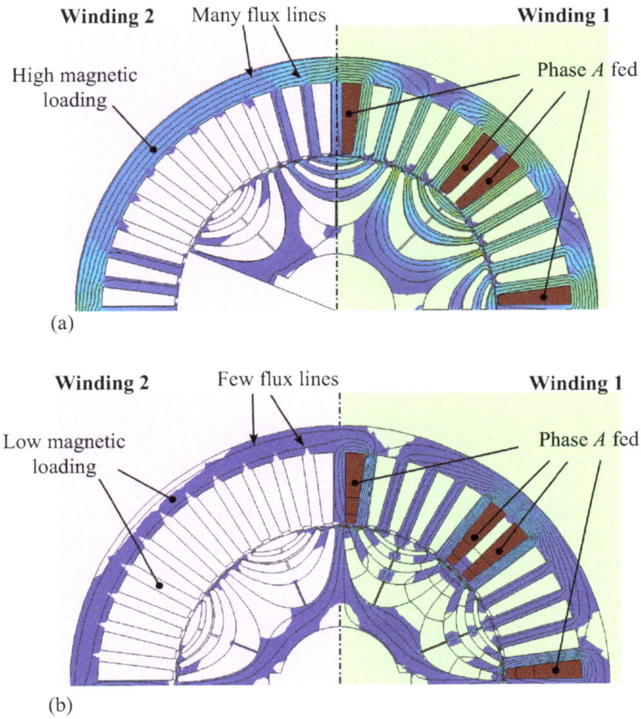

Figure 18.7 Flux density plot for different rotor positions when only the A_1-phase is supplied: (a) maximum magnetic coupling and (b) minimum magnetic coupling

is aligned to the A_1-axis, the relative flux linkage is minimum as well as the winding 2 flux linkages. This is also confirmed by the flux-density plot in Figure 18.7(b) in which it is evident the low magnetic loading of the winding 2 stator region and the low number of flux lines that flow through the stator yoke. In general, the ratio between the flux linkage of A_1-phase and the others ranges between 10 and 70, depending on the rotor position.

18.7.2 Experimental results

The mutual magnetic coupling between the winding 1 and 2 has been verified experimentally by means of the test bench setup reported in Figure 18.8. In such an experiment, the rotor of the motor under test (MUT) is kept in rotation by a prime mover at constant speed (i.e. 37.5 rpm) while only one phase (i.e. A_1) is fed by AC voltage supply with an electric frequency much greater than the motor electric one (i.e. 50 Hz). Then, the voltages of the disconnected phases and the current of the supplied phase are measured by means of voltage and current probes.

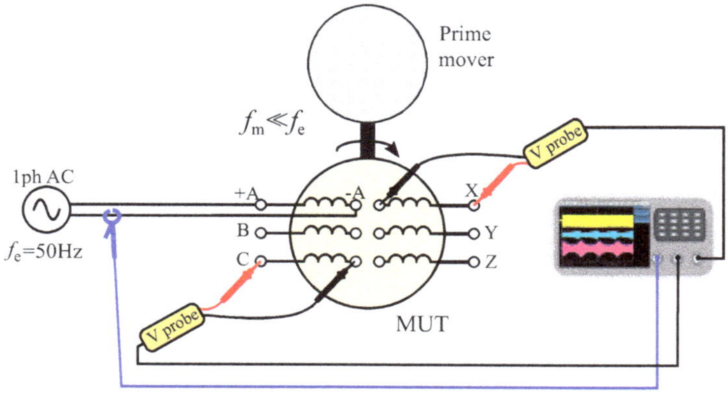

Figure 18.8 Test bench setup when only the phase A is fed

Their waveforms are finally displayed on the oscilloscope and are reported in Figure 18.9. The first plot shows the current behavior of the supplied phase. Its oscillation is due to the change of the motor inductance during the rotation. Its value is maximum when the rotor d-axis is aligned to the phase axis and is minimum after 90 electric degrees. The voltage of the supplied phase is reported in the second plot, its amplitude is constant since it is imposed by the power supply. The second and third plots show the voltage of the phase B_1 and C_1, respectively. It can be observed that the relative position of their waveforms reflects the FE results in Figure 18.6. Moreover, their maximum amplitude is about half of V_{A1} since in the three-phase winding system the mutual inductance is an half of the self-inductance. The last three plots show the phase voltages of winding 2. Their maximum amplitude is about 10% of the supplied phase that confirms the low magnetic coupling between the winding 1 and 2.

18.8 Three-phase short circuit test

This section reports the machine performance when the winding 1 is regularly supplied and a three-phase short circuit is applied to the winding 2 (SC-HWM). The aim of such an analysis is evaluating the maximum amplitude of the short-circuit currents and their effect on the average torque. The effect on the torque ripple has not been verified experimentally because of the test bench limitation.

18.8.1 FE analysis results

The simulations are carried out feeding the winding 1 with a given current vector and short circuiting the terminals of the winding 2. The results obtained at rated current in MTPA condition are reported in the following.

Figure 18.10 reports the torque waveforms in OC-HWM and SC-HWM. This analysis is carried out neglecting the skewing effect for simplicity. It can be noted that

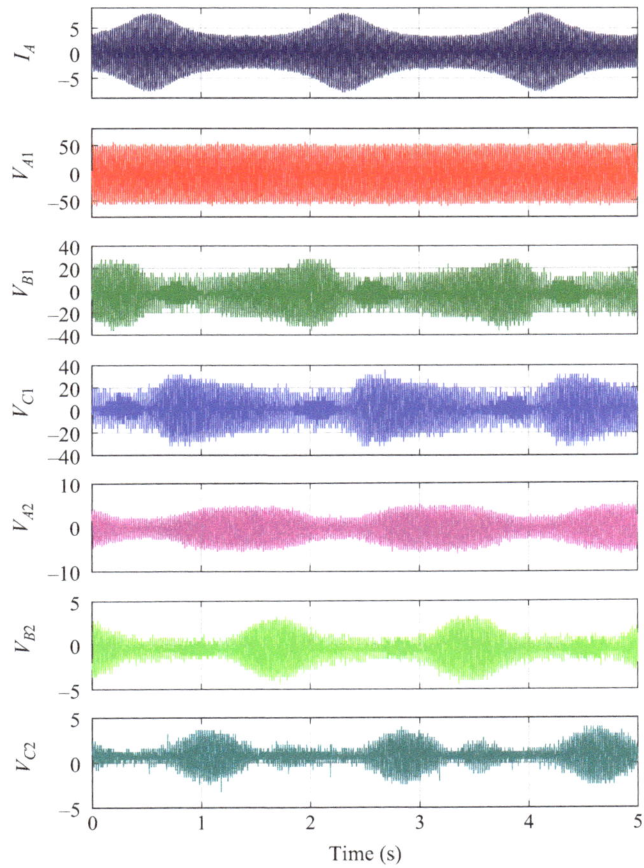

Figure 18.9 Voltages and current behaviors of the mutual coupling test. The voltage V_A is imposed by the power supply while the other voltages are induced. The maximum current amplitude is equal to the rated one (i.e. 8A)

the short-circuit condition involves a lower torque oscillation and slightly increases the average torque.

These results can be explained considering the flux density plots in Figure 18.11(a) (winding 2 open circuited) and (b) (winding 2 short-circuited). It is evident that the magnetic loading in short-circuit condition is lower as well as the flux leakage that flow in the winding 2 region. This is due to the demagnetization effect of the short-circuit currents.

The short-circuit current behaviors have been simulated with rated and overload conditions (i.e. 16 A) at the mechanical speed of 500 rpm, and the results are plotted in Figure 18.12(b) and (c), respectively. The maximum amplitude is always lower than 4 A, which means that such an operating condition does not represent a risk for the

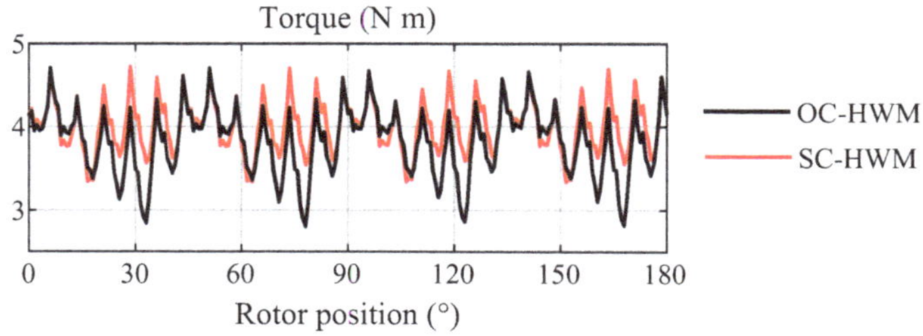

Figure 18.10 Torque versus rotor position in OC-HWM and SC-HWM in MTPA condition

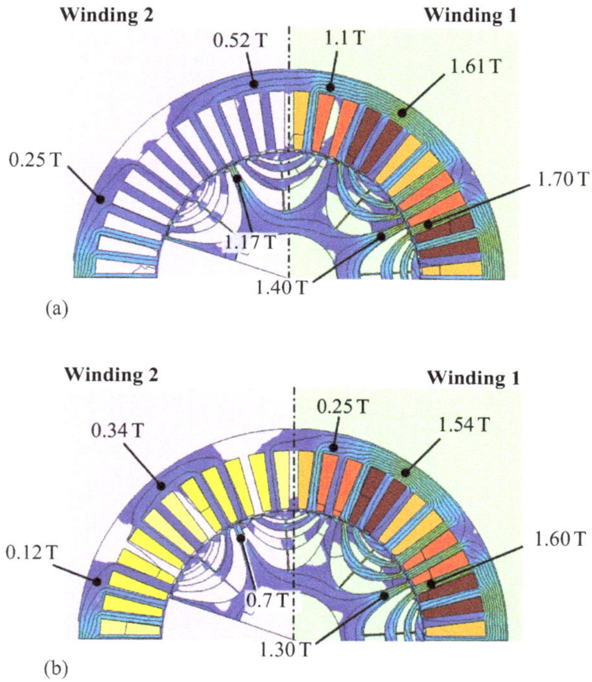

Figure 18.11 Flux density and current density plots in (a) OC-HWM and (b) SC-HWM. The winding 1 is normally fed with the rated current in Maximum Torque Per Ampere (MTPA) condition

inverter and the machine itself. Of course, the presence of the short-circuit currents affects the efficiency of the machine since additional Joule losses are present.

18.8.2 Experimental results

The test bench setup is shown in Figure 18.12(a). The winding 1 of the MUT is regularly fed with the current vector imposed by the user by means of the Host PC,

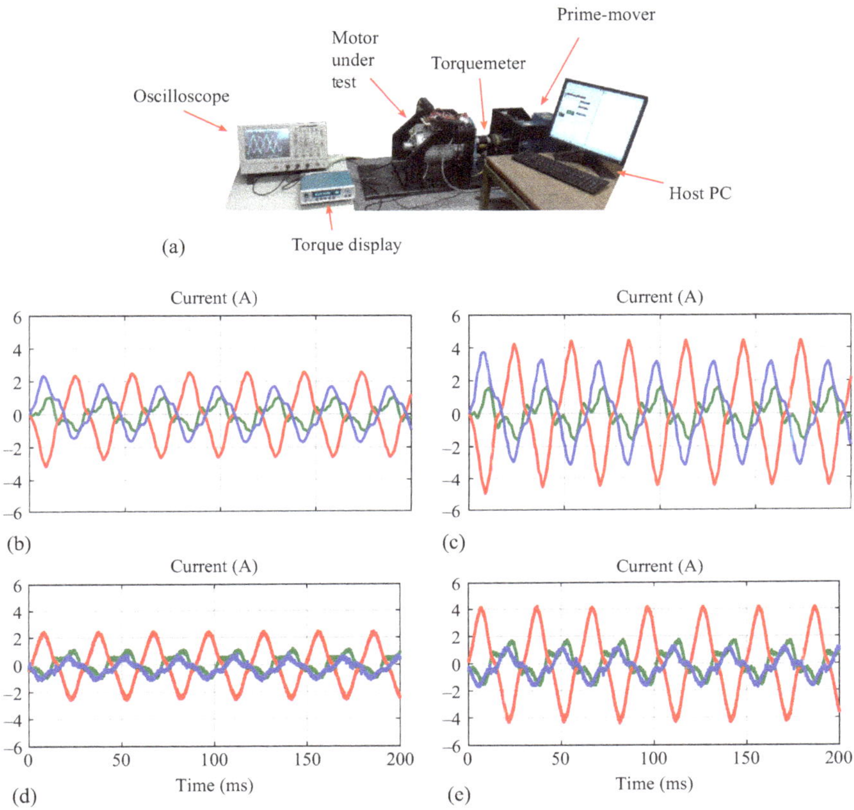

Figure 18.12 Short-circuit current behaviors for different current amplitudes in the winding 1: (a) test bench setup; (b) 8 A at 500 rpm, FEA; (c) 16 A at 500 rpm, FEA; (d) 8 A at 500 rpm, measure and (e) 16 A at 500 rpm, measure

and the mechanical speed is set by the user as well. The MUT is connected to the prime mover with a torquemeter, and its signal is then visualized by the torque display. The short-circuit currents are measured by means of three different current probes and then visualized on the oscilloscope screen.

The short-circuit currents waveform measured at different speed and current supply conditions are shown in Figure 18.12(d) and (e). It can be observed that their behavior and maximum amplitude are in good agreement with the FE results discussed before. The small difference between the waveform is due to the skewing effect of the experimental test.

Different working conditions have been tested varying the mechanical speed and the current amplitude of the winding 1, the results are reported in Table 18.3. In general, the average torque in SC-HWM is higher than the OC-HWM, which confirms

Table 18.3 Average torque at different mechanical speeds in OC-HWM and SC-HWM

I	HWM	Mechanical speed [rpm]				
[A]		100	200	300	400	500
8	OC	6.96	6.91	6.88	6.84	6.8
	SC	7.14	7.15	7.15	7.12	7.11
12	OC	11.49	11.45	11.4	11.36	11.34
	SC	11.94	12.08	12.1	12.11	12.1
16	OC	15.29	15.24	15.21	15.16	15.12
	SC	15.93	16.15	16.22	16.24	16.25

the observations made on the FE results. In particular, this increment is amplified by the mechanical speed. For example, with the rated current at 100 rpm, the torque increment is about 2.6% while, at 500 rpm, the increment is about 4.6%. In overload condition, the torque increment, in the same mechanical speeds, passes from 4.2% to 7.5%.

18.9 Discussion

This chapter dealt with the optimization and analysis of an SyR motor with dual three-phase winding for FT applications.

The stator and rotor have been optimized by means of a multi-objective differential evolution algorithm in order to obtain good performance both in full-winding and half-winding modes. Two stator geometries have been investigated: the standard one and the custom one. The optimization results revealed that only the first one is suitable both in terms of time consumption and results.

A prototype of the most promising individual provided by the optimization has been manufactured and tested to prove the FE results. The torque waveform provided good results in terms of average torque and torque ripple both in full-winding and half-winding modes. The mutual magnetic coupling between the phases of the dual three-phase winding has been tested in order to verify the amplitude of the induced voltages in the disconnected phases in half-winding mode. The experimental results confirmed the FE predictions since very low induced voltages have been observed. A three-phase short circuit of one of the two three-phase winding has been finally tested to verify the amplitude of the short-circuit currents and their effect on the motor performance. From the measurements and FE analysis as well, it has been observed that the maximum short-circuit current is always lower than the rated one and, in addition, the demagnetizing effect of the short-circuit currents has a beneficial effect on the average torque and torque ripple.

Chapter 19

Self-sensing-oriented optimization of synchronous reluctance machine design

This chapter deals with self-sensing-oriented optimization of Synchronous Reluctance Machines (SYRMs). This kind of machine is among the most challenging ones to control without the position sensor at low speed. In fact, typical position estimations adopt High-Frequency (HF) voltage injection that heavily relies on the intrinsic machine saliency. However, both at low and high currents, such saliency is not guaranteed due to the presence of the iron ribs and to the saturation of the iron material, respectively. Furthermore, the estimation algorithm could also become unstable due to the absence of convergence points. The aim of the chapter is to tackle this issue, embedding proper sensorless-control capability into the design through multi-objective optimizations.

Several works in the past tried to enhance the self-sensing capability of interior permanent magnet motors [195–197], focusing on avoiding that the Maximum Torque Per Ampere (MTPA) trajectory crosses the point where the current error signal is zero. Despite that, few or no research is available for SYRM, where the convergence issue is different in nature. In this chapter, the non-convergence issue is tackled from early electromagnetic design phase, trying to embed the solution into a complete optimization of the motor. The sensorless technique considered hereafter is the injection of HF pulsating voltage along the estimated d-axis, referred to as $\tilde{d}$, and the response is the $\tilde{q}$-axis current (Figure 19.1(a)). A scheme of the electric drive with the HF injection is shown in (Figure 19.1(b)). So the applied HF voltage vector is

$$u_{\tilde{d}h}(t) = U_{\tilde{d}h}\cos\omega_h t\,, \qquad u_{\tilde{q}h}(t) = 0 \tag{19.1}$$

where ω_h is the HF signal angular frequency and $U_{\tilde{d}h}$ is the injected voltage amplitude. The resulting current error signal is

$$\begin{aligned} i_{\tilde{q}h}(t) &= I_{\tilde{q}h}\sin\omega_h t \\ I_{\tilde{q}h} &= \frac{-U_{\tilde{d}h}}{\omega_h[\ell_d\ell_q - \ell_{dq}^2]}[\ell_\Delta \sin 2\Delta\vartheta + \ell_{dq}\cos 2\Delta\vartheta] \end{aligned} \tag{19.2}$$

where $\Delta\vartheta$ is the electrical position estimation error, and ℓ_x are the differential inductances.* If there was no ℓ_{dq}, such response would be zero when the position estimation

*See for the derivation.

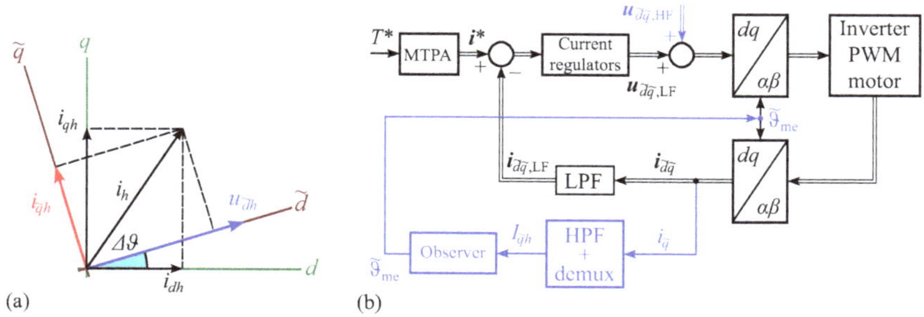

Figure 19.1 Operating principle and control scheme implementation of the high-frequency signal injection for sensorless rotor position estimation: (a) position estimation error and current error signal and (b) control scheme of the electric drive with high-frequency signal injection

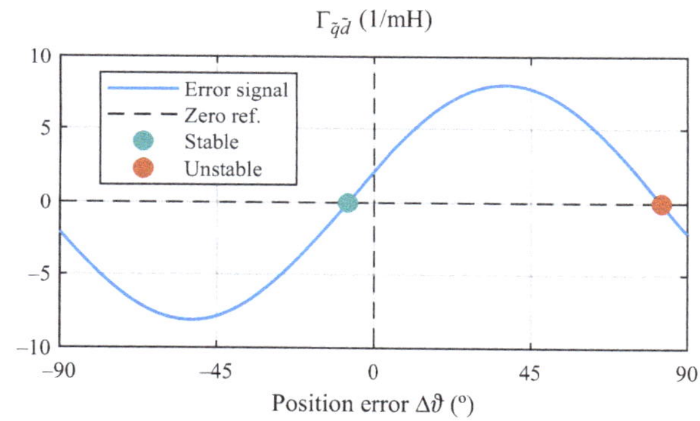

Figure 19.2 Ideal error signal, $\Gamma_{\tilde{q}\tilde{d}}$, with constant differential inductances

error is zero. Figure 19.2 reports the ideal error signal, $\Gamma_{\tilde{q}\tilde{d}}$. It is a pure sinusoid that crosses zero twice in a semi-period. One of the zeros is stable, the green one, while the other one is unstable, the red one. The stable convergence point is slightly leading the correct rotor position (zero estimating error $\Delta\vartheta$) due to the presence of the cross-saturation differential inductance, ℓ_{dq}.

The chapter is organized as follows: at first, classical optimization is performed on the rotor of the machine to maximize the output torque and minimize the torque ripple. One optimal geometry is then investigated to determine its HF sensorless-control capability through error signal maps. Next, the stator and rotor are optimized jointly with the additional objective of minimizing the cross-saturation differential inductance. Finally, one more objective is added to the multi-objective optimization to further improve the self-sensing capability of the motor.

19.1 Example of typical optimization

The specifications and the constraints of the machine under analysis are reported in Table 19.1. In particular, the requirement is to obtain about 90 N m with the highest possible efficiency within the given dimensions. At first, also the stator geometry is a constraint, and its data are also reported in Table 19.1.

State-of-the-art optimization coupled to Finite Element Analysis (FEA) has been performed in order to assess the feasibility of the design and to get the overall machine performance. The main objectives for SYRMS are average torque and torque ripple. Since most of the losses are located in the stator and in this optimization the stator is kept the same, the efficiency depends only upon the average torque.

The objectives plane is reported in Figure 19.3. Every dot represents a simulated individual. Each evaluation took about 150 seconds. The color of the dot represents the age of the individual, while the black diamonds are the Pareto points that are those individuals who are not dominated by others: this means that no other individual has both objectives better. Thanks to the fact that almost all Pareto points exhibit proper average torque, and the individual with the lowest ripple at 92.3 N m has been selected for further investigations. The corresponding efficiency is 93.1%.

A complete mapping of the (i_d, i_q) plane has been carried out on this individual with FEA. In particular, the flux linkages $\lambda_d(i_d, i_q)$ and $\lambda_q(i_d, i_q)$ are computed and they are reported in Figure 19.4. These two maps are the fundamental elements for the following analysis. In fact, the electromagnetic torque can be readily computed, neglecting the dependence on the rotor position, through

$$T_m(i_d, i_q) = \frac{3}{2}p\left[\lambda_d(i_d, i_q)i_q - \lambda_q(i_d, i_q)i_d\right] \tag{19.3}$$

Table 19.1 Specifications of the motor

Quantity	Value	u.m.
Torque	90	N m
Speed	2 000	rpm
Efficiency	92	%
Stator outer diameter	180	mm
Stack length	288	mm
Number of slots	36	
Number of poles	4	
Stator inner diameter	110	mm
Tooth width	4.83	mm
Slot height	22.4	mm
Slot opening width	2.5	mm
Slot opening height	1	mm
Slot wedge height	0.5	mm

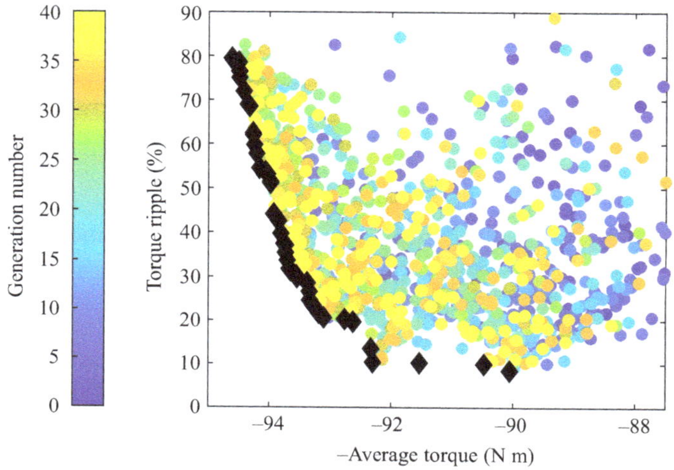

Figure 19.3 Objectives plane for the first rotor optimization

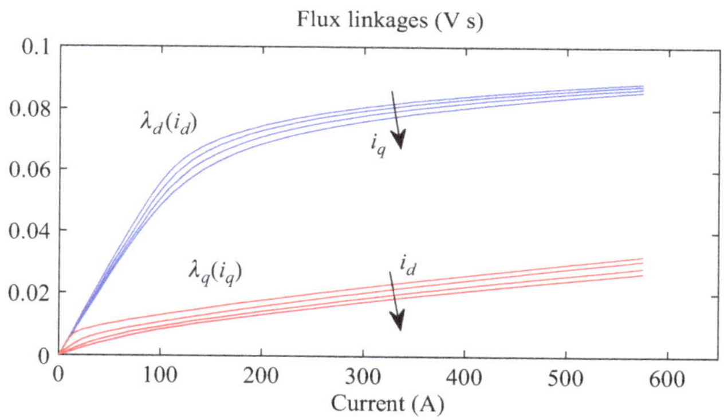

Figure 19.4 Flux linkages as functions of the two axes currents

Then, the differential inductances are defined as

$$\ell_d = \frac{\partial \lambda_d}{\partial i_d}, \quad \ell_{dq} = \frac{\partial \lambda_d}{\partial i_q} = \frac{\partial \lambda_q}{\partial i_d} = \ell_{qd}, \quad \ell_q = \frac{\partial \lambda_q}{\partial i_q} \tag{19.4}$$

These inductances are defined locally and strongly depend on the machine saturation. The half-difference between ℓ_q and ℓ_d is defined as $\ell_\Delta = (\ell_q - \ell_d)/2$ and this quantity displays the available saliency for HF signals. Also ℓ_Δ results to be a map of the (i_d, i_q) plane. In Figure 19.5 such a map is reported. The contour $\ell_\Delta = 0$ is highlighted in red, the MTPA trajectory with black dots, while the rated current circle is the black line.

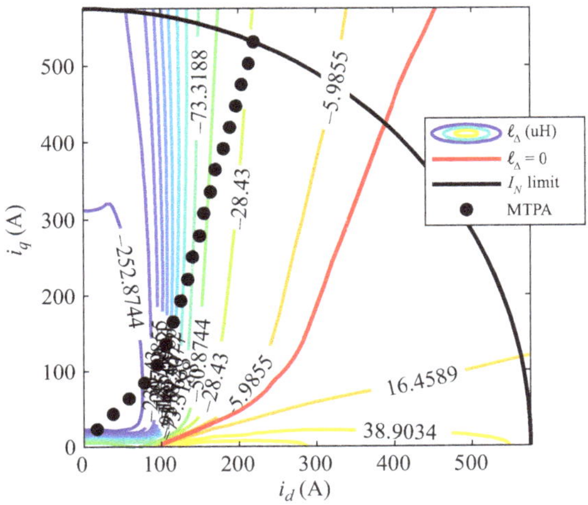

Figure 19.5 Map of the differential inductance ℓ_Δ in the (i_d, i_q) plane

Often in literature the region where $\ell_\Delta = 0$ has been reported as an unfeasible region for HF-injection sensorless control due to the loss of error signal proportional to ℓ_Δ. However, thanks to the presence of ℓ_{dq}, the sensorless estimation would still converge to a value of $\Delta\vartheta = \pi/2$. In the specific machine under analysis, the MTPA trajectory is quite far from the problematic $\ell_\Delta = 0$ contour. Even though this fact seems to guarantee proper convergence of the estimating algorithm, when the position is not known, the control may apply a current vector that is very different from the desired one. For instance, let us consider the base point that is the MTPA point at the rated current. That point will be the reference from which the estimation error $\Delta\vartheta$ will be measured. And since the control may apply any current vector with that amplitude, the estimation error signal is a complicated function of $\Delta\vartheta$ (the dependency of the quantities on the current references (i_d^*, i_q^*) is omitted for convenience):

$$I_{\tilde{q}h}(\Delta\vartheta) = \frac{U_{\tilde{d}h}}{\omega_h} \underbrace{\left[-\frac{\ell_\Delta(\Delta\vartheta)\sin 2\Delta\vartheta + \ell_{dq}(\Delta\vartheta)\cos 2\Delta\vartheta}{\ell_d(\Delta\vartheta)\ell_q(\Delta\vartheta) - \ell_{dq}^2(\Delta\vartheta)} \right]}_{\Gamma_{\tilde{q}\tilde{d}}} \tag{19.5}$$

The term between brackets in (19.5) is referred to as $\Gamma_{\tilde{q}\tilde{d}}$ in the following, because it acts like the inverse of an inductance.

In Figure 19.6(a) the behavior of $\Gamma_{\tilde{q}\tilde{d}}$ as a function of $\Delta\vartheta$ is reported. The most notable thing is that this waveform does not cross the x-axis, which means that the observer would not be able to converge to any working point, not even a wrong one. Recent works tried to tackle this issue with a current, angle, or current and angle compensations [198,199]. The scheme of the drive with the HF sensorless rotor position estimation with current and angle compensations is shown in Figure 19.7.

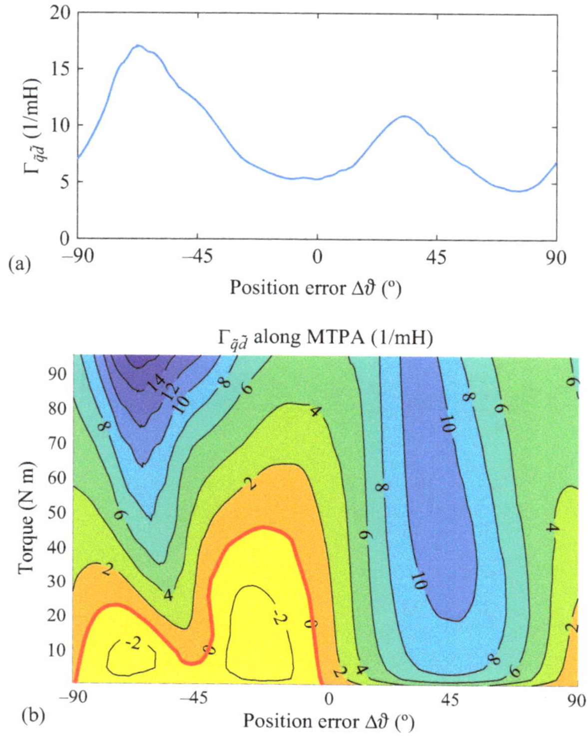

Figure 19.6 $\Gamma_{\tilde{q}\tilde{d}}$ plots for the base point and for every point along the MTPA trajectory: (a) $\Gamma_{\tilde{q}\tilde{d}}$ as a function of $\Delta\vartheta$ for the base point at the rated current and (b) $\Gamma_{\tilde{q}\tilde{d}}$ as a function of the output torque and position estimation error

Repeating the same computation of $\Gamma_{\tilde{q}\tilde{d}}$ for all the possible working points along the MTPA, a map of $\Gamma_{\tilde{q}\tilde{d}}$ as a function of the output torque and estimation error can be obtained. The map for the motor under analysis is shown in Figure 19.6(b). The contour $\Gamma_{\tilde{q}\tilde{d}} = 0$ identifies the operational limit of the HF-injection sensorless control. In fact, each horizontal slice of the map in Figure 19.6(b) is the estimation error signal (Figure 19.6(a)) used by the observer for the selected torque level. It can be noted that for this machine, the last achievable torque level without any compensation is at about 45N m (49% of the rated one).

Since $\Gamma_{\tilde{q}\tilde{d}}(\Delta\vartheta)$ stays always positive when convergence problems appear (see Figure 19.6(a)), the convergence region can also be found in the whole (i_d, i_q) plane for every working point looking at the minimum value of the $\Gamma_{\tilde{q}\tilde{d}}$ waveform: if this value is lower than zero, then there is a zero-crossing and therefore the estimating algorithm converges. Thus, the contour $\min \Gamma_{\tilde{q}\tilde{d}} = 0$ delimits the unfeasible region. In Figure 19.8, it can be observed that such a region is quite wide. The cause of this

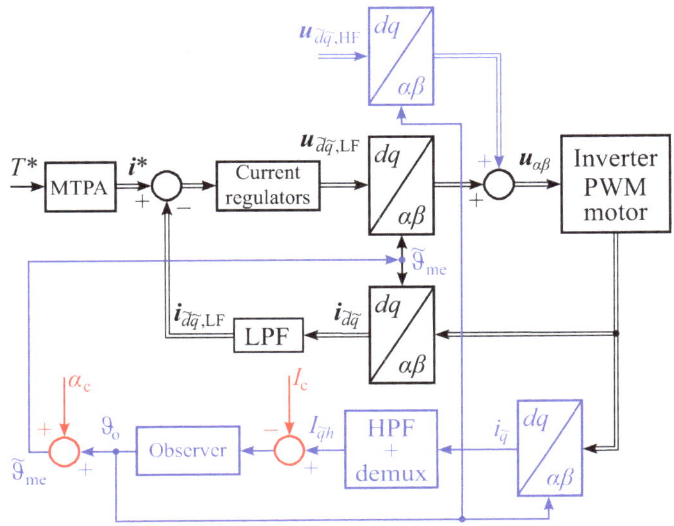

Figure 19.7 Control scheme of the electric drive with high-frequency signal injection and current and angle compensation

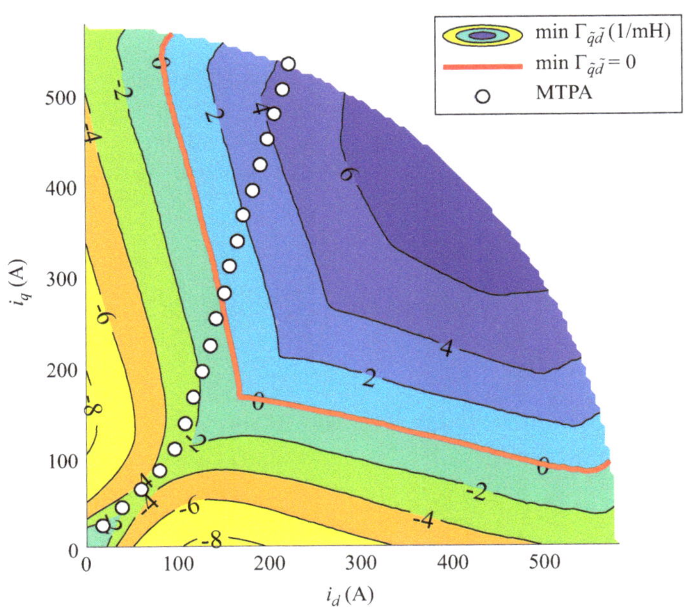

Figure 19.8 $\min \Gamma_{\tilde{q}\tilde{d}}(\Delta\vartheta)$ *map in the* (i_d, i_q) *plane for every working point*

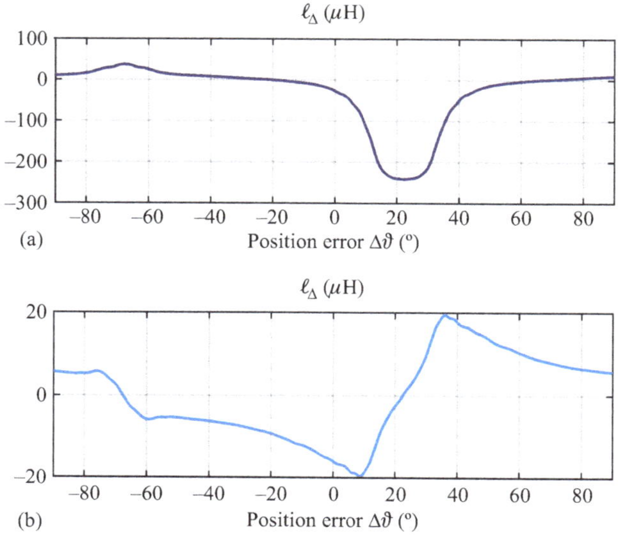

Figure 19.9 The differential inductances (a) ℓ_Δ and (b) ℓ_{dq} as functions of position estimation error at rated current

issue is the presence of harmonics in both ℓ_Δ and ℓ_{dq} along the current circles. The ℓ_Δ and ℓ_{dq} behaviors are reported in Figure 19.9 as functions of the position estimation error $\Delta\vartheta$ for $\alpha_i^e = \alpha_{i\,\mathrm{MTPA}}^e$.

The presence of a second-order harmonic is evident (in the figure only a semi-period is shown). The error signal can be computed substituting ℓ_Δ and ℓ_{dq} harmonics into the numerator of (19.5). Defining ϑ_i the angle starting from the d-axis, the error angle $\Delta\vartheta$ is defined through $\vartheta_i = \alpha_i^e + \Delta\vartheta$, where α_i^e is the reference current angle. So

$$\begin{aligned}
\ell_\Delta(\vartheta_i) &\approx \ell_\Delta^{(0)} + \ell_\Delta^{(2)} \cos 2\vartheta_i \\
&\approx \ell_\Delta^{(0)} + \ell_\Delta^{(2)} \cos 2\alpha_i^e \cos 2\Delta\vartheta - \ell_\Delta^{(2)} \sin 2\alpha_i^e \sin 2\Delta\vartheta \\
\ell_{dq}(\vartheta_i) &\approx \ell_{dq}^{(2)} \sin 2\vartheta_i \\
&\approx \ell_{dq}^{(2)} \sin 2\alpha_i^e \cos 2\Delta\vartheta + \ell_{dq}^{(2)} \cos 2\alpha_i^e \sin 2\Delta\vartheta
\end{aligned} \tag{19.6}$$

and then

$$\begin{aligned}
\Gamma_{\tilde{q}\tilde{d}} \sim\ & \underbrace{\left(\frac{\ell_{dq}^{(2)} - \ell_\Delta^{(2)}}{2}\right) \sin 2\alpha_i^e}_{\text{bias}} + \underbrace{\ell_\Delta^{(0)} \sin 2\Delta\vartheta}_{\text{useful signal}} \\
&+ \left(\frac{\ell_\Delta^{(2)} + \ell_{dq}^{(2)}}{2}\right) [\cos 2\alpha_i^e \sin 4\Delta\vartheta + \sin 2\alpha_i^e \cos 4\Delta\vartheta]
\end{aligned} \tag{19.7}$$

It can be seen that the second-order harmonics in ℓ_Δ and ℓ_{dq} are responsible for a bias that shifts vertically the $\Gamma_{\tilde{q}\tilde{d}}$ waveform. It should be noted that ℓ_{dq} is also the source of steady-state position estimation error, which is defined as

$$\varepsilon = \frac{1}{2}\arctan_2(\ell_{dq}, -\ell_\Delta) \tag{19.8}$$

In fact, if there was no ℓ_{dq}, the estimation error signal $\Gamma_{\tilde{q}\tilde{d}}$ would always go to zero when $\Delta\vartheta = 0$, regardless of the ℓ_Δ behavior. On the other hand, a large ℓ_Δ value is desirable to get HF current.

19.2 Self-sensing-oriented optimization

Two different optimizations have been carried out, with the aim of extending the HF-injection self-sensing torque range. The first optimization tried to minimize the cross-saturation differential inductance ℓ_{dq} for an intermediate current angle $\alpha_i^e = 45°$ at the rated current. The second optimization also aims at minimizing the ratio R_Δ, defined as

$$R_\Delta = \frac{\ell_\Delta^{(0)}}{\left|\ell_\Delta^{(2)}\right|} \approx \frac{2\langle\ell_\Delta\rangle}{|\ell_\Delta^{90°} - \ell_\Delta^{0°}|} \tag{19.9}$$

which is equal to maximize the absolute value of the average of ℓ_Δ (numerator) along the rated current circle and, at the same time, to minimize its second-order harmonic (denominator). To compute all these quantities, some linearized incremental permeability simulations are performed, which do not add too much computational load to the optimization procedure.

19.2.1 First self-sensing-oriented optimization

In this optimization, there were three objectives: torque ripple, efficiency and cross-saturation differential inductance. The parameters (degrees of freedom) were the three flux-barrier angles, the magnetic insulation ratio, the stator split ratio, the stator tooth width and slot height.

The results of the first optimization are reported in Figure 19.10. It can be observed that the cross-saturation differential inductance ℓ_Δ is proportional to the efficiency. So the higher the efficiency, the higher ℓ_{dq}. Such quantity strongly depends on the saturation of the machine, which is affected by the geometrical parameters (and so is the torque, thus the efficiency). Larger iron areas improve ℓ_{dq} but negatively affect the motor torque, and so the efficiency. In fact the rotor diameter decreases to leave more space for the stator yoke, and the slots shrink to obtain larger teeth. As a consequence, the slot area has to decrease, and so the copper losses increase.

In this optimization the Pareto front is a 3D surface. In Figure 19.10, the efficiency and ℓ_{dq} are reported in the two axes, while the torque ripple is displayed through the size of the markers: the bigger the marker, the lower the ripple. The average torque is not an objective now, but it is implicitly taken into account in the torque ripple and in

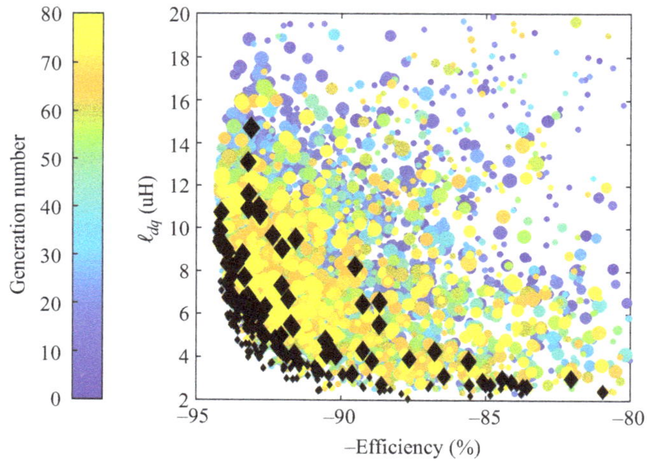

Figure 19.10 Objectives plane for the first self-sensing-oriented optimization. The dimension of the marker is inversely proportional to the torque ripple. The black diamonds are the Pareto points.

the efficiency. Some individuals on the Pareto front were selected for the complete mapping of the (i_d, i_q) plane to test their HF sensorless-control capabilities.

The map of $\Gamma_{\tilde{q}\tilde{d}}$ for one individual is shown in Figure 19.11. It can be seen that the highest torque that guarantees the convergence of the observer is now about 58N m. So the minimization of ℓ_{dq} improved the torque range of the sensorless control, even though the error signal still has a large negative bias: the negative values of the map are larger than the positive ones in absolute value. Also the $\Gamma_{\tilde{q}\tilde{d}}$ behavior at high torque for position error values between $-90°$ and $0°$ is improved (Figure 19.11), because now the highest values are closer to zero (the x-axis), requiring lower current amplitude compensation [199].

19.2.2 Second self-sensing-oriented optimization

This time two cascaded optimizations were performed: the first one considered three objectives (efficiency, cross-saturation differential inductance and the ratio R_Δ introduced in (19.9)), while the second one focused on the torque ripple and average torque only. This is due to the fact that the torque ripple is mainly due to the rotor geometry, while the self-sensing capability mainly depends on the stator parameters. The R_Δ numerator, $\langle \ell_\Delta \rangle$, is computed using four samples along the current limit circle: at 0°, 45°, 90° and at 67.5°, where also ℓ_{dq} was sampled. So the R_Δ numerator results in

$$\langle \ell_\Delta \rangle = \frac{2\ell_\Delta^{0°} + 3\ell_\Delta^{45°} + 2\ell_\Delta^{67.5°} + \ell_\Delta^{90°}}{8} \tag{19.10}$$

from a stepped average of the quantity.

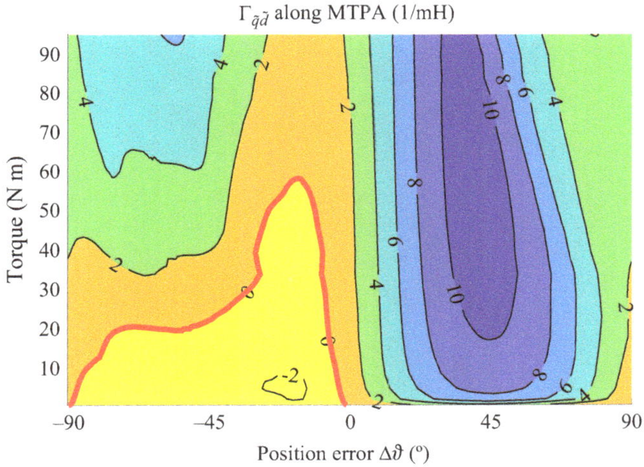

Figure 19.11 $\Gamma_{\tilde{q}\tilde{d}}$ *as a function of the output torque and estimation error*

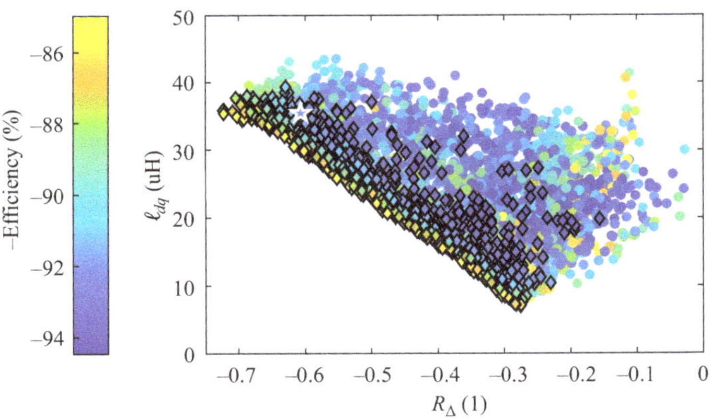

Figure 19.12 Objectives plane for the second self-sensing-oriented optimization. The color is proportional to the efficiency. The diamonds are the Pareto points, the star the solution that has been further investigated.

The results of the second self-sensing-oriented optimization are summarized in Figure 19.12. The figure displays ℓ_{dq} and R_{Δ} of the simulated motors along the two axes, while the color is proportional to the value of the efficiency, so a darker color means a higher efficiency. The diamonds are the individual on the Pareto front. A very clear relationship between ℓ_{dq} and R_{Δ} can be noticed, since there seems to be a linear limit on the values that these quantities may assume. In addition, the efficiency

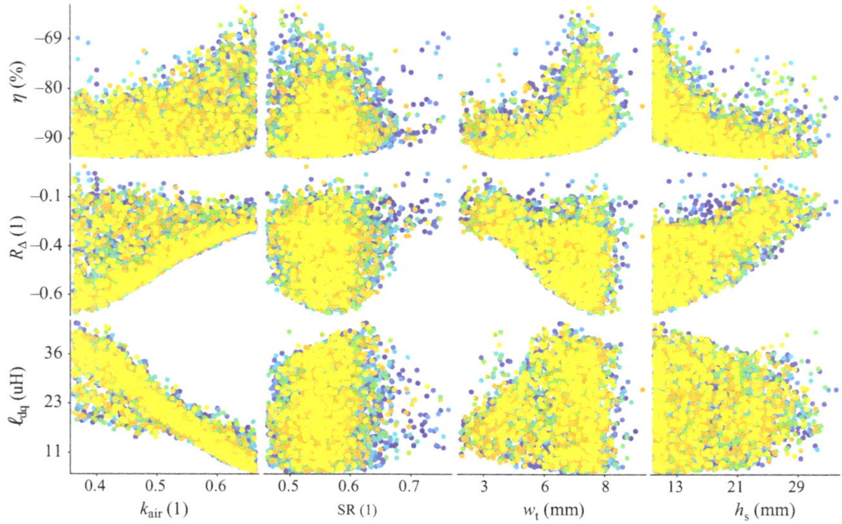

Figure 19.13 Objectives versus parameters plots for the third self-sensing-oriented optimization

increases on a direction almost perpendicular to this fictitious line, and this proves that efficiency is conflicting with the other two objectives.

The reason why ℓ_{dq} and R_Δ are conflicting objectives can be understood from Figure 19.13, looking at the second and third plots in the first column. In fact, there is a lower limit on the value of R_Δ as a function of k_{air}, and this limit increases with the increase of k_{air}. On the other hand, ℓ_{dq} decreases with the increase of k_{air}, which is beneficial. This means that increasing the amount of air in the rotor increases the saturation of the rotor iron, obviously, and improves the cross-saturation, which is mostly due to the saturation of the stator yoke. But it also leads to lower differential saliency, so a lower difference between the d- and q-axes inductances, and therefore to a lower absolute value of R_Δ.

Also in this case some individuals on the Pareto front have been selected for the complete mapping of the currents plane. In particular, the one marked by the star in Figure 19.12 is the most promising one, because it exhibits a high R_Δ ratio and proper efficiency. In Figure 19.14, the corresponding $\Gamma_{\tilde{q}\tilde{d}}$ map is shown. It can be observed that the maximum torque that guarantees the convergence is about 80N m which corresponds to 87% of the rated torque. This has been possible only thanks to the combined optimization of the stator and rotor geometries. Of course, this improvement comes at a cost: the efficiency drops to about 91.4%. The torque ripple has been minimized through a second optimization to a value of about 5%. It has been found that a low-torque ripple is fundamental to maintain the same self-sensing performance during the rotor rotation. The extension of the self-sensing range is

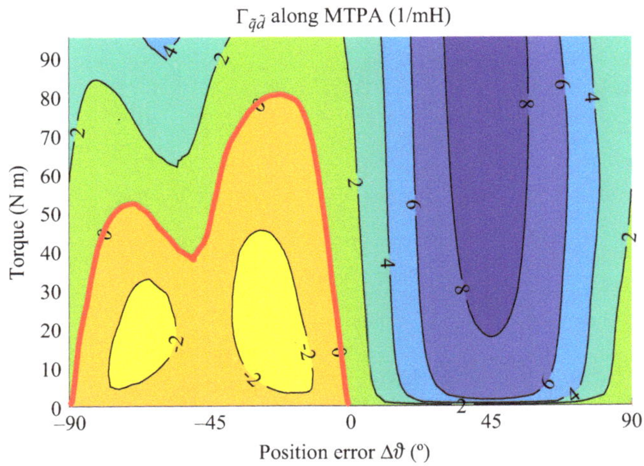

Figure 19.14 $\Gamma_{\tilde{q}\tilde{d}}$ *as a function of the output torque and estimation error*

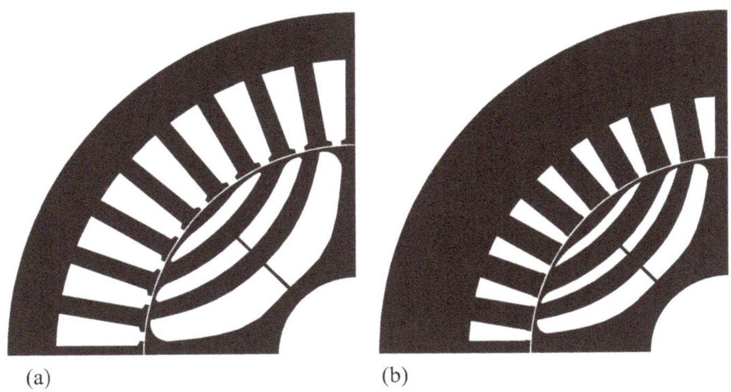

Figure 19.15 Investigated geometries resulting from the two optimizations: (a) typical optimization with given stator and (b) second self-sensing-oriented optimization

evident also looking at Figure 19.16(b) and comparing it to Figure 19.8. The unfeasible region now is greatly reduced and it presents lower absolute values.

In Figure 19.15, the geometry analyzed is shown and compared to the one first studied. The differences are evident. In particular, the rotor diameter is smaller, but that does not affect the torque capability of the machine. Overall, there is now more iron in the stator, both in the yoke and in the teeth. As a result, the slot area becomes smaller, which explains the decrease in the efficiency.

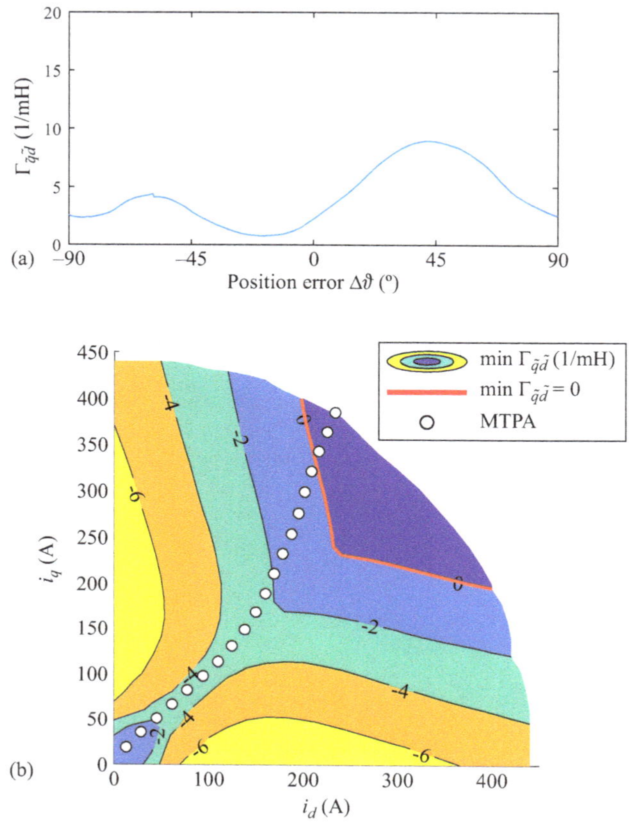

Figure 19.16 $\Gamma_{\tilde{q}\tilde{d}}$ *plots for the selected individual of the second self-sensing oriented optimization: (a)* $\Gamma_{\tilde{q}\tilde{d}}$ *as a function of* $\Delta\vartheta$ *for the base point and (b)* max $\Gamma_{\tilde{q}\tilde{d}}(\Delta\vartheta)$ *map in the* (i_d, i_q) *plane for every working point*

Finally, Figure 19.16(a) reports the waveform of $\Gamma_{\tilde{q}\tilde{d}}$ as a function of the estimation error for the base point. Comparing it with Figure 19.16(a), it can be noted that the behavior is improved even though it still does not cross the x-axis. In fact, it has less negative offset, and also the peak before $\Delta\vartheta = 0°$ is more pronounced and closer to zero. This means that smaller current and angle compensations are needed.

19.3 Discussion

In this chapter, an HF-injection self-sensing-oriented optimization scheme has been presented. A heavily saturated synchronous reluctance motor has been selected as a case study. At first, typical optimization has been performed on the rotor of the machine. The torque range for sensorless control was rather limited. After that, the source of the issue has been highlighted, together with the possible mitigating solutions. These solutions were embedded into the optimization algorithm, in two different ways. Even though the solution found was not able to exhibit self-sensing capability at the maximum torque, the torque range has been greatly extended to about 87% of the rated torque.

It has been demonstrated that the factor that mainly affects the converge of the observer is R_{Δ}, due to the stator saturation. However, the rotor should be designed accordingly to guarantee a high average torque, high efficiency and low-torque ripple. Finally, it has been deduced that a better self-sensing capability comes at the cost of lower efficiency. This fact highlights the importance of considering HF-injection sensorless control at early design phases.

Chapter 20
Conclusions

Due to the rapid interest in designing electric machines with no rare–earth materials, the synchronous reluctance machine is a perfect candidate. It can be without magnets at all or even assisted by low-energy magnets as ferrite magnets. The synchronous reluctance machine reaches performance slightly lower than the permanent magnet machines, but it is characterized by other interesting advantages. Among the others, a cheap and robust structure, capacity to sustain high overload, a high fault-tolerant capability. The possible drawbacks of such a machine, such as torque ripple, low-power factor or low power in flux-weakening operations, have been deeply analyzed and several remedies have been proposed.

In this monograph, a deep investigation of the performance of the synchronous reluctance machine has been carried out. Besides the common studies about such a machine, as found in literature, special applications have been analyzed and the performance of the machine has been highlighted. In addition, unusual approaches in the analysis have been proposed, which allow a better understanding of the machine characteristics.

A nonlinear analytical model of the machine is described to derive both average and torque harmonics as a function of the rotor geometry. Applying a proper saturation coefficient at the air gap, the analytical model results to be fast and accurate in any calculation. The torque maps allow the proper combination of barrier angles to be found to obtain maximum average torque and minimum torque ripple. In addition, the procedure to achieve the efficiency map is clearly described, so as to help the reader to apply to its projects.

The modern approaches to the design of the electric machines are often based on analyses based on finite-element method and optimization algorithms. Therefore, the multi-objective optimization is described and applied to the design of the synchronous reluctance machine. Since the rotor design is a more complex task, the typical approach is to determine the flux-barrier angles of the rotor in order to achieve the maximum average torque and a properly low-torque ripple.

A special application that has been deeply analyzed is that requiring high-speed operations. It is shown that the performance of the motor rapidly worsens with the speed due to the mechanical requirement to sustain the centrifugal forces on the rotor islands. Even if some advantages can be found as far as the increase of the electromechanical power with the speed is concerned, the power factor results quite low.

Therefore, effective advantages can be achieved only by small machines with low rotor diameters.

A peculiar characteristic of the synchronous reluctance machine is the absence of permanent magnet, whose flux linkage could induce electromotive force on the stator windings. This feature makes the machine suitable for fault-tolerant applications. A dual three–phase motor has been designed, manufactured and tested, showing that such a motor is able to operate even when a part of the winding is short-circuited. A minimum current value is induced in the short circuit so that the motor can continue to operate.

Finally, the anisotropic rotor allows a zero-speed sensorless control of the motor. The current response to a high-frequency voltage injection is adopted to identify the rotor position. However, it is shown that the rotor saliency is not always guaranteed due to the presence of the iron ribs and to the saturation of the iron material at high currents. The estimation algorithm could become unstable due to the absence of convergence points.

Although there are other possible studies that can be addressed about such a kind of machine, the Authors guess that this book is a proper starting point to extend the knowledge on the synchronous reluctance machine. From these chapters the active reader can get the proper tools to use in designing the machine for his specific application.

References

[1] Bianchi N, Bolognani S, and Chalmers BJ. Salient-rotor PM synchronous motors for an extended flux-weakening operation range. *IEEE Transactions on Industry Applications*. 2000;4(36):1118–1125.

[2] Taghavi S and Pillay P. A sizing methodology of the synchronous reluctance motor for traction applications. *IEEE Journal of Emerging and Selected Topics in Power Electronics*. 2014;2(2):329–340.

[3] Castagnaro E, Bacco G, and Bianchi N. Impact of geometry on the rotor iron losses in synchronous reluctance motors. *IEEE Transactions on Industry Applications*. 2019;55(6):5865–5872.

[4] Park J, Park S, Lee M, *et al.* Rotor design on torque ripple reduction for a synchronous reluctance motor with concentrated winding using response surface methodology. In: INTERMAG 2006 – IEEE International Magnetics Conference; 2006. p. 984–984.

[5] Sanada M, Hiramoto K, Morimoto S, *et al.* Torque ripple improvement for synchronous reluctance motor using an asymmetric flux barrier arrangement. *IEEE Transactions on Industry Applications*. 2004;40(4): 1076–1082.

[6] Vagati A, Boazzo B, Guglielmi P, *et al.* Ferrite assisted synchronous reluctance machines: A general approach. In: Proceedings – 2012 20th International Conference on Electrical Machines, ICEM 2012; 2012.

[7] Bianchi N, Degano M, and Fornasiero E. Sensitivity analysis of torque ripple reduction of synchronous reluctance and interior PM motors. *IEEE Transactions on Industry Applications*. 2015;51(1):187–195.

[8] Almeida AD, Ferreira F, and Baoming G. Beyond induction motors—Technology trends to move up efficiency. *IEEE Transactions on Industry Applications*. 2013;50(3):2103–2114.

[9] Kostko JK. Polyphase reaction synchronous motors. *Journal of the American Institute of Electrical Engineers*. 1923;42(11):1162–1168.

[10] Lamme BG. The story of the induction motor. *Journal of the American Institute of Electrical Engineers*. 1921;40(3):203–223.

[11] Matsuo T and Lipo TA. Field oriented control of synchronous reluctance machine. In: Proceedings of IEEE Power Electronics Specialist Conference-PESC'93; 1993. p. 425–431.

[12] Matsuo T and Lipo TA. Rotor design optimization of synchronous reluctance machine. *IEEE Transactions on Energy Conversion*. 1994;9(2):359–365.
[13] Sagawa M, Fujimura S, Togawa N, *et al.* New material for permanent magnets on a base of Nd and Fe. *Journal of Applied Physics*. 1984;55(6):2083–2087.
[14] Cui J, Kramer M, Zhou L, *et al.* Current progress and future challenges in rare-earth-free permanent magnets. *Acta Materials*. 2018;158:118–137.
[15] Mohri K and Yamasaky J. Permanent magnet and method for producing same. In: U.S. Patent, 4,756,848; 1988.
[16] Viswanathan P. Rare-earth iron-boron magnet containing Cerium and Lanthanum. In: U.S. Patent, 6,261,387 B1; 2001.
[17] Hofmann H and Sanders SR. High-speed synchronous reluctance machine with minimized rotor losses. *IEEE Transactions on Industry Applications*. 2000;36(2):531–539.
[18] Rawlings RD. *Materials Science and Engineering – Volume II*. Encyclopedia of Life Support Systems; Physical Sciences, Eng. Singapore: EOLSS Publishers Company Limited; 2009.
[19] Miller TJE. Brushless Permanent-magnet and Reluctance Motor Drives. Monographs in electrical and electronic engineering. Glasgow: Clarendon; 1989. Available from: https://books.google.it/books?id=2AmxHwAACAAJ.
[20] Hadjipanayis GC. Bonded Magnets: Proceedings of the NATO Advanced Research Workshop on Science and Technology of Bonded Magnets Newark, U.S.A. 22–25 August 2002. NATO Science Series II: Springer Netherlands; 2012. Available from: https://books.google.it/books?id=0L7tCAAAQBAJ.
[21] Poskovic E, Babetto C, Bianchi N, *et al.* The study of permanent magnet assisted reluctance machine with the adoption of NdFeB bonded magnets. In: 2018 International Symposium on Power Electronics, Electrical Drives, Automation and Motion (SPEEDAM); 2018. p. 274–279.
[22] Pošković E, Babetto C, Bianchi N, *et al.* Bonded magnets in PM-assisted synchronous reluctance machines: Performance dependence on the production technology. In: 2019 IEEE International Electric Machines Drives Conference (IEMDC); 2019. p. 442–448.
[23] Bai S-x, Zhang H, Lv L, *et al.* Progress of research on bonded Nd-Fe-B magnets. *Journal of Iron and Steel Research, International*. 2006;13: 489–493. Available from: http://www.sciencedirect.com/science/article/pii/S1006706X08602368.
[24] Ferraris L, Ferraris P, Poskovic E, *et al.* Theoretic and experimental approach to the adoption of bonded magnets in fractional machines for automotive applications. *IEEE Transactions on Industrial Electronics*. 2012;59(5):2309–2318.
[25] Ferraris L and Poškovic E. Bonded magnets for brushless fractional machines: Process parameters effects evaluation. In: IECON 2013 – 39th Annual Conference of the IEEE Industrial Electronics Society; 2013. p. 2632–2637.
[26] Moghaddam RR. Synchronous Reluctance Machine for VSD Application [PhD Thesis]. KTH School of Electrical Engineering; 2011.

[27] Wang Y, Bacco G, and Bianchi N. Geometry analysis and optimization of PM-assisted reluctance motors. *IEEE Transactions on Industry Applications*. 2017;53(5):1. Available from: http://ieeexplore.ieee.org/document/7921587/.

[28] Babetto C, Bacco G, and Bianchi N. Design methodology for high speed synchronous reluctance machines. *IET Electric Power Applications*. 2018;12(8).

[29] Vagati A, Franceschini G, Marongiu I, *et al.* Design criteria of high performance synchronous reluctance motors. In: Conference Record of the 1992 IEEE Industry Applications Society Annual Meeting. vol. 1; 1992. p. 66–73.

[30] López C, Michalski T, Espinosa A, *et al.* Rotor of synchronous reluctance motor optimization by means reluctance network and genetic algorithm. In: 2016 XXII International Conference on Electrical Machines (ICEM); 2016. p. 2052–2058.

[31] Torres CL, Michalski T, Espinosa AG, *et al.* Fast optimization of the magnetic model by means of reluctance network for PMa-SynRM. In: IECON 2016 – 42nd Annual Conference of the IEEE Industrial Electronics Society; 2016. p. 1642–1647.

[32] Babetto C, Bianchi N, López C, *et al.* High-speed synchronous reluctance motors: Computation of the power limits by means of reluctance networks. In: 2018 IEEE 18th International Power Electronics and Motion Control Conference (PEMC); 2018. p. 556–561.

[33] López-Torres C, Espinosa AG, Riba JR, *et al.* Design and optimization for vehicle driving cycle of rare-earth-free SynRM based on coupled lumped thermal and magnetic networks. *IEEE Transactions on Vehicular Technology*. 2018 1;67(1):196–205.

[34] Bianchi N, Bolognani S, Bon D, *et al.* Rotor flux-barrier design for torque ripple reduction in synchronous reluctance and PM-assisted synchronous reluctance motors. *IEEE Transactions on Industry Applications*. 2009;45(3):921–928.

[35] Bianchi N, Bolognani S, Bon D, *et al.* Torque ripple of a synchronous reluctance motor: Effect of the position of the flux-barriers and rotor design for its reduction; 2007.

[36] Barcaro M. Design and Analysis of Interior Permanent Magnet Synchronous Machines for Electric Vehicles [PhD Thesis]. University of Padova; 2011. Available from: http://paduaresearch.cab.unipd.it/3497/.

[37] Chua LO, Desoer CA, and Kuh ES. *Linear and Nonlinear Circuits*. Rao S, Elken AE, editors. New York, NY: McGraw-Hill; 1987.

[38] Bianchi N. Analysis of the IPM motor – Part I. In: Bianchi N, Jahns TM, editors. *Design, Analysis, and Control of Interior PM Synchronous Machines, Tutorial Course Notes IAS*. 1st ed. Seattle, WA: CLEUP Padova; 2006.

[39] Bianchi N and Barcaro M. Iron losses reduction in synchronous motors with anisotropic rotor. In: Proceedings – 34th Annual Conference of the IEEE Industrial Electronics Society, IECON 2008. 1. Orlando, FL; 2008. p. 1258–1263.

[40] Han SH, Soong WL, and Jahns TM. An analytical design approach for reducing stator iron losses in interior PM synchronous machines during flux-weakening operation. In: 2007 IEEE Industry Applications Annual Meeting. New Orleans; 2007. p. 103–110.

[41] Pellegrino G, Guglielmi P, Vagati A, *et al.* Core losses and torque ripple in IPM machines: Dedicated modeling and design tradeoff. *IEEE Transactions on Industry Applications*. 2010;46(6):2381–2391.

[42] Barcaro M, Meneghetti G, and Bianchi N. Structural analysis of the interior PM rotor considering both static and fatigue loading. *IEEE Transactions on Industry Applications*. 2014;50(1):253–260.

[43] Babetto C, Bacco G, and Bianchi N. Analytical power limits curves of high-speed synchronous reluctance machines. *IEEE Transactions on Industry Applications*. 2018;55(2):1–8.

[44] Lipo TA, Vagati A, Miller TJ, *et al.* *Synchronous Reluctance Motors and Drives – A New Alternative*. Denver, CO: IEEE Industry Applications Society; 1994.

[45] Barcaro M and Bianchi N. Interior PM machines using ferrite to replace rare-earth surface PM machines. *IEEE Transactions on Industry Applications*. 2014;50(2):979–985.

[46] Silvester PP and Ferrari RL. *Finite Elements for Electrical Engineers*. Cambridge, UK: Cambridge University Press; 1996.

[47] Bianchi N. *Electrical Machine Analysis Using Finite Elements. Power Electronics and Applications*. Boca Raton, FL: CRC Press; 2005.

[48] Salon S. *Finite Element Analysis of Electrical Machines. Power Electronics and Power Systems*. New York, NY: Springer US; 2012.

[49] Bianchi N, Fornasiero E, Carraro E, *et al.* Electric vehicle traction based on a PM assisted synchronous reluctance motor. In: 2014 IEEE International Electric Vehicle Conference, IEVC 2014. Florence; 2014. p. 6–11.

[50] Bianchi N, Fornasiero E, Ferrari M, *et al.* Experimental comparison of PM assisted synchronous reluctance motors. *IEEE Transactions on Industry Applications*. 2016;52(1):163–171.

[51] Ferrari M, Bianchi N, and Fornasiero E. Analysis of rotor saturation in synchronous reluctance and PM-assisted reluctance motors. *IEEE Transactions on Industry Applications*. 2015;51(1):169–177.

[52] Vagati A, Pastorelli M, Franceschini G, *et al.* Design of low-torque-ripple synchronous reluctance motors. *IEEE Transactions on Industry Applications*. 1998;34(4):758–765.

[53] Moghaddam RR. Synchronous Reluctance Machine (SynRM) in Variable Speed Drives (VSD) Applications [PhD Thesis]. The Royal Institute of Technology, Stockholm; 2011. Available from: http://www.diva-portal.org/smash/record.jsf?pid=diva2:417890.

[54] Gamba M. Design of Non Conventional Synchronous Reluctance Machine [PhD Thesis]; 2017.

[55] Bacco G and Bianchi N. Choice of flux-barriers position in synchronous reluctance machines. In: 2017 IEEE Energy Conversion Congress and Exposition (ECCE). Cincinnati, OH; 2017. p. 1872–1879.

[56] Bianchi N, Bolognani S, Bon D, *et al.* Rotor flux-barrier design for torque ripple reduction in synchronous reluctance motors. In: Conference Record of the 2006 IEEE 41st Industry Application Annual Meeting. vol. 3. Tampa, FL; 2006. p. 1193 –1200.

[57] Barcaro M, Bianchi N, and Magnussen F. Rotor flux-barrier geometry design to reduce iron losses in synchronous IPM motors under FW operations. In: IEEE International Electric Machines and Drives Conference, 2009. IEMDC '09; 2009. p. 928 –935.

[58] Bianchi N, Bolognani S, Bon D, *et al.* Torque harmonic compensation in a synchronous reluctance motor. *IEEE Transactions on Energy Conversion*. 2008;23(2).

[59] Barcaro M and Bianchi N. Torque components in integral- and fractional-slot IPM machines. In: 2011 IEEE International Electric Machines Drives Conference (IEMDC); 2011. p. 1340–1345.

[60] Dular P and Geuzaine C. GetDP reference manual: The documentation for GetDP, a general environment for the treatment of discrete problems. Available from: https://getdp.info/doc/texinfo/getdp.html

[61] Dular P, Geuzaine C, Henrotte F, *et al.* A general environment for the treatment of discrete problems and its application to the finite element method. *IEEE Transactions on Magnetics*. 1998;34(5):3395–3398.

[62] Slaughter WS. The Linearized Theory of Elasticity. Birkhäuser Boston; 2012. Available from: https://books.google.it/books?id=TxvnBwAAQBAJ.

[63] Dell'Acqua LC. Meccanica delle strutture. No. v. 1 in Collana di Istruzione Scientifica.Serie di Ingegneria Civile, 3 vol. McGraw-Hill Companies; 1992. Available from: https://books.google.it/books?id=b25uAAAACAAJ.

[64] Meneghetti G, Manzolaro M, and Quaresimin M. *Introduction to the Structural Analysis with ANSYS Numerical Code*. Progetto Libreria; 2014. Available from: https://books.google.it/books?id=Uv3voQEACAAJ.

[65] Geuzaine C and Remacle J. Gmsh: A three-dimensional finite element mesh generator with built-in pre- and post-processing facilities. *International Journal for Numerical Methods in Engineering*. 2009;79:1309–1331.

[66] Lancaster PR and Mitchell D. *Advanced Solid Mechanics: Theory, Worked Examples and Problems*. Macmillan; 1980. Available from: https://books.google.it/books?id=EoFUNAAACAAJ.

[67] Lazzarin P. Esercizi di costruzione di macchine. CUSL Nuova vita; 1995. Available from: https://books.google.it/books?id=VWhGMwEACAAJ.

[68] López C, Michalski T, Espinosa A, *et al.* New SynRM design approach based on behaviour maps analysis. In: 2016 XXII International Conference on Electrical Machines (ICEM); 2016. p. 1915–1921.

[69] Goss J, Popescu M, Staton D, *et al.* A comparison between maximum torque/ampere and maximum efficiency control strategies in IPM synchronous machines. In: 2014 IEEE Energy Conversion Congress and Exposition (ECCE); 2014. p. 2403–2410.

[70] Mahmoudi A, Soong WL, Pellegrino G, *et al.* Loss function modeling of efficiency maps of electrical machines. *IEEE Transactions on Industry Applications*. 2017;53(5):4221–4231.

[71] Shah SB and Arkkio A. Efficiency map prediction of flux switching machine. In: 2015 18th International Conference on Electrical Machines and Systems (ICEMS); 2015. p. 1490–1493.

[72] Bojoi R, Armando E, Pastorelli MA, *et al.* Efficiency and loss mapping of AC motors using advanced testing tools. In: 2016 XXII International Conference on Electrical Machines (ICEM); 2016. p. 1043–1049.

[73] Stipetic S, Goss J, Zarko D, *et al.* Calculation of efficiency maps using a scalable saturated model of synchronous permanent magnet machines. *IEEE Transactions on Industry Applications*. 2018;54(5):4257–4267.

[74] Pellegrino G, Vagati A, Guglielmi P, *et al.* Performance comparison between surface-mounted and interior PM motor drives for electric vehicle application. *IEEE Transactions on Industrial Electronics*. 2012 2;59(2): 803–811.

[75] Carbonieri M, Bianchi N, and Alberti L. Direct analysis of induction motor using finite element. In: 2018 IEEE Energy Conversion Congress and Exposition (ECCE); 2018. p. 277–283.

[76] Carbonieri M, Bianchi N, and Alberti L. A fast and direct analysis of three-phase induction motors using finite element. In: 2018 XIII International Conference on Electrical Machines (ICEM); 2018. p. 18–24.

[77] Boglietti A, Cavagnino A, Lazzari M, *et al.* A simplified thermal model for variable-speed self-cooled industrial induction motor. *IEEE Transactions on Industry Applications*. 2003;39(4):945–952.

[78] Cavazzuti M. *Optimization Methods: From Theory to Design*. Modena, Italy: Springer; 2013.

[79] Storn R and Price K. Differential evolution – A simple and efficient heuristic for global optimization over continuous spaces. *Journal of Global Optimization*. 1997;11(4):341–359. Available from: http://dx.doi.org/10.1023/A:1008202821328.

[80] Narzisi G. *Multi-Objective Optimization*. New York City; 2008.

[81] Bramerdorfer G, Tapia JA, Pyrhonen JJ, *et al.* Modern electrical machine design optimization: Techniques, trends, and best practices. *IEEE Transactions on Industrial Electronics*. 2018;65(10):7672–7684.

[82] Mohammadi MH. Rotor Design Optimization of Permanent Magnet—Assisted Synchronous Reluctance Machines for Traction Applications [Master Thesis]. McGill University, Montreal, Quebec, Canada; 2015.

[83] D'Amato G, Cecchi L, D'amato M, *et al.* Urban air pollution and climate change as environmental risk factors of respiratory allergy: An update. *Journal of Investigational Allergology and Clinical Immunology*. 2010;20(2):95–102.

[84] Lave LB and Seskin EP. *Air Pollution and Human Health*. New York, NY: RFF Press; 2013.

[85] Gurjar B, Jain A, Sharma A, *et al.* Human health risks in megacities due to air pollution. *Atmospheric Environment*. 2010;44(36):4606–4613.

[86] Zhang X, Xie J, Rao R, *et al.* Policy incentives for the adoption of electric vehicles across countries. *Sustainability*. 2014;6(11):8056–8078.

[87] de Santiago J, Bernhoff H, Ekergard B, *et al.* Electrical motor drivelines in commercial all-electric vehicles: A review. *IEEE Transactions on Vehicular Technology*. 2012 2;61(2):475–484.

[88] Frieske B, Kloetzke M, and Mauser F. Trends in vehicle concept and key technology development for hybrid and battery electric vehicles. In: 2013 World Electric Vehicle Symposium and Exhibition (EVS27). Barcelona, Spain; 2013. p. 1–12.

[89] Bianchi N, Bolognani S, and Zigliotto M. High-performance PM synchronous motor drive for an electrical scooter. *IEEE Transactions on Industry Applications*. 2001 9;37(5):1348–1355.

[90] Ravi N, Ekram S, and Mahajan D. Design and development of a in-wheel brushless D.C. motor drive for an electric scooter. In: 2006 International Conference on Power Electronic, Drives and Energy Systems. New Delhi, India; 2006. p. 1–4.

[91] Song B, Chang K, and Choi J. Design of an outer-rotor-type permanent magnet motor for electric scooter propulsion systems. In: The 2010 International Power Electronics Conference – ECCE ASIA. Sapporo, Japan; 2010. p. 2763–2742.

[92] Gruber W, Bäck W, and Amrhein W. Design and implementation of a wheel hub motor for an electric scooter. In: 2011 IEEE Vehicle Power and Propulsion Conference. Chicago, IL, USA; 2011. p. 1–6.

[93] Fodorean D, Idoumghar L, and Szabo L. Motorization for an electric scooter by using permanent-magnet machines optimized based on a hybrid metaheuristic algorithm. *IEEE Transactions on Vehicular Technology*. 2013; 62(1):39–49.

[94] Omekanda AM. Switched reluctance machines for EV and HEV propulsion: State-of-the-art. In: 2013 IEEE Workshop on Electrical Machines Design, Control and Diagnosis (WEMDCD). Paris, France; 2013. p. 70–74.

[95] Moghaddam RR, Magnussen F, and Sadarangani C. Novel rotor design optimization of synchronous reluctance machine for high torque density. In: 6th IET International Conference on Power Electronics, Machines and Drives (PEMD 2012); 2012. p. B32–B32. Available from: http://digital-library.theiet.org/content/conferences/10.1049/cp.2012.0256.

[96] Pellegrino G, Cupertino F, and Gerada C. Automatic design of synchronous reluctance motors focusing on barrier shape optimization. *IEEE Transactions on Industry Applications*. 2015;51(2):1465–1474.

[97] Pellegrino G, Jahns TM, Bianchi N, *et al. The Rediscovery of Synchronous Reluctance and Ferrite Permanent Magnet Motors: Tutorial Course Notes*. New York, NY: Springer; 2016.

[98] Cai H, Guan B, and Xu L. Low-cost ferrite PM-assisted synchronous reluctance machine for electric vehicles. *IEEE Transactions on Industrial Electronics*. 2014;61(10):5741–5748.

[99] Obata M, Morimoto S, Sanada M, *et al.* Performance of PMASynRM with ferrite magnets for EV/HEV applications considering productivity. *IEEE Transactions on Industry Applications*. 2014;50(4):2427–2435.

[100] Ehsani M, Gao Y, Longo S, *et al. Modern Electric, Hybrid Electric, and Fuel Cell Vehicles*. Boca Raton, FL: CRC Press; 2018.

[101] Pellegrino G and Cupertino F. FEA-based multi-objective optimization of IPM motor design including rotor losses. In: 2010 IEEE Energy Conversion Congress and Exposition. Atlanta, GA, USA; 2010. p. 3659–3666.

[102] Ooi S, Morimoto S, Sanada M, *et al.* Performance evaluation of a high-power-density PMASynRM with ferrite magnets. *IEEE Transactions on Industry Applications*. 2013;49(3):1308–1315.

[103] Babetto C, Bianchi N, and Benedetti G. Design and optimization of a PMASR motor for low-voltage E-scooter applications. In: 2019 IEEE International Electric Machines Drives Conference (IEMDC); 2019. p. 1016–1023.

[104] Pyrhonen J, Jokinen T, and Hrabovcova V. *Design of Rotating Electrical Machines*. Hoboken, NJ: John Wiley & Sons; 2013.

[105] Liwschitz-Garik M and Whipple CC. *Electric Machinery. No. v. 2 in Electric Machinery*. New York, NY: D. Van Nostrand Company, Inc.; 1947.

[106] Meeker D. *Finite Element Method Magnetics*; 2015. Available from: https://www.femm.info/wiki/HomePage

[107] Grauers A. Directly driven wind turbine generators. In: International Conference on Electrical Machines (ICEM). Vigo , Spain; 1996. p. 417–422.

[108] Lampola P. Electromagnetic design of an unconventional directly driven permanent-magnet wind generator. In: International Conference on Electrical Machines (ICEM)1. Istanbul, Turkey; 1998. p. 1705–1710.

[109] Popescu M, Mitulet A, Chihaia R, *et al.* Finite element analysis of a low speed permanent magnets synchronous generator with direct drive. In: 2013 – 8th International Symposium on Advanced Topics in Electrical Engineering, ATEE 2013; 2013.

[110] Zhou G, Xu X, Xiong Y, *et al.* Design and analysis of low-speed high torque direct-driven permanent magnet synchronous machines (PMSM) with fractional-slot concentrated winding used in coal mine belt conveyor system. In: 2017 20th International Conference on Electrical Machines and Systems, ICEMS 2017; 2017.

[111] Mohanarajah T, Rizk J, Hellany A, *et al.* Torque ripple improvement in synchronous reluctance machines. In: 2018 2nd International Conference On Electrical Engineering (EECon): November 2018. p. 44–50. Available from: https://ieeexplore.ieee.org/document/8541021/.

[112] Carraro E, Morandin M, and Bianchi N. Traction PMASR motor optimization according to a given driving cycle. *IEEE Transactions on Industry Applications*. 2016;52(1):209–216.

[113] Guglielmi P, Pastorelli MA, Carrer A, *et al.* An IPM-PMASR motor for home appliance washing machines. In: IECON 2013 – 39th Annual Conference of the IEEE Industrial Electronics Society; 2013. p. 2608–2613.

[114] Guglielmi P, Boazzo B, Armando E, *et al.* Magnet minimization in IPM-PMASR motor design for wide speed range application. In: 2011 IEEE Energy Conversion Congress and Exposition; 2011. p. 4201–4207.

[115] Bianchi N and Bolognani S. Design procedure of a synchronous motor for flux-weakening applications. In: Proceedings of EPE Symposium. Nancy, France; 1996. p. 235–240.

[116] Bianchi N, Fornasiero E, and Soong W. Selection of PM flux linkage for maximum low-speed torque rating in a PM-assisted synchronous reluctance machine. *IEEE Transactions on Industry Applications*. 2015;51(5):3600–3608.

[117] Babetto C, Bacco G, and Bianchi N. Analytical approach to determine the power limit of high-speed synchronous reluctance machines. In: 2017 IEEE International Electric Machines and Drives Conference, IEMDC 2017. Miami, FL; 2017. p. 2306–2312.

[118] Gerada D, Mebarki A, Brown NL, *et al.* High-speed electrical machines: Technologies, trends, and developments. *IEEE Transactions on Industrial Electronics*. 2014 6;61(6):2946–2959.

[119] Leuzzi R, Monopoli VG, Cupertino F, *et al.* High-speed machines: Typologies, standards, and operation under PWM supply. In: 2018 AEIT International Annual Conference; 2018. p. 1–6.

[120] Lahne H and Gerling D. Comparison of state-of-the-art high-speed high-power machines: Research study including a design example of a 50000 rpm induction machine. In: IECON 2015 – 41st Annual Conference of the IEEE Industrial Electronics Society; 2015. p. 3519–3524.

[121] Terzic MV, Mihic DS, and Vukosavić SN. Design of high-speed, low-inertia induction machines with drag-cup rotor. *IEEE Transactions on Energy Conversion*. 2014;29(1):169–177.

[122] Larsson M, Johansson M, Naslund L, *et al.* Design and evaluation of high-speed induction machine. In: IEEE International Electric Machines and Drives Conference, 2003. IEMDC'03. vol. 1; 2003. p. 77–82.

[123] Hao Z and Fengxiang W. Comparative study on high speed induction machine with different rotor structures. In: 2007 International Conference on Electrical Machines and Systems (ICEMS); 2007. p. 1009–1012.

[124] Gerada D, Mebarki A, Brown NL, *et al.* Design aspects of high-speed high-power-density laminated-rotor induction machines. *IEEE Transactions on Industrial Electronics*. 2011;58(9):4039–4047.

[125] Klima J and Vitek O. The influence of the rotor slits on the performance of high-speed solid-rotor induction machine. In: 2016 17th International Conference on Mechatronics – Mechatronika (ME); 2016. p. 1–5.

[126] Lateb R, Enon J, and Durantay L. High speed, high power electrical induction motor technologies for integrated compressors. In: 2009 International Conference on Electrical Machines and Systems; 2009. p. 1–5.

[127] Boules N. Impact of slot harmonics on losses of high-speed permanent magnet machines with a magnet retaining ring. *Electric Machines & Power Systems*. 1981;6(6):527–539. Available from: https://doi.org/10.1080/03616968108960089.

[128] Zhu ZQ, Ng K, and Howe D. Design and analysis of high-speed brushless permanent magnet motors. In: 1997 Eighth International Conference on

Electrical Machines and Drives (Conf. Publ. No. 444). Cambridge, UK; 1997. p. 381–385.

[129] Binder A, Schneider T, and Klohr M. Fixation of buried and surface-mounted magnets in high-speed permanent-magnet synchronous machines. *IEEE Transactions on Industry Applications*. 2006 7;42(4):1031–1037.

[130] Zhang F, Du G, Wang T, *et al.* Rotor retaining sleeve design for a 1.12-MW high-speed PM machine. *IEEE Transactions on Industry Applications*. 2015;51(5):3675–3685.

[131] Alberti L, Fornasiero E, Bianchi N, *et al.* Impact of rotor losses in a 12-slot 10-pole axial flux PM machine. In: 2008 IEEE Industry Applications Society Annual Meeting; 2008. p. 1–8.

[132] Berardi G and Bianchi N. Rotor losses reduction in high speed PM generators for organic Rankine cycle systems. In: 2017 IEEE Energy Conversion Congress and Exposition (ECCE); 2017. p. 5396–5402.

[133] Bianchi N, Bolognani S, and Luise F. Potentials and limits of high-speed PM motors. *IEEE Transactions on Industry Applications*. 2004;40(6): 1570–1578.

[134] Xing J, Wang F, Wang T, *et al.* Study on anti-demagnetization of magnet for high speed permanent magnet machine. *IEEE Transactions on Applied Superconductivity*. 2010;20(3):856–860.

[135] Berardi G, Bianchi N, and Gasperini D. A high speed PM generator for an Organic Rankine Cycle system. In: 2017 IEEE International Electric Machines and Drives Conference (IEMDC); 2017. p. 1–8.

[136] Rotor for a permanent magnet synchronous machine. US2003173853 (A1); 2003. Available form: https://patents.google.com/patent/US6768238

[137] Gan J, Chau KT, Chan CC, *et al.* A new surface-inset, permanent-magnet, brushless DC motor drive for electric vehicles. *IEEE Transactions on Magnetics*. 2000;36(5):3810–3818.

[138] Marino M. Analytical modeling and optimization of a radial permanent magnets synchronous machine. In: International Conference on Energy Efficiency in Motor Driven Systems; 2009.

[139] Kolehmainen J. Dovetail Rotor Poles in Synchronous Permanent Magnet and Reluctance Machines [PhD Thesis]. Aalto University; 2011.

[140] Vagati A, Pellegrino G, and Guglielmi P. Comparison between SPM and IPM motor drives for EV application. In: The XIX International Conference on Electrical Machines – ICEM 2010; 2010. p. 1–6.

[141] Fairall EW, Bilgin B, and Emadi A. State-of-the-art high-speed switched reluctance machines. In: 2015 IEEE International Electric Machines Drives Conference (IEMDC); 2015. p. 1621–1627.

[142] Raminosoa T, Blunier B, Fodorean D, *et al.* Design and optimization of a switched reluctance motor driving a compressor for a PEM fuel-cell system for automotive applications. *IEEE Transactions on Industrial Electronics*. 2010;57(9):2988–2997.

[143] Won SH, Ahn J S, Kim KC, *et al.* Performance prediction of switched reluctance motor considering windage loss for high speed application. In: 2006

12th Biennial IEEE Conference on Electromagnetic Field Computation; 2006. p. 417.

[144] Doherty RE and Nickle CA. Synchronous machines: An extension of Blondel's two-reaction theory. *Journal of the AIEE*. 1926;45(10):974–987.

[145] Hofer M and Schroedl M. Comparison of a flux barrier and a salient pole synchronous reluctance machine for high rotational speeds in electric traction applications. In: 2017 20th International Conference on Electrical Machines and Systems (ICEMS); 2017. p. 1–6.

[146] Hadi Zaim ME. High-speed solid rotor synchronous reluctance machine design and optimization. *IEEE Transactions on Magnetics*. 2009;45(3): 1796–1799.

[147] Ikäheimo J, Kolehmainen J, Känsäkangas T, Kivelä V, and Moghaddam RR. Synchronous high-speed reluctance machine with novel rotor construction. *IEEE Transactions on Industrial Electronics*. 2014 6;61(6):2969–2975.

[148] Cupertino F, Palmieri M, and Pellegrino G. Design of high-speed synchronous reluctance machines. In: 2015 IEEE Energy Conversion Congress and Exposition (ECCE); 2015. p. 4828–4834.

[149] Di Nardo M, Galea M, Gerada C, *et al.* Multi-physics optimization strategies for high speed synchronous reluctance machines. In: 2015 IEEE Energy Conversion Congress and Exposition (ECCE); 2015. p. 2813–2820.

[150] Di Nardo M, Calzo GL, Galea M, *et al.* Structural design optimization of a high speed synchronous reluctance machine. In: 2016 XXII International Conference on Electrical Machines (ICEM); 2016. p. 2073–2079.

[151] Jahns TM and Soong WL. Pulsating torque minimization techniques for permanent magnet AC motor drives – A review. *IEEE Transactions on Industrial Electronics*. 1996;43(2):321–330.

[152] Geuzaine C, Dular P, and Legros W. A general environment for the treatment of discrete problems and its application to coupled finite element and boundary integral methods. In: Proceedings of the 8th International IGTE Symposium on Numerical Field Calculation in Electrical Engineering. Graz, Austria; 1998.

[153] Babetto C, Bacco G, and Bianchi N. Analytical power limits curves of high-speed synchronous reluctance machines. *IEEE Transactions on Industry Applications*. 2018:1–8.

[154] Geuzaine C. GetDP: A general finite-element solver for the de Rham complex. In: PAMM Volume 7 Issue 1. Special Issue: Sixth International Congress on Industrial Applied Mathematics (ICIAM07) and GAMM Annual Meeting, Zürich 2007. vol. 7. Wiley; 2008. p. 1010603–1010604.

[155] Vagati A, Pastorelli M, Francheschini G, *et al.* Design of low-torque-ripple synchronous reluctance motors. *IEEE Transactions on Industry Applications*. 1998;34(4):758–765.

[156] Consoli A. Special section on robust operation of electrical drives. *IEEE Transactions on Power Electronics*. 2012;27(2):476–478.

[157] Bolvashenkov I, Herzog HG, Ismagilov F, *et al.* *Fault-Tolerant Traction Electric Drives: Reliability, Topologies and Components Design*. SpringerBriefs

in Electrical and Computer Engineering Series. Singapore: Springer Verlag; 2019. Available from: https://books.google.it/books?id=jRegDwAAQBAJ.

[158] Ruba M and Szabo L. Fault tolerant electrical machines. State of the art and future directions. *Journal of Computer Science and Control Systems*. 2008;1.

[159] Toliyat HA, Nandi S, Choi S, *et al. Electric Machines: Modeling, Condition Monitoring, and Fault Diagnosis*. CRC Press; 2017. Available from: https://books.google.it/books?id=HwPOBQAAQBAJ.

[160] El Hachemi Benbouzid M. A review of induction motors signature analysis as a medium for faults detection. *IEEE Transactions on Industrial Electronics*. 2000;47(5):984–993.

[161] Jack AG, Mecrow BC, and Haylock JA. A comparative study of permanent magnet and switched reluctance motors for high-performance fault-tolerant applications. *IEEE Transactions on Industry Applications*. 1996; 32(4):889–895.

[162] Fodorean D, Ruba M, Szabo L, *et al.* Comparison of the main types of fault-tolerant electrical drives used in vehicle applications. In: 2008 International Symposium on Power Electronics, Electrical Drives, Automation and Motion; 2008. p. 895–900.

[163] Williamson S, Smith S, and Hodge CG. Fault tolerance in multiphase propulsion motors. *Proceedings of IMarEST – Part A – Journal of Marine Engineering and Technology*. 2004;2004:3–7.

[164] Apsley JM, Williamson S, Smith AC, *et al.* Induction motor performance as a function of phase number. *IEE Proceedings – Electric Power Applications*. 2006;153(6).

[165] Jawad F, Vahid G, and Gojko J. *Fault Diagnosis of Induction Motors*. Energy Engineering. London: Institution of Engineering and Technology; 2017. Available from: https://digital-library.theiet.org/content/books/po/pbpo108e.

[166] Elch-Heb T and Hautier JP. Remedial strategy for inverter-induction machine system faults using two-phase operation. In: 1993 Fifth European Conference on Power Electronics and Applications. vol. 5; 1993. p. 151–156.

[167] Liu T-H, Fu J-R, and TA L. A strategy for improving reliability of field-oriented controlled induction motor drives. *IEEE Transactions on Industry Applications*. 1993;29(5):910–918.

[168] Beltrao de Rossiter Correa M, Brandao Jacobina C, Cabral da Silva ER, *et al.* An induction motor drive system with improved fault tolerance. *IEEE Transactions on Industry Applications*. 2001;37(3):873–879.

[169] Mendes AMS and Marques Cardoso AJ. Fault-tolerant operating strategies applied to three-phase induction-motor drives. *IEEE Transactions on Industrial Electronics*. 2006;53(6):1807–1817.

[170] Kastha D and Bose BK. Fault mode single-phase operation of a variable frequency induction motor drive and improvement of pulsating torque characteristics. *IEEE Transactions on Industrial Electronics*. 1994;41(4): 426–433.

[171] Alberti L and Bianchi N. Impact of winding arrangement in dual 3-phase induction motor for fault tolerant applications. In: The XIX International Conference on Electrical Machines – ICEM 2010; 2010. p. 1–6.

[172] Zhong Peng, Zheng Z, Li Y, *et al.* Fault-tolerant control of multiphase induction machine drives based on virtual winding method. In: 2017 IEEE Transportation Electrification Conference and Expo (ITEC); 2017. p. 252–256.

[173] Chen Q, Liu G, Gong W, *et al.* A new fault-tolerant permanent-magnet machine for electric vehicle applications. *IEEE Transactions on Magnetics*. 2011;47(10):4183–4186.

[174] Parsa L and Toliyat HA. Five-phase permanent-magnet motor drives. *IEEE Transactions on Industry Applications*. 2005;41(1):30–37.

[175] Alberti L and Bianchi N. Experimental tests of dual three-phase induction motor under faulty operating condition. *IEEE Transactions on Industrial Electronics*. 2012;59(5):2041–2048.

[176] Ruba M. Fault-Tolerant Electrical Machines and Drives. In: Demetgul M, Ünal M, editors. *Fault Diagnosis and Detection*. Rijeka: IntechOpen; 2017. Available from: https://doi.org/10.5772/67354.

[177] Switched reluctance machine having unbalance forces compensation coils. US5422525A; 1994. Available from: https://patentscope.wipo.int/search/en/detail.jsf?docId=WO1996000882; https://patents.google.com/patent/EP0769133A1/en

[178] Mecrow BC, El-Kharashi EA, Finch JW, *et al.* Segmental rotor switched reluctance motors with single-tooth windings. *IEE Proceedings – Electric Power Applications*. 2003;150(5):591–599.

[179] Xu Z, Lee D, and Ahn J. Design of a novel 6/5 segmental rotor type switched reluctance motor. In: 2014 IEEE Industry Application Society Annual Meeting; 2014. p. 1–7.

[180] Welchko BA, Jahns TM, Soong WL, *et al.* IPM synchronous machine drive response to symmetrical and asymmetrical short circuit faults. *IEEE Transactions on Energy Conversion*. 2003;18(2):291–298.

[181] Bianchi N, Bolognani S, and Dai Prè M. Design of a fault-tolerant IPM motor for electric power steering. *IEEE Transactions on Vehicular Technology*. 2006;55(4):2873–2880.

[182] Jahns TM. Improved reliability in solid-state AC drives by means of multiple independent phase drive units. *IEEE Transactions on Industry Applications*. 1980;IA-16(3):321–331.

[183] Bianchi N, Pre MD, Grezzani G, *et al.* Design considerations on fractional-slot fault-tolerant synchronous motors. In: IEEE International Conference on Electric Machines and Drives, 2005; 2005. p. 902–909.

[184] Bianchi N, Prè MD, and Alberti L. Theory and Design of Fractional-slot Pm Machines. IEEE IAS Tutorial Course Notes. CLEUP; 2007.

[185] Parsa L and Toliyat HA. Fault-tolerant five-phase permanent magnet motor drives. In: Conference Record of the 2004 IEEE Industry Applications Conference, 2004. 39th IAS Annual Meeting. vol. 2; 2004. p. 1048–1054.

[186] Bianchi N, Bolognani S, and Dai Pré M. Design and tests of a fault-tolerant five-phase permanent magnet motor. In: 2006 37th IEEE Power Electronics Specialists Conference; 2006. p. 1–8.
[187] Bianchi N, Bolognani S, and Dai Pre M. Strategies for the fault-tolerant current control of a five-phase permanent-magnet motor. *IEEE Transactions on Industry Applications*. 2007;43(4):960–970.
[188] Barcaro M, Bianchi N, and Magnussen F. Analysis and tests of a dual three-phase 12-slot 10-pole permanent-magnet motor. *IEEE Transactions on Industry Applications*. 2010;46(6):2355–2362.
[189] Barcaro M, Bianchi N, and Magnussen F. Faulty operations of a PM fractional-slot machine with a dual three-phase winding. *IEEE Transactions on Industrial Electronics*. 2011;58(9):3825–3832.
[190] Qin D, Luo X, and Lipo TA. Reluctance motor control for fault-tolerant capability. In: 1997 IEEE International Electric Machines and Drives Conference Record; 1997, pp. WA1/1.1–WA1/1.6, doi: 10.1109/IEMDC.1997.604282.
[191] Matos DB, Estima JO, and Cardoso AJM. Performance of a synchronous reluctance motor drive with a fault-tolerant three-level neutral point clamped inverter. In: 2016 XXII International Conference on Electrical Machines (ICEM); 2016. p. 1152–1159.
[192] Babetto C and Bianchi N. Synchronous reluctance motor with dual three-phase winding for fault-tolerant applications. In: 2018 XIII International Conference on Electrical Machines (ICEM); 2018. p. 2297–2303.
[193] Bianchi N and Dai Pré M. Use of the star of slots in designing fractional-slot single-layer synchronous motors. *IEE Proceedings – Electric Power Applications*. 2006;153(3):459–466.
[194] Babetto C, Bianchi N, Torreggiani A, *et al.* Optimal design and experimental validation of a synchronous reluctance machine for fault-tolerant applications. In: 2019 IEEE Energy Conversion Congress and Exposition (ECCE). Baltimore; 2019.
[195] Bianchi N and Bolognani S. Sensorless-oriented design of PM motors. *IEEE Transactions on Industry Applications*. 2009;45(4):1249–1257.
[196] Bianchi N, Fornasiero E, and Bolognani S. Effect of stator and rotor saturation on sensorless rotor position detection. *IEEE Transactions on Industry Applications*. 2013;49(3):1333–1342.
[197] Kano Y. Sensorless-oriented design of IPMSM. In: 2014 International Power Electronics Conference, IPEC-Hiroshima – ECCE Asia 2014; 2014. p. 2457–2464.
[198] Kwon YC, Lee J, and Sul SK. Full torque-range low-speed sensorless drive for heavily saturated IPMSMs by manipulation of convergence point. In: 2017 IEEE Energy Conversion Congress and Exposition (ECCE). Cincinnati; 2017. p. 865–872.
[199] Manzolini V and Bolognani S. On the rotor position self-sensing capability of IPM and reluctance synchronous motors. In: 2018 IEEE 9th International Symposium on Sensorless Control for Electrical Drives (SLED). 2. Helsinki, Finland: IEEE; 2018. p. 108–113.

Appendix A
Iron losses insights

In this chapter, some insights on iron losses are investigated.

A.1 Eddy currents iron losses coefficient

Consider an infinitely long conductive sheet of thickness d along the y-axis, depicted in Figure A.1. Let $B(t)$ be the forcing time-varying magnetic flux density, parallel to the z-axis:

$$B(t) = \{0, 0, B_z(t)\}^{\mathrm{T}}$$

Recalling Faraday–Neumann–Henry–Lenz's law,

$$\nabla \times E = -\frac{\partial B}{\partial t}$$

Since B has z-component only, $\nabla \times E$ must be parallel to it:

$$(\nabla \times E) \cdot \hat{z} = \partial_x E_y - \partial_y E_x$$

Looking at Figure A.1, we can see that eddy currents are directed along the x-axis only. And because of the constitutive equation (Ohm's law)*

$$J = \sigma E$$

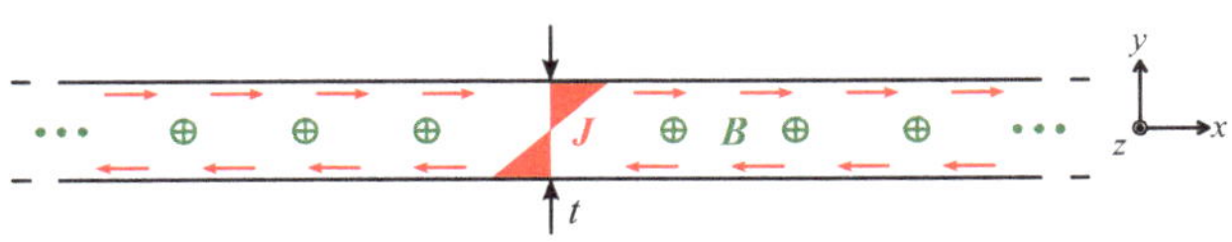

Figure A.1 Conductive lamination sheet of thickness d and chosen reference frame

*Assuming a uniform, homogeneous and isotropic material.

the E field must have x-component only. Therefore,

$$-\partial_y E_x = -\partial_t B_z$$
$$\frac{\partial E_x}{\partial y} = \frac{\partial B_z}{\partial t} \tag{A.1}$$

This is a first-order one-dimensional partial differential equation. Employing Steinmetz (or Fourier) transform, we can transform it into a complex first-order ordinary differential equation†:

$$\frac{\mathrm{d}\underline{E}_x}{\mathrm{d}y} = \mathrm{j}\omega \underline{B}_z \tag{A.2}$$

Since $\underline{B}_z$ is supposed to be imposed and uniform, this differential equation is trivial to solve, and thanks to the proper choice of the reference frame, the integration constant is zero:

$$\underline{E}_x = \mathrm{j}\omega \underline{B}_z y$$

As shown in Figure A.1, both the electric field and the current density field have a linear distribution along the y-coordinate. In fact,

$$\underline{J}_x = \mathrm{j}\omega\sigma \underline{B}_z y \tag{A.3}$$

Now we can compute the associated power losses due to eddy currents. Also the losses will result a function of the y-coordinate‡:

$$p_J = \frac{1}{2}\rho \left|\underline{J}_x\right|^2 = \frac{1}{2}\omega^2\sigma \underline{B}_z^2 y^2$$

At this point, we can compute the space average of these power losses, doing

$$\langle p_J \rangle = \frac{1}{d}\int_{-d/2}^{d/2} \frac{1}{2}\omega^2\sigma \underline{B}_z^2 y^2 \,\mathrm{d}y$$
$$= \frac{\omega^2\sigma \underline{B}_z^2}{2d} \left.\frac{y^3}{3}\right|_{-d/2}^{d/2}$$
$$= \frac{\omega^2\sigma \underline{B}_z^2 d^2}{24} = \frac{4\pi^2 f^2 \sigma \underline{B}_z^2 d^2}{24} = \frac{\pi^2\sigma d^2}{6}\underline{B}_z^2 f^2$$

At this point, we can define the eddy currents coefficient, g_{ec}, as

$$g_{\mathrm{ec}} = \frac{\pi^2\sigma d^2}{6\gamma} \qquad [\mathrm{W/(kgT^2Hz^2)}] \tag{A.4}$$

where γ is the volumetric mass density of the material.

†With "_" we denote a phasor quantity. In this case, the phasor expresses the peak value of the underlying quantity.

‡We need the factor 1/2 because we are using phasors with peak values.

A.2 Element-by-element iron losses

The partial differential equation that governs low-frequency electromagnetic problems is the following:

$$\nabla \times \nu \nabla \times A = J \tag{A.5}$$

The current density vector may be due to many causes, in particular, to the supply or to eddy currents:

$$\nabla \times \nu \nabla \times A = J_s - \sigma \nabla V - \sigma \frac{\partial A}{\partial t} \qquad \text{with e.c. and voltage supply}$$

$$-\nabla \cdot \nu \nabla A_z = J_s - \sigma \frac{\partial A_z}{\partial t} \qquad \text{in 2D}$$

$$-\nabla \cdot \nu \nabla \underline{A}_z = \underline{J}_s - j\omega\sigma \underline{A}_z \qquad \text{in 2D and in frequency}$$

$$\nabla \cdot \nu \nabla \underline{A}_z - j\omega\sigma \underline{A}_z + \underline{J}_s = 0 \qquad \text{standard form}$$

A.2.1 Permanent magnets eddy currents

In magnets, we have to guarantee that only zero-sum currents flow:

$$\begin{aligned} &\int_{\Omega_{PM}} J_{eddy}\, dS = 0 \\ &J_{eddy}(x, y, t) = -\frac{\partial A_z(x, y, t)}{\partial t} \end{aligned} \tag{A.6}$$

But if we were to compute the eddy currents with the derivative, errors due to noise may occur. We want to show that imposing (A.6) in time is the same as imposing it in frequency. In fact, we can write

$$J_{eddy}(x, y, t) = \sum_{h=1}^{+\infty} j_h(x, y)\, e^{jh\omega t}$$

and substitute in (A.6)

$$\int_{\Omega_{PM}} \sum_{h=1}^{+\infty} j_h(x, y)\, e^{jh\omega t}\, dS = 0$$

$$\sum_{h=1}^{+\infty} e^{jh\omega t} \int_{\Omega_{PM}} j_h(x, y)\, dS = 0$$

Since the basis functions $e^{jh\omega t}$ are never zero, we need

$$\int_{\Omega_{PM}} j_h(x, y)\, dS = 0$$

which is valid for every time instant, as requested.

A.2.2 Harmonic iron losses

Iron losses in ferromagnetic material can be estimated knowing the time variation of the flux density in each part of the machine. In finite-element analysis, the flux density components $(B_x(t), B_y(t))$ are known in every mesh element. Referring to them as $(B_x^e(t), B_y^e(t))$, we can express them in Fourier series as:

$$B_x^e(t) = \sum_{h=0}^{\infty} B_{xh}^e \cos(h\omega t - \phi_h)$$

$$B_y^e(t) = \sum_{h=0}^{\infty} B_{yh}^e \cos(h\omega t - \psi_h)$$

Since iron losses depend on the square of the element flux densities, we can write

$$B_{h,e}^2 = (B_{xh}^e)^2 + (B_{yh}^e)^2$$

Determining appropriately the hysteresis and iron losses coefficients, one could express the iron losses due to all the harmonics as:

$$p_{\text{Fe}}^e = \sum_{h=1}^{+\infty} \left(c_{\text{hy}} B_{h,e}^2 hf + c_{\text{ec}} B_{h,e}^2 h^2 f^2\right) \qquad [\text{W/kg}]$$

To get the total iron losses, we multiply the element iron losses with the respective element weight:

$$P_{\text{Fe}} = \sum_{e=1}^{N_{\text{e}}} \sum_{h=1}^{+\infty} \left(c_{\text{hy}} B_{h,e}^2 hf + c_{\text{ec}} B_{h,e}^2 h^2 f^2\right) \gamma V_e \qquad [\text{W}] \tag{A.7}$$

where N_{e} is the number of mesh elements of the iron, γ is the material volumetric mass density and V_e is the volume of the element e.

Appendix B
MMF distribution along the stator periphery

Suppose to modulate a spatial sinusoid

$$f(\vartheta) = \sin\vartheta \tag{B.1}$$

with a cosinusoidal time signal of angular frequency ω:

$$\tilde{f}(\vartheta, t) = \sin\vartheta\cos\omega t = \tfrac{1}{2}(\sin(\vartheta + \omega t) + \sin(\vartheta - \omega t)) \tag{B.2}$$

The obtained signal is a spatial sinusoid that oscillates in time with frequency ω, just like a standing wave in a stretched rope. Moreover, this very same signal can be reproduced by superposition of two traveling waves, as shown in the right-hand side (RHS) term of (B.2).

In order to better understand the characteristics of these traveling waves, let us consider just the second addendum of the RHS:

$$\sin(\vartheta - \omega t) = \sin\left(\frac{2\pi}{\lambda}x - \omega t\right) = \sin(k(x - ct))$$

where λ is the wavelength of the wave, equal to twice the pole pitch in an electrical machine:

$$2\tau = \frac{\pi D}{p}$$

Here k is the wavenumber, spatial analogous to the angular frequency:

$$k = \frac{2\pi}{\lambda}$$

and $c = \lambda f$ is the speed of propagation. The speed c can also be derived considering to "ride the wave," so to move synchronously with the wave on top of the same point. So

$$\frac{2\pi}{\lambda}x - \omega t = \text{const.}$$

and taking the derivative and expliciting the speed,

$$c = \frac{\mathrm{d}x}{\mathrm{d}t} = \frac{\omega}{\frac{2\pi}{\lambda}} = \frac{2\pi f}{2\pi}\lambda = f\lambda$$

B.1 Spatial harmonic

Let us suppose now that the spatial sinusoid has higher wavenumber, so smaller wavelength:

$$\tilde{f}^{\nu} = \sin \nu\vartheta \cos \omega t \;, \qquad k^{\nu} = \nu k \qquad \lambda^{\nu} = \frac{\lambda}{\nu} \;, \qquad \nu \in \mathbb{Z}_0 \tag{B.3}$$

Repeating the same step as before, we can easily find out that the speed of the associated waves must be lower than before and it is

$$c^{\nu} = \frac{c}{\nu} \tag{B.4}$$

This is especially true if one thinks that in the same period (ω is the same) the wave travels for one oscillation, the wavelength, which is much smaller, so the speed must be smaller too.

In electrical machines, we are interested in angular velocity, obtainable through

$$\Omega^{\nu} = \frac{c^{\nu}}{\frac{D}{2}} = \frac{f\lambda^{\nu}}{\frac{D}{2}} = \frac{2f\,2\tau}{D} = \frac{2f\,\pi D}{\nu p\,D} = \frac{2\pi f}{\nu p} = \frac{\omega_{\mathrm{m}}}{\nu}$$

so the angular velocity of the νth harmonic is $1/\nu$ times the speed of the fundamental. However, the quantities shown before were all mechanical quantities, and we are interested in the electrical ones. So what happens is that every harmonic speed has to be multiplied by the correct number of pole, which is $p^{\nu} = \nu p$. For instance, a fifth harmonic has an equivalent number of poles that is $5p$ due to the five wavelengths contained in one fundamental wavelength.

B.2 Rotating MMF

Let us now suppose to have three balanced and symmetrical Magnetomotive Force (MMF) waves, displaced in space *and* time:

$$\tilde{f}_1(\vartheta, t) = F \sin \vartheta \cos \omega t$$

$$\tilde{f}_2(\vartheta, t) = F \sin\left(\vartheta - \frac{2}{3}\pi\right) \cos\left(\omega t - \frac{2}{3}\pi\right)$$

$$\tilde{f}_3(\vartheta, t) = F \sin\left(\vartheta - \frac{4}{3}\pi\right) \cos\left(\omega t - \frac{4}{3}\pi\right)$$

Decomposing each of them in the same way as before, we can see that they are sum of a progressive wave and a regressive one, just like before. Summing them together though, the regressive terms cancel each other and what we obtain is a progressive term equal to

$$f(\vartheta, t) = \frac{3}{2}F \sin(\vartheta - \omega t) \tag{B.5}$$

This is the rotating fundamental MMF due to a three-phase system of symmetrical currents flowing in three sets of coils displaced in space by 120°. This MMF causes a rotating flux density distribution along the air gap.

B.2.1 Rotating harmonics

Repeating the same steps we can find that for every harmonic one type of traveling wave cancels and the other "survives." For example, the fifth harmonic travels in the opposite way of the fundamental, with the speed computed before. So it is convenient to assign the harmonic order a sign to express the direction of motion of the associated MMF. All the harmonics present in the described three-phase system are

$$\nu = 6k + 1\,, \quad k \in \mathbb{Z} \qquad (\text{e.g. } \nu = +1, -5, +7, -11, +13, \ldots) \tag{B.6}$$

In order to get the correct result one must remember that a spatial displacement of a signal results in a different displacement of the harmonics proportional to the harmonic order:

$$\begin{aligned} y(\vartheta) &\overset{\mathscr{F}\mathscr{S}}{\longleftrightarrow} c_\nu \\ y(\vartheta + \beta) &\overset{\mathscr{F}\mathscr{S}}{\longleftrightarrow} \mathrm{e}^{j\nu\beta}\, c_\nu \end{aligned} \tag{B.7}$$

Resuming the fifth harmonic, the set of single-phase MMF equations is

$$\begin{aligned} f_1^{(-5)}(\vartheta, t) &= F^{(-5)} \sin\left[-5\vartheta\right] \cos(\omega t) \\ f_2^{(-5)}(\vartheta, t) &= F^{(-5)} \sin\left[-5\left(\vartheta - \frac{2}{3}\pi\right)\right] \cos\left(\omega t - \frac{2}{3}\pi\right) \\ f_3^{(-5)}(\vartheta, t) &= F^{(-5)} \sin\left[-5\left(\vartheta - \frac{4}{3}\pi\right)\right] \cos\left(\omega t - \frac{4}{3}\pi\right) \end{aligned}$$

Then the generic rotating MMF harmonic can be expressed as

$$f^{\nu}(\vartheta, t) = \frac{3}{2} F^{\nu} \sin(\nu\vartheta - \omega t) \tag{B.8}$$

Notice that if ν is negative, we can collect the minus sign and bring it out of the sine, so what remains inside the operator is a regressive wave with negative amplitude where the fundamental is positive.

Appendix C
HS-SyR analytical model constants

The constants used in the analytical model presented in Chapter 5 are here reported. The coefficients ρ_1, ρ_2, ρ_3, b, e, f, r, s, z and other missing variables can be found in [36]:

$$X = AI_dI_q + (B + C + D + E)I_d$$

$$A = \frac{9}{2\pi} p\mu_0 \left(\frac{k_w N_s}{2p}\right)^2 \frac{D_s L_{stk}}{g''}$$

$$B = \frac{3}{16} k_w N_s \mu_0 L_{stk} \frac{D_s^2 \hat{K}_s}{pg''} \left(\sin\left(\pi - \alpha_i^e\right) - \sin\left(-\alpha_i^e\right)\right)$$

$$C = -\frac{3}{4} k_w N_s \mu_0 L_{stk} \frac{D_s^2}{g''} \left[\frac{\hat{K}_{sq}}{p} \rho_1\right] [\vartheta_{b3} - \vartheta_{b2}]$$

$$D = -\frac{3}{4} k_w N_s \mu_0 L_{stk} \frac{D_s^2}{g''} \left[\frac{\hat{K}_{sq}}{p} \rho_2\right] [\vartheta_{b2} - \vartheta_{b1}]$$

$$E = -\frac{3}{4} k_w N_s \mu_0 L_{stk} \frac{D_s^2}{g''} \left[\frac{\hat{K}_{sq}}{p} \rho_1\right] \vartheta_{b1}$$

$$Y = (F + G + M)I_d$$

$$F = \frac{3}{4} p k_w N_s \mu_0 L_{stk}^2 \frac{D_s}{g''} \frac{D_r^3}{8} B_{sat} \frac{\nu_\sigma \gamma_{Lam}}{\sigma_r} [\vartheta_{b3} - \vartheta_{b2}] \cdot \Big\{ s\Re_{b1} 2\vartheta_{b1} - \sin 2\vartheta_{b1}) + y\Re_{b2} k_2 (2\vartheta_{b2} - \sin 2\vartheta_{b2}) - r\Re_{b3} k_3 (2\vartheta_{b3} - \sin 2\vartheta_{b3}) \Big\}$$

$$G = \frac{3}{8} p k_w N_s \mu_0 L_{stk}^2 \frac{D_r^3}{8} B_{sat} \frac{D_s}{g''} \frac{\nu_\sigma \gamma_{Lam}}{\sigma_r} [\vartheta_{b2} - \vartheta_{b1}] \cdot \Big\{ (2\vartheta_{b1} - \sin 2\vartheta_{b1})[\Re_{b1} zs + e)] + (2\vartheta_{b2} - \sin 2\vartheta_{b2})[\Re_{b2} k_2 (y - z)] + (2\vartheta_{b3} - \sin 2\vartheta_{b3})(r\Re_{b3} k_3) \Big\}$$

$$M = \frac{3}{8} p k_\mathrm{w} N_\mathrm{s} \mu_0 L_\mathrm{stk}^2 \frac{D_\mathrm{r}^3}{8} B_\mathrm{sat} \frac{D_\mathrm{s}}{g''} \frac{\nu_\sigma \gamma_\mathrm{Lam}}{\sigma_\mathrm{r}} [2\vartheta_\mathrm{b1}] \cdot \Big\{ (2\vartheta_\mathrm{b1} - \sin 2\vartheta_\mathrm{b1})[b\Re_\mathrm{b1}(zs + e - 1)] + (2\vartheta_\mathrm{b2} - \sin 2\vartheta_\mathrm{b2}) bz\Re_\mathrm{b2} k_2 (y-1) - (2\vartheta_\mathrm{b3} - \sin 2\vartheta_\mathrm{b3}) r\Re_\mathrm{b3} k_3 \Big\}$$

$$f_\mathrm{r1} = b\Big[z\,(sU_\mathrm{sat1} + yU_\mathrm{sat2} - rU_\mathrm{sat3}) + [(e-1)U_\mathrm{sat1} - zU_\mathrm{sat2}] \Big]$$

$$f_\mathrm{r2} = z\Big[(s + e/z)U_\mathrm{sat1} + (y-1)U_\mathrm{sat2} - rU_\mathrm{sat3} \Big]$$

$$f_\mathrm{r3} = sU_\mathrm{sat1} + yU_\mathrm{sat2} - rU_\mathrm{sat3}$$

$$g_\mathrm{r1} = b\big(rzU_\mathrm{sat3} - (zs + e - 1)U_\mathrm{sat1} - z(y-1)U_\mathrm{sat2} \big)$$

$$g_\mathrm{r2} = z\big(rU_\mathrm{sat3} - (s + e/z)U_\mathrm{sat1} - (y-1)U_\mathrm{sat2} \big)$$

$$g_\mathrm{r3} = rU_\mathrm{sat3} - sU_\mathrm{sat1} - yU_\mathrm{sat2}$$

Appendix D

GetDP elasticity formulation

The GetDP formulation code (content in the file `Elasticity_2D.pro`) used to solve the 2D elasticity problems in Chapter 7 and in all the FE structural analyses of the present work is here reported. For more details see [60,61].

```
Group{
  DefineGroup[Domain_Disp, Domain_Force, Domain_Force_Sur,
    Domain_Force_Lin] ;
  DomainTot = Region[ {Domain_Disp, Domain_Force} ];
}


Group{
  DefineGroup[ DomainInf ] ;
  DefineVariable[ Val_Rint, Val_Rext ] ;
}


Jacobian{
  { Name Vol ;
    Case {
      { Region DomainInf ; Jacobian VolSphShell {Val_Rint,
    Val_Rext} ; }
      { Region All ; Jacobian Vol ; }
    }
  }
  { Name Sur ;
    Case { { Region All ; Jacobian Sur ; }
    }
  }
  { Name Lin ;
    Case { { Region All ; Jacobian Lin ; }
    }
  }
  { Name SurLinVol ;
    Case {
      { Region Domain_Force_Sur ; Jacobian Sur ; }
      { Region Domain_Force_Lin ; Jacobian Lin ; }
```

```
        { Region All ; Jacobian Vol ; }
      }
    }
  }

  Integration{
    { Name GradGrad ;
      Case {
        { Type Gauss ;
          Case {
            { GeoElement Point        ; NumberOfPoints  1 ; }
            { GeoElement Line         ; NumberOfPoints  2 ; }
            //          { GeoElement Line2        ;
      NumberOfPoints  2 ;
            }
            { GeoElement Triangle     ; NumberOfPoints  4 ; }
            //          { GeoElement Triangle2    ;
      NumberOfPoints  6 ;
            }
            { GeoElement Quadrangle   ; NumberOfPoints  4 ; }
          }
        }
      }
    }
  }

  FunctionSpace{
    { Name H_u_Mec2D ; Type Vector ;
      BasisFunction {
        { Name sxn ; NameOfCoef uxn ; Function BF_NodeX ;
          dFunction {BF_NodeX_D12, BF_Zero} ;
          Support DomainTot ; Entity NodesOf[ All ] ; }
        { Name syn ; NameOfCoef uyn ; Function BF_NodeY ;
          dFunction {BF_NodeY_D12, BF_Zero} ;
          Support DomainTot ; Entity NodesOf[ All ] ; }
      }
      Constraint{
        { NameOfCoef uxn ;
          EntityType NodesOf ; NameOfConstraint DeplacementX ; }
        { NameOfCoef uyn ;
          EntityType NodesOf ; NameOfConstraint DeplacementY ; }
      }
    }
  }

  Formulation{
    { Name Mec2D_u ; Type FemEquation ;
      Quantity {
        { Name u ; Type Local ; NameOfSpace H_u_Mec2D ; }
```

```
      }
      Equation {
        Galerkin { [ C_m[] * Dof{D1 u}, {D1 u} ] ;
          In Domain_Disp ; Jacobian Vol ; Integration GradGrad ;
       }
        Galerkin { [ -F[] , {u} ] ;
          In Domain_Force; Jacobian SurLinVol; Integration
     GradGrad; }
      }
    }
}

Resolution{
  { Name Mec2D_u ;
    System {
      { Name Sys_Mec ; NameOfFormulation Mec2D_u ; }
    }
    Operation{ Generate Sys_Mec ; Solve Sys_Mec ; SaveSolution
     Sys_Mec ; }
  }
}

PostProcessing{
  { Name Mec2D_u ; NameOfFormulation Mec2D_u ;
    PostQuantity {
      { Name u ; Value { Term { [ {u} ]   ; In Domain_Disp ; }
     } }
      { Name um ; Value { Term { [Norm[{u}]] ; In Domain_Disp
    ; } } }
      { Name F ; Value { Term { [F[]]   ; In Domain_Force ; }
    } }
      { Name Vm ; Value { Term {
            [Sqrt[ CompX[C_m[]*{D1 u}]#1^2 + CompY[C_m[]*{D1
            u}]#2^2 - #1*#2
            +  3*(CompZ[C_m[]*{D1 u}])^2   ] ] ; In
    Domain_Disp ; }
            } }
    }
  }
}
```

Appendix E
Maxwell stress tensor derivation

First, recall the set of differential Maxwell's equations

$$\nabla \cdot \boldsymbol{D} = \rho$$

$$\nabla \times \boldsymbol{E} = \frac{\partial \boldsymbol{B}}{\partial t}$$

$$\nabla \cdot \boldsymbol{B} = 0$$

$$\nabla \times \boldsymbol{H} = \boldsymbol{J} + \frac{\partial \boldsymbol{D}}{\partial t}$$

and the equivalent integral formulation

$$\oint_{\partial\Omega} \boldsymbol{D} \cdot \hat{\boldsymbol{n}} \, \mathrm{d}S = \int_{\Omega} \rho \, \mathrm{d}V$$

$$\oint_{\partial\Sigma} \boldsymbol{E} \cdot \hat{\boldsymbol{t}} \, \mathrm{d}l = \frac{\mathrm{d}}{\mathrm{d}t} \int_{\Sigma} \boldsymbol{B} \cdot \hat{\boldsymbol{n}} \, \mathrm{d}S$$

$$\oint_{\partial\Omega} \boldsymbol{B} \cdot \hat{\boldsymbol{n}} \, \mathrm{d}S = 0$$

$$\oint_{\partial\Sigma} \boldsymbol{H} \cdot \hat{\boldsymbol{t}} \, \mathrm{d}l = \int_{\Sigma} \boldsymbol{J} \cdot \hat{\boldsymbol{n}} \, \mathrm{d}S + \frac{\mathrm{d}}{\mathrm{d}t} \int_{\Sigma} \boldsymbol{D} \cdot \hat{\boldsymbol{n}} \, \mathrm{d}S$$

The most used ones in electrical machines are (E), (E), respectively, Faraday's and Ampère's laws. Usually the former is in the form

$$e = \frac{\mathrm{d}\lambda}{\mathrm{d}t}$$

where e is the back electromotive force (BEMF) while λ is the flux linkage of the considered winding. In the latter, the time-derivative is usually neglected because it has almost no effect at low frequencies.

In order to derive the force acting on bodies subjected to electromagnetic fields, we have to start from the Lorentz force on a charged particle:

$$\boldsymbol{F} = q\,(\boldsymbol{E} + \boldsymbol{v} \times \boldsymbol{B}) \qquad [\mathrm{N}] \tag{E.1}$$

where q is the charge of the particle and $\boldsymbol{v}$ is its speed. This formula can be generalized to the force density on a continuum media, so

$$\begin{aligned} \boldsymbol{f} &= \rho\boldsymbol{E} + \boldsymbol{J}\times\boldsymbol{B} \qquad [\mathrm{N/m^3}] \\ &= (\nabla\cdot\boldsymbol{D})\boldsymbol{E} + (\nabla\times\boldsymbol{H})\times\boldsymbol{B} - \frac{\partial\boldsymbol{D}}{\partial t}\times\boldsymbol{B} \end{aligned}$$

The time derivative can be rewritten as

$$\frac{\partial}{\partial t}(\boldsymbol{D}\times\boldsymbol{B}) = \frac{\partial\boldsymbol{D}}{\partial t}\times\boldsymbol{B} + \boldsymbol{D}\times\frac{\partial\boldsymbol{B}}{\partial t}$$

Therefore,

$$\frac{\partial\boldsymbol{D}}{\partial t}\times\boldsymbol{B} = \frac{\partial}{\partial t}(\boldsymbol{D}\times\boldsymbol{B}) + \boldsymbol{D}\times(\nabla\times\boldsymbol{E})$$

$$\begin{aligned} \boldsymbol{f} &= [(\nabla\cdot\boldsymbol{D})\boldsymbol{E} - \boldsymbol{D}\times(\nabla\times\boldsymbol{E})] + [-\boldsymbol{B}\times(\nabla\times\boldsymbol{H})] + \frac{\partial}{\partial t}(\boldsymbol{D}\times\boldsymbol{B}) \\ &= [(\nabla\cdot\boldsymbol{D})\boldsymbol{E} - \boldsymbol{D}\times(\nabla\times\boldsymbol{E})] + [(\nabla\cdot\boldsymbol{B})\boldsymbol{H} - \boldsymbol{B}\times(\nabla\times\boldsymbol{H})] - \frac{\partial}{\partial t}(\boldsymbol{D}\times\boldsymbol{B}) \end{aligned}$$

Through the vector calculus identity,

$$\boldsymbol{A}\times(\nabla\times\boldsymbol{A}) = \tfrac{1}{2}\nabla(\boldsymbol{A}\cdot\boldsymbol{A}) - (\boldsymbol{A}\cdot\nabla)\boldsymbol{A}$$

$$\begin{aligned} \boldsymbol{f} &= [(\nabla\cdot\boldsymbol{D})\boldsymbol{E} + (\boldsymbol{D}\cdot\nabla)\boldsymbol{E}] + [(\nabla\cdot\boldsymbol{B})\boldsymbol{H} + (\boldsymbol{B}\cdot\nabla)\boldsymbol{H}] \\ &\qquad - \tfrac{1}{2}\nabla(\boldsymbol{D}\cdot\boldsymbol{E} + \boldsymbol{B}\cdot\boldsymbol{H}) - \frac{\partial}{\partial t}(\boldsymbol{D}\times\boldsymbol{B}) \\ &= \nabla\cdot(\boldsymbol{D}\otimes\boldsymbol{E} + \boldsymbol{B}\otimes\boldsymbol{H}) - \tfrac{1}{2}\nabla(\boldsymbol{D}\cdot\boldsymbol{E} + \boldsymbol{B}\cdot\boldsymbol{H}) - \frac{\partial}{\partial t}(\boldsymbol{D}\times\boldsymbol{B}) \end{aligned}$$

where $\otimes$ is the dyadic product, or tensor product.* It can be shown that

$$\nabla(\boldsymbol{D}\cdot\boldsymbol{E}) = \nabla\cdot((\boldsymbol{D}\cdot\boldsymbol{E})\boldsymbol{I})$$

*Let $\boldsymbol{a}$ and $\boldsymbol{b}$ be two column vectors: the dot (scalar) product can be defined as

$$\boldsymbol{a}\cdot\boldsymbol{b} = \boldsymbol{a}^{\mathrm{T}}\boldsymbol{b} = \sum_i a_i b_i$$

while the dyadic product is defined as

$$\boldsymbol{a}\otimes\boldsymbol{b} = \boldsymbol{a}\,\boldsymbol{b}^{\mathrm{T}} = \begin{bmatrix} a_1b_1 & \cdots & a_1b_n \\ \vdots & \ddots & \vdots \\ a_nb_1 & \cdots & a_nb_n \end{bmatrix}$$

where $\boldsymbol{I}$ is the identity matrix. Therefore,

$$\boldsymbol{f} = \nabla \cdot \underbrace{\left(\boldsymbol{D} \otimes \boldsymbol{E} + \boldsymbol{B} \otimes \boldsymbol{H} - \tfrac{1}{2}(\boldsymbol{D} \cdot \boldsymbol{E} + \boldsymbol{B} \cdot \boldsymbol{H})\boldsymbol{I}\right)}_{\overleftrightarrow{\boldsymbol{M}}} - \frac{\partial}{\partial t}\big(\varepsilon\mu \underbrace{\boldsymbol{E} \times \boldsymbol{H}}_{\boldsymbol{S}}\big)$$

$$\boldsymbol{f} = \nabla \cdot \overleftrightarrow{\boldsymbol{M}} - \varepsilon\mu \frac{\partial \boldsymbol{S}}{\partial t}$$

This equation fully shows the force due to the presence of the electromagnetic fields. $\overleftrightarrow{\boldsymbol{M}}$ is the Maxwell Stress Tensor (MST), while $\boldsymbol{S}$ is the Poynting vector, representing either a sort of electromagnetic momentum density or alternatively the directional energy flux density. Noticing that the last term on the right-hand side (RHS) derives from the rate of change in time of the displacement field, in the quasi-magnetostatic case (usual for electrical machines) it can be neglected. Thus

$$\boldsymbol{f} = \nabla \cdot \overleftrightarrow{\boldsymbol{M}} \tag{E.2}$$

In order to get the total force acting on a body contained in the volume Ω, we integrate

$$\boldsymbol{F} = \int_{\Omega} \nabla \cdot \overleftrightarrow{\boldsymbol{M}}\, \mathrm{d}V = \oint_{\Sigma} \overleftrightarrow{\boldsymbol{M}} \cdot \hat{\boldsymbol{n}}\, \mathrm{d}S \tag{E.3}$$

Therefore, in order to compute the total electromagnetic force acting on a body enclosed by the surface Σ (where $\Sigma = \partial\Omega$ is the boundary of Ω), it is sufficient to compute the flux of the MST through it.

The MST can be written also as

$$\sigma_{ij} = D_i E_j - \tfrac{1}{2}\delta_{ij} \sum_k D_k E_k + B_i H_j - \tfrac{1}{2}\delta_{ij} \sum_k B_k H_k$$

where δ_{ij} is Kronecker's delta.[†]

Even though electrical machines have the adjective electrical, the electromechanical energy conversion is realized through the interaction of magnetic fields. So the electrical terms in the MST are not needed in the following:

$$\overleftrightarrow{\boldsymbol{M}} = \boldsymbol{B} \otimes \boldsymbol{H} - \tfrac{1}{2}(\boldsymbol{B} \cdot \boldsymbol{H})\boldsymbol{I} \tag{E.4}$$

The second term on the RHS resembles the magnetic pressure (which is also the energy density), and it contributes to the MST only in the diagonal terms. Hence in the Cartesian coordinate system

$$\overleftrightarrow{\boldsymbol{M}} = \begin{bmatrix} \tfrac{1}{2}B_x H_x - \tfrac{1}{2}B_y H_y - \tfrac{1}{2}B_z H_z & B_x H_y & B_x H_z \\ B_y H_x & \tfrac{1}{2}B_y H_y - \tfrac{1}{2}B_x H_x - \tfrac{1}{2}B_z H_z & B_y H_z \\ B_z H_x & B_z H_y & \tfrac{1}{2}B_y H_y - \tfrac{1}{2}B_x H_x - \tfrac{1}{2}B_y H_y \end{bmatrix}$$

[†]It is simply

$$\delta_{ij} = \begin{cases} 1 & \text{if } i = j \\ 0 & \text{if } i \neq j \end{cases}.$$

However, typical electrical machines have a cylindrical rotor, so it is rather convenient to express the MST in cylindrical coordinates and in terms of stresses:

$$\overleftrightarrow{\boldsymbol{M}} = \begin{bmatrix} \sigma_{rr} & \sigma_{r\theta} & \sigma_{rz} \\ \sigma_{\theta r} & \sigma_{\theta\theta} & \sigma_{\theta z} \\ \sigma_{zr} & \sigma_{z\theta} & \sigma_{zz} \end{bmatrix} \tag{E.5}$$

At this point, the electromagnetic torque generated by the machine will be the electromagnetic torque times the lever arm. If R is the rotor radius, the lever arm vector (in cylindrical coordinates) is

$$\boldsymbol{b} = \{R, 0, 0\}^{\mathrm{T}}$$

and the torque

$$\boldsymbol{T}_{\mathrm{em}} = \boldsymbol{F} \times \boldsymbol{b}$$

where $\times$ is the vector (cross) product of vectors. After some simple steps, we would get

$$\boldsymbol{T}_{\mathrm{em}} = \{0, RF_z, -RF_\theta\}^{\mathrm{T}}$$

What we are interested in is the last term, so $T_{\mathrm{m},z} = RF_\theta$, regardless of the sign.

Since we are adopting cylindrical coordinates, the normal versor in (E.3) is simply $\hat{\boldsymbol{r}} = \{1, 0, 0\}^{\mathrm{T}}$, and so

$$\overleftrightarrow{\boldsymbol{M}} \cdot \hat{\boldsymbol{n}} = \overleftrightarrow{\boldsymbol{M}} \cdot \hat{\boldsymbol{r}} = \begin{bmatrix} \sigma_{rr} & \sigma_{r\theta} & \sigma_{rz} \\ \sigma_{\theta r} & \sigma_{\theta\theta} & \sigma_{\theta z} \\ \sigma_{zr} & \sigma_{z\theta} & \sigma_{zz} \end{bmatrix} \cdot \begin{Bmatrix} 1 \\ 0 \\ 0 \end{Bmatrix} = \begin{Bmatrix} \sigma_{rr} \\ \sigma_{\theta r} \\ \sigma_{zr} \end{Bmatrix}$$

Therefore, choosing a cylinder as the integration surface right at the rotor periphery, we can look at the tangential component of the force only, expressed as

$$\begin{aligned} F_\theta &= \oint_{\Sigma} \sigma_{\theta r} \; \mathrm{d}S \\ &= L \oint_{0}^{2\pi} \sigma_{\theta r} R \; \mathrm{d}\theta \\ &= 2\pi RL \langle \sigma_{\theta r} \rangle \end{aligned} \tag{E.6}$$

where L is the active length of the machine and $\langle \sigma_{\theta r} \rangle$ is the average air-gap stress. So the torque results simply in

$$T_{\mathrm{em}} = 2\pi R^2 L \langle \sigma_{\theta r} \rangle = 2V_{\mathrm{r}} \langle \sigma_{\theta r} \rangle \tag{E.7}$$

where V_{r} is the rotor volume, and

$$\langle \sigma_{\theta r} \rangle = \langle B_\theta H_r \rangle = \langle B_r H_\theta \rangle = \langle \sigma_{r\theta} \rangle$$

because of the symmetry of the MST. If the stator current is distributed on an infinitely thin sheet $H_\theta = K_s$, where K_s is the electric loading. If this quantity and the air-gap magnetic flux density are sinusoidally distributed quantities

$$\langle \sigma_{r\theta} \rangle = \frac{\hat{B}_r \hat{H}_\theta}{2} = \frac{\hat{B}_g \hat{K}_s}{2}$$

Doing the last step

$$T_{em} = V_r \hat{B}_g \hat{K}_s \tag{E.8}$$

we would get Esson's rule.

Appendix F
High-frequency signal injection mathematical model

Start from the general voltage equation of any passive-rotor* synchronous electric machine:

$$\begin{cases} u_d = Ri_d + \dfrac{\mathrm{d}\lambda_d}{\mathrm{d}t} - \omega_{\mathrm{me}}\lambda_q \\ u_q = Ri_q + \dfrac{\mathrm{d}\lambda_q}{\mathrm{d}t} + \omega_{\mathrm{me}}\lambda_d \end{cases} \tag{F.1}$$

At low speed

$$\begin{cases} u_d = Ri_d + \dfrac{\mathrm{d}\lambda_d}{\mathrm{d}t} \\ u_q = Ri_q + \dfrac{\mathrm{d}\lambda_q}{\mathrm{d}t} \end{cases} \tag{F.2}$$

Neglecting the resistive terms, considering the small-signal model of the motor and using phasors for these signals:

$$\underline{U} = \mathrm{j}\omega \boldsymbol{L}\underline{\boldsymbol{I}} \tag{F.3}$$

Let $\boldsymbol{T}$ be the transformation matrix from the estimated reference frame to the correct one, and $\Delta\vartheta$ the estimation error angle:

$$\boldsymbol{T} = \begin{bmatrix} \cos\Delta\vartheta & -\sin\Delta\vartheta \\ \sin\Delta\vartheta & \cos\Delta\vartheta \end{bmatrix}$$

*No winding on the rotor.

Defining $\boldsymbol{\Gamma} = \boldsymbol{L}^{-1}$, we could write

$$\underline{\boldsymbol{I}} = \frac{1}{\mathrm{j}\omega}\boldsymbol{\Gamma}\underline{\boldsymbol{U}}$$

$$\boldsymbol{T}\underline{\tilde{\boldsymbol{I}}} = \frac{1}{\mathrm{j}\omega}\boldsymbol{\Gamma}\boldsymbol{T}\underline{\tilde{\boldsymbol{U}}}$$

$$\boldsymbol{T}^{-1}\boldsymbol{T}\underline{\tilde{\boldsymbol{I}}} = \frac{1}{\mathrm{j}\omega}\boldsymbol{T}^{-1}\boldsymbol{\Gamma}\boldsymbol{T}\underline{\tilde{\boldsymbol{U}}}$$

$$\underline{\tilde{\boldsymbol{I}}} = \frac{1}{\mathrm{j}\omega}\tilde{\boldsymbol{\Gamma}}\underline{\tilde{\boldsymbol{U}}}$$

where

$$\boldsymbol{\Gamma} = \frac{1}{\ell_d\ell_q - \ell_{dq}^2}\begin{bmatrix} \ell_q & -\ell_{qd} \\ -\ell_{dq} & \ell_q \end{bmatrix}$$

$$\tilde{\boldsymbol{\Gamma}} = \frac{1}{\ell_d\ell_q - \ell_{dq}^2}\begin{bmatrix} \ell_\Sigma + \ell_\Delta\cos 2\Delta\vartheta - \ell_{dq}\sin 2\Delta\vartheta & -\ell_\Delta\sin 2\Delta\vartheta - \ell_{dq}\cos 2\Delta\vartheta \\ -\ell_\Delta\sin 2\Delta\vartheta - \ell_{dq}\cos 2\Delta\vartheta & \ell_\Sigma - \ell_\Delta\cos 2\Delta\vartheta + \ell_{dq}\sin 2\Delta\vartheta \end{bmatrix}$$

F.1 Induction motor

The induction motor requires double the voltage equations needed for its mathematical model:

$$\begin{cases} u_{sd} = R_s i_{sd} + \dfrac{\mathrm{d}\lambda_{sd}}{\mathrm{d}t} - \omega_s\lambda_{sq} \\ u_{sq} = R_s i_{sq} + \dfrac{\mathrm{d}\lambda_{sq}}{\mathrm{d}t} + \omega_s\lambda_{sd} \\ 0 = R_r i_{rd} + \dfrac{\mathrm{d}\lambda_{rd}}{\mathrm{d}t} - \omega_{\mathrm{sl}}\lambda_{rq} \\ 0 = R_r i_{rq} + \dfrac{\mathrm{d}\lambda_{rq}}{\mathrm{d}t} + \omega_{\mathrm{sl}}\lambda_{rd} \end{cases} \tag{F.4}$$

At low speeds and slips,

$$\begin{cases} u_{sd} = R_s i_{sd} + \dfrac{\mathrm{d}\lambda_{sd}}{\mathrm{d}t} \\ u_{sq} = R_s i_{sq} + \dfrac{\mathrm{d}\lambda_{sq}}{\mathrm{d}t} \\ 0 = R_r i_{rd} + \dfrac{\mathrm{d}\lambda_{rd}}{\mathrm{d}t} \\ 0 = R_r i_{rq} + \dfrac{\mathrm{d}\lambda_{rq}}{\mathrm{d}t} \end{cases} \tag{F.5}$$

The fluxes–currents relationships for the induction machine are more complex, since we have more windings and mutual couplings. Let us consider the small-signal model, so the linearized machine model around a specific working point. We are going to refer to (differential) inductances of the same set as ℓ, while to inductances of different set as m. Then, the cause–effect relationship is going to be expressed as $\ell^{\text{cause}}_{\text{effect}}$.

In general,

$$\boldsymbol{\lambda} = \begin{Bmatrix} \lambda_{sd} \\ \lambda_{sq} \\ \lambda_{rd} \\ \lambda_{rq} \end{Bmatrix} = \begin{bmatrix} \ell^{sd}_{sd} & \ell^{sq}_{sd} & m^{rd}_{sd} & m^{rq}_{sd} \\ \ell^{sd}_{sq} & \ell^{sq}_{sq} & m^{rd}_{sq} & m^{rq}_{sq} \\ m^{sd}_{rd} & m^{sq}_{rd} & \ell^{rd}_{rd} & \ell^{rq}_{rd} \\ m^{sd}_{rq} & m^{sq}_{rq} & \ell^{rd}_{rq} & \ell^{rq}_{rq} \end{bmatrix} \begin{Bmatrix} i_{sd} \\ i_{sq} \\ i_{rd} \\ i_{rq} \end{Bmatrix} = \boldsymbol{L}\boldsymbol{i} \tag{F.6}$$

$$\begin{Bmatrix} \lambda_s \\ \lambda_r \end{Bmatrix} = \begin{bmatrix} \boldsymbol{\ell}^s_s & \boldsymbol{m}^r_s \\ \boldsymbol{m}^s_r & \boldsymbol{\ell}^r_r \end{bmatrix} \begin{Bmatrix} i_s \\ i_r \end{Bmatrix}$$

Adopting the same hypotheses as before, we would end up with the same expression of (F.3).

Appendix G

Incremental permeability simulations for differential inductances computation

In this chapter incremental permeability simulations are used to estimate the differential inductances of the machine around a specific working point.

The considered motor is a synchronous reluctance motor, and its geometry is depicted in Figure G.1.

It resulted from a multi-objective optimization, aiming at a high average torque and a low-torque ripple. Only the rotor was modified during the optimization.

G.1 Differential inductance computation

The fluxes–currents relationships are the following:

$$\begin{aligned} \delta\lambda_d &= \ell_d \delta i_d + \ell_{dq} \delta i_q \\ \delta\lambda_q &= \ell_{qd} \delta i_d + \ell_q \delta i_q \end{aligned} \tag{G.1}$$

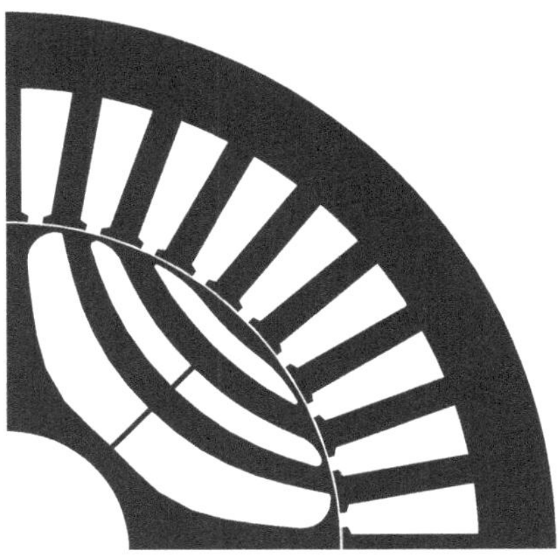

Figure G.1 Geometry of the Synchronous Reluctance Synchronous Reluctance (SYR) motor analyzed

and to get the differential inductance, we can just do

$$\begin{aligned} \ell_d &= \left.\frac{\delta\lambda_d}{\delta i_d}\right|_{\delta i_q=0} & \ell_{qd} &= \left.\frac{\delta\lambda_q}{\delta i_d}\right|_{\delta i_q=0} \\ \ell_{dq} &= \left.\frac{\delta\lambda_d}{\delta i_q}\right|_{\delta i_d=0} & \ell_q &= \left.\frac{\delta\lambda_q}{\delta i_q}\right|_{\delta i_d=0} \end{aligned} \tag{G.2}$$

and for the reciprocity of the linear mutual inductor $\ell_{dq} = \ell_{qd}$.

To estimate these differential inductances, we can do in two ways:

- magnetostatic nonlinear simulations around the working point;
- magnetostatic linear simulations around the working point using the previous element permeability (incremental permeability).

In any case, the number of magnetostatic simulations required is three. However, for the former case, they are "full" nonlinear magnetostatic simulations, while for the latter only one "full" nonlinear simulation is required, and the following increments are just linear simulations, which are very quick.

In the following, we will compare the two methods.

G.2 Plain magnetostatic only simulations

The most basic idea to compute differential inductances is to look in the neighborhood of the working point, slightly moving the d- and q-axes currents individually.

Let I_N and α_i^e be the rated current vector amplitude and angle. Then

$$\begin{aligned} I_d &= I_N \cos\alpha_i^e \\ I_q &= I_N \sin\alpha_i^e \end{aligned}$$

The first simulations is done with these currents.

The second one

$$\begin{aligned} I_d &= I_N \cos\alpha_i^e + \delta I \\ I_q &= I_N \sin\alpha_i^e \end{aligned}$$

and the third one

$$\begin{aligned} I_d &= I_N \cos\alpha_i^e \\ I_q &= I_N \sin\alpha_i^e + \delta I \end{aligned}$$

From the post-processed results, it is easy then to derive the increments of the quantities (flux linkages, energies, etc.) and the differential inductances.

G.3 Incremental permeability simulations

For this method, the first simulation required is the same as in the previous case, because we have to extract the incremental and cross-coupling permeability for every mesh element.

The following simulations are just incremental, in the sense that the imposed currents are only the increments. So, for the second simulation

$$I_d = \delta I$$
$$I_q = 0$$

while for the third

$$I_d = 0$$
$$I_q = \delta I$$

The obtained flux linkages are then directly the incremental ones.

G.4 Results comparison

For the previous motor, considering $n_{cs} = 1$, $I_N = 575$ A, $\alpha_i^e = 64°$ and $\delta I = I_N/1\,000$, the results are the following:

	Nonlinear	**Incremental**
T **[s]**	**10.5**	**6.5**
ℓ_d	$6.9223 \cdot 10^{-5}$	$6.9359 \cdot 10^{-5}$
ℓ_{qd}	$-1.4570 \cdot 10^{-5}$	$-1.4582 \cdot 10^{-5}$
ℓ_{dq}	$-1.4578 \cdot 10^{-5}$	$-1.4582 \cdot 10^{-5}$
ℓ_q	$3.6114 \cdot 10^{-5}$	$3.6121 \cdot 10^{-5}$

It can be noted that the "full" nonlinear simulation takes about 60% longer time to complete. While this seems quite a lot, in an optimization function evaluation, which lasts about 100 s for this motor example, it reduces to a 3% saving time.

These results were obtained before the April 21, 2019 release of FEMM, which introduced the possibility to start the linearized system from the previous nonlinear solution. So the use of the incremental permeability technique is less justified when computing the differential inductances from a computational time point of view.

G.5 Meaning of incremental permeability

At first, consider a unidimensional problem, where it is true that

$$B = \mu H$$

Taking the derivative with respect to H, we obtain the incremental permeability definition

$$\mu_{\mathrm{inc}}(H) = \frac{\mathrm{d}B}{\mathrm{d}H} = \frac{\mathrm{d}\mu}{\mathrm{d}H}H + \mu \tag{G.3}$$

In a two-dimensional Cartesian coordinate system we may write

$$\boldsymbol{B} = \mu(\boldsymbol{H})\boldsymbol{H} = \mu(|\boldsymbol{H}|)\boldsymbol{H}$$

where

$$|\boldsymbol{H}| = \sqrt{H_x^2 + H_y^2}$$

Explicitly, this means

$$\begin{Bmatrix} B_x \\ B_y \end{Bmatrix} = \mu(|\boldsymbol{H}|)\begin{Bmatrix} H_x \\ H_y \end{Bmatrix}$$

provided that the material is isotropic in the considered plane. We could write the linearized problem using Taylor first order approximation around a working point, identified by the magnetic field $\boldsymbol{H}_0$ and the magnetic flux density $\boldsymbol{B}_0$:

$$\boldsymbol{B} \approx \boldsymbol{B}_0 + \left.\frac{\partial \boldsymbol{B}}{\partial \boldsymbol{H}}\right|_{\boldsymbol{H}_0} \cdot (\boldsymbol{H} - \boldsymbol{H}_0) \tag{G.4}$$

where $\cdot$ is the scalar product between the matrix and the vector. The term $\partial \boldsymbol{B}/\partial \boldsymbol{H}$ is the 2-by-2 Jacobian matrix, and it is actually the definition of the incremental permeability $\overleftrightarrow{\boldsymbol{\mu}}$ tensor. In 2D Cartesian coordinates, it results

$$\mu_{xx} = \frac{\partial B_x}{\partial H_x} = \frac{\partial \mu}{\partial H}\frac{H_x^2}{\sqrt{H_x^2 + H_y^2}} + \mu$$

$$\mu_{xy} = \frac{\partial B_x}{\partial H_y} = \frac{\partial \mu}{\partial H}\frac{H_x H_y}{\sqrt{H_x^2 + H_y^2}}$$

$$\mu_{yy} = \frac{\partial B_y}{\partial H_y} = \frac{\partial \mu}{\partial H}\frac{H_y^2}{\sqrt{H_x^2 + H_y^2}} + \mu$$

$$\mu_{yx} = \frac{\partial B_y}{\partial H_x} = \mu_{xy}$$

When we are interested in the increment of $\boldsymbol{B}$, we may write (G.4) simply with

$$\delta\boldsymbol{B} = \overleftrightarrow{\boldsymbol{\mu}} \cdot \delta\boldsymbol{H}$$

Now, let us consider the case of a horizontal magnetic field, $\boldsymbol{H}_0$, so

$$\boldsymbol{H}_0 = \begin{Bmatrix} H_0 \\ 0 \end{Bmatrix}, \qquad \boldsymbol{B}_0 = \begin{Bmatrix} B_0 \\ 0 \end{Bmatrix} = \begin{Bmatrix} \mu(H_0)H_0 \\ 0 \end{Bmatrix}$$

Then the incremental magnetic permeability tensor becomes

$$\mu_{xx} = \left.\frac{\partial B_x}{\partial H_x}\right|_{\boldsymbol{H}_0} = \left.\frac{\partial \mu}{\partial H}\right|_{\boldsymbol{H}_0} H_0 + \mu(H_0)$$

$$\mu_{xy} = \left.\frac{\partial B_x}{\partial H_y}\right|_{\boldsymbol{H}_0} = \left.\frac{\partial B_y}{\partial H_x}\right|_{\boldsymbol{H}_0} = \mu_{yx} = 0$$

$$\mu_{yy} = \left.\frac{\partial B_y}{\partial H_y}\right|_{\boldsymbol{H}_0} = \mu(H_0)$$

Looking at the first of these equations, we see that the μ_{xx} expression coincides with the definition of incremental permeability given in (G.3).

Expressing the increments of the quantities, we may write

$$\delta B_x = \mu_{xx}\delta H_x = \mu_{\text{inc}}(H_0)\delta H_x$$

$$\delta B_y = \mu_{yy}\delta H_y = \mu(H_0)\delta H_y$$

It can be noted that the magnetic permeability acting along the main flux lines is the incremental one, while the apparent permeability is acting on the perpendicular direction.*

*The incremental permeability represents the slope of the tangent to the *B–H* characteristic. The apparent permeability is the slope of the line connecting the working point along the *B–H* curve to the origin.

Index

www.ingramcontent.com/pod-product-compliance
Lightning Source LLC
LaVergne TN
LVHW051049210926
848486LV00001B/53